T0328904

Spectroscopy, Diffraction and Tomography in Art and Heritage Science

Spectroscopy, Diffraction and Tomography in Art and Heritage Science

Edited by

Mieke Adriaens

Department of Chemistry, Ghent University, Ghent, Belgium

Mark Dowsett

Department of Physics, The University of Warwick, Coventry, Warwickshire, United Kingdom
Department of Chemistry, Ghent University, Ghent, Belgium

ELSEVIER

Elsevier
Radarweg 29, PO Box 211, 1000 AE Amsterdam, Netherlands
The Boulevard, Langford Lane, Kidlington, Oxford OX5 1GB, United Kingdom
50 Hampshire Street, 5th Floor, Cambridge, MA 02139, United States

Notices
Knowledge and best practice in this field are constantly changing. As new research and experience broaden
our understanding, changes in research methods, professional practices, or medical treatment may
become necessary.

Practitioners and researchers must always rely on their own experience and knowledge in evaluating
and using any information, methods, compounds, or experiments described herein. In using such
information or methods they should be mindful of their own safety and the safety of others, including
parties for whom they have a professional responsibility.

To the fullest extent of the law, neither the Publisher nor the authors, contributors, or editors, assume
any liability for any injury and/or damage to persons or property as a matter of products liability,
negligence or otherwise, or from any use or operation of any methods, products, instructions, or
ideas contained in the material herein.

Library of Congress Cataloging-in-Publication Data
A catalog record for this book is available from the Library of Congress

British Library Cataloguing-in-Publication Data
A catalogue record for this book is available from the British Library

ISBN: 978-0-12-818860-6

For information on all Elsevier publications
visit our website at https://www.elsevier.com/books-and-journals

Publisher: Susan Dennis
Acquisitions Editor: Kathryn Eryilmaz
Editorial Project Manager: Andrea Dulberger
Production Project Manager: Sreejith Viswanathan
Cover Designer: Alan Studholme

Typeset by SPi Global, India

Contents

CHAPTER 7 **Laser-induced breakdown spectroscopy in cultural heritage science 209**

Rosalba Gaudiuso

Contributors

Mieke Adriaens
Department of Chemistry, Ghent University, Ghent, Belgium

Stephen Bauters
European Synchrotron Radiation Facility (ESRF), Helmholtz-Zentrum Dresden-Rossendorf CRG-Beamline, Grenoble Cedex 9, France

Andrea Denker
Protons for Therapy, Helmholtz-Zentrum Berlin, Berlin, Germany; University of Applied Sciences, Beuth Hochschule für Technik Berlin, Berlin, Germany

Mark Dowsett
Department of Physics, The University of Warwick, Coventry, Warwickshire, United Kingdom; Department of Chemistry, Ghent University, Ghent, Belgium

Rosalba Gaudiuso
Université du Québec à Montréal, Montréal, QC, Canada

Marie Godet
IPANEMA, CNRS USR 3461, MCC, UVSQ, Paris-Saclay University, Gif-sur-Yvette, France

Evelyne Godfrey
Uffington Heritage Watch Ltd, Faringdon, Oxfordshire, United Kingdom

Chris Jeynes
University of Surrey Ion Beam Centre, Guildford, United Kingdom

Winfried Kockelmann
STFC Rutherford Appleton Laboratory, ISIS Neutron Facility, Oxfordshire, United Kingdom

Brecht Laforce
Department of Chemistry, Ghent University, Ghent, Belgium

Eberhard Lehmann
Paul Scherrer Institute, Villigen, Switzerland

David Mannes
Paul Scherrer Institute, Villigen, Switzerland

Ella De Pauw
Department of Chemistry, Ghent University, Ghent, Belgium

Paola Ricciardi
The Fitzwilliam Museum, University of Cambridge, Cambridge, United Kingdom

Anastasia Rousaki
Raman Spectroscopy Research Group, Department of Chemistry, Ghent University, Ghent, Belgium

Philippe Sciau
CEMES, CNRS, Toulouse University, Toulouse, France

Geert Silversmit
Research Department, Belgian Cancer Registry, Koningsstraat, Brussels, Belgium

Pieter Tack
Department of Chemistry, Ghent University, Ghent, Belgium

Peter Vandenabeele
Raman Spectroscopy Research Group, Department of Chemistry; Archaeometry Research Group, Department of Archaeology, Ghent University, Ghent, Belgium

Bart Vekemans
Department of Chemistry, Ghent University, Ghent, Belgium

Laszlo Vincze
Department of Chemistry, Ghent University, Ghent, Belgium

Rita Wiesinger
Institute of Science and Technology in Art, Academy of Fine Arts Vienna, Vienna, Austria

Origins and fundamentals

1

Mark Dowsett[a,b] and Mieke Adriaens[b]

[a]*Department of Physics, The University of Warwick, Coventry, Warwickshire, United Kingdom*
[b]*Department of Chemistry, Ghent University, Ghent, Belgium*

1 Introduction

This book describes some of the principal spectroscopic, diffraction, and tomography techniques used in the chemical and structural analysis of artworks and historical artifacts. By chemical analysis we mean the identification of some or all of the elements and molecules comprising a sample, while the determination of structure has two principal objectives: (i) to map the spatial arrangement of the atoms and molecules, e.g., to indicate whether a material is crystalline, amorphous, or a mixture of both, to describe how the chemistry varies in three dimensions etc.; and (ii) to investigate larger-scale organization such as stratigraphy, stress patterns, joins, and cracks.

In the search for contributors, our objective has been to find authors for the specialist chapters who can present the techniques in which they are expert in an accessible way. Since their expertise derives both from extensive heritage applications experience and from pioneering developments in the instrumentation and techniques, they can open the "black box" which so many techniques are becoming and provide the background required for properly informed analysis.

All the techniques in this book involve the exposure of the sample to one kind of radiation (primary radiation) which is usually directed at the sample in a beam, and the study of radiation emitted by, transmitted through, or scattered from the sample as a consequence. The word "beam" may mean that the incoming radiation has a well-defined direction (parallel beam), or radiates from a small source (divergent beam), or is focused into a spot (convergent or focused beam). The emitted radiation might not be the same as the primary radiation although scattered and transmitted radiations will be. Typical examples of the incoming and outgoing radiations might be light, X-rays, electrons, protons, heavy ions, and neutrons, but this is not an exhaustive list. Depending on the physics underlying a particular analytical technique it might be convenient to treat the incident and exiting radiations as a stream of particles, or as a wave or as both (in different parts of the interaction) when

Spectroscopy, Diffraction and Tomography in Art and Heritage Science. https://doi.org/10.1016/B978-0-12-818860-6.00010-6

describing the effects initiated in the sample by the beam and when analyzing the data. We discuss the origins of this dual approach—a cultural phenomenon thousands of years old—in Section 2.

Whatever the radiation used, great care and technique expertise are essential to ensure that irreplaceable samples (actual museum objects and artworks) do not suffer measurable damage either during analysis, or later as a result of some process initiated by the bombardment (bleaching or darkening of a particular pigment in a painting, for example) [1]. However, many studies can be carried out on simulants (e.g., corrosion and corrosion inhibition studies), fragments, or objects in reserve collections, and such studies can pave the way for the safe analysis of a unique object or provide the required information directly. Of course, where an artifact is at risk from uncontrolled degradation which must be understood to save the object (e.g., bacterial attack in conserved wood [2]), some of the techniques described herein may play an essential role in its preservation, even if they cause a small amount of damage or require a sample to be removed. At the end of this chapter in Section 6 we discuss some of the terminology relating to analytical risk in heritage science, specifically "nondestructive," "microdestructive," and "noninvasive." We do this after describing some of the energies and interactions involved in preceding sections.

These descriptions will unavoidably make use of scientific terminology, as does the rest of this book. Later in this chapter we will, therefore, describe some of the basic parameters such as wavelength, frequency, energy, wave number, particle mass, and so on, as well as their relationships and importance to the data.

What then are spectroscopy, diffraction, and tomography, and what can they tell us about objects such as paintings, glassware, textiles, frescos, sculpture, musical instruments, historical artifacts, and even buildings?

Various spectroscopies are discussed in Chapters 2–4, 6, 7, and 9–11. Spectroscopy is the science of acquiring and interpreting a spectrum which is a measurement of the intensity of some radiation versus its wavelength (color in the case of light) or a related parameter such as frequency, energy, or particle mass. (The relationship between particle mass and wavelength is discussed in Section 4.2.) To give two contrasting examples: The combination of the human eye and the visual cortex in the brain, performs a kind of spectroscopic analysis on white light scattered from a painting and this results in the perception of differently colored regions. Tailored magnetic and electric fields in an electron microscope can both focus the electron beam and disperse electrons whose energy has been changed by interaction with a sample across an area detector to obtain a spectrum which can reveal details of sample chemistry on the nano-scale (see Chapter 3 for example).

Some diffraction techniques are described in Chapter 3 (electron diffraction), Chapter 6 (X-ray diffraction), and Chapter 8 (neutron diffraction). Diffraction (as an analytical method rather than a physical phenomenon) is, similarly, the acquisition and interpretation of the intensity pattern as a function of angle for coherent scattering of incident radiation by the object being studied. Coherent scattering means that the scattered radiation has the same wavelength as the incident radiation. Such a pattern can reveal crystalline structure in a sample, identify the specific crystalline compounds present, and give information on microstructure which can reveal methods of manufacture (e.g., the composition of an alloy and whether it was cast or forged).

Tomography (Chapter 5) is the reconstruction of a 3D density (or other) map from 2D maps of absorption (as with a medical X-ray) or emission (as with an MRI scan) of radiation in a volume or microvolume of sample. Tomography offers a potentially nondestructive and noninvasive way of observing the internal structure of an object—i.e., without cutting it open.

Taken together, these methodologies offer means for complete study of the chemistry and structure of their subjects on the centimeter to nanometer scale. They can be used to identify corrosion, evaluate countermeasures, compare pigments, determine methods of manufacture, help ascertain age, provenance, and use, find internal damage, indicate the original appearance, identify fakes, and much more.

At another level, they are cultural phenomena in their own right and just as dependent on human creativity, perception, and aspiration as an art object. So, let us begin by looking at the (at least) 8-millennia-long cultural heritage of optics and spectroscopy themselves.

2 A brief cultural history of optics and spectroscopy

2.1 Optics

With one exception, the analytical techniques described in this book grew out of the huge strides made in atomic and particle physics in the hundred years between 1850 and 1950. The exception is optical spectroscopy, i.e., the spectroscopy of visible light. From the perspective of the history and application of science it is interesting to note that all the techniques were originally conceived in order to explore fundamentals—to determine the nature of light, to prove the existence of the atom, then to understand its structure, to probe the chemistry of stars, to investigate the nature of subatomic particles, and to provide the experimental proofs for new theories of quantum physics and relativity—still an ongoing process today. Many of the instruments we now use for routine chemical and structural analysis, instruments which can be bought off the shelf and used with minimal understanding of their underlying principles (at the user's peril, of course), are the descendants of the large hadron colliders of their time: once fantastically taxing to build with the materials available, temperamental and difficult to use, and built as prototypes to investigate the building blocks of matter and the universe.

Optical spectroscopy and diffraction are the exception as their origins are far older. (As we shall see, the science of optics, which underlies everything in the book and pretty much the whole of modern physics and technology, goes back at least 8000 years.) Perhaps this is unsurprising as we have both the inbuilt detectors—the cone cells in the retina of the eye—and processing power in the visual cortex to pursue spectroscopy with nothing more than a light source. In common with other animals, we instinctively use spectroscopy to judge the time of day, estimate the likely weather ahead, and identify whether fruit is ripe, among many other tasks. Diffraction effects are common in nature, in sound and water waves as well as light, and will have been observed for millennia, even if not understood. In a heritage context, the spectroscopic power of the eye-brain combination used in visual comparison both with the memory of color and between an artifact and a set of reference colors is a powerful tool for setting safe limits on illumination [3, 4].

It was probably Joseph von Fraunhofer in the early 1800s who assembled various optical elements into the first true spectroscope and who experimented with the detailed optical properties of glasses needed to optimize the device [5]. He used sunlight for his experiments to get a sufficiently intense source, and consequently discovered the lines in the solar spectrum named after him. These lines were identified as being due to the Sun's chemistry following the invention of the chemical spectroscope by Gustav Kirchhoff and Robert Bunsen who, in 1859, made a prism-based instrument [6] to study the light emitted from elements heated in the flame of a Bunsen burner (also invented for that specific purpose) (Fig. 1). By 1861 they had discovered two new alkali metals, rubidium and cesium, from the unique pattern of colored lines in their spectra. Although the experimenter's eye was their detector, the color of a line was defined by its refracted angle in the spectroscope and therefore absolutely related to wavelength rather than contextually dependent on the observer's vision.

Fraunhofer, Bunsen, and Kirchhoff drew on around 8000 years of optical research and development in building their device, but it is impossible to say when the study of optics really began. The first optical devices known at the time of writing are polished obsidian mirrors excavated from a Neolithic settlement in what is now Turkey [7]. These have been dated to around 6000 BCE. Lenses may have arrived somewhat later in human civilization with one of the earliest examples being the Nimrud lens [8, 9] (on display in the British Museum) found in modern-day Iraq and dated to

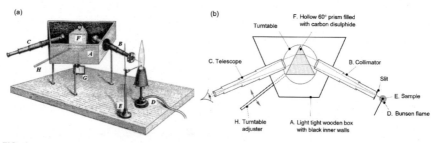

FIG. 1

Depiction of the spectroscope invented by Bunsen and Kirchhoff. (A) From the original paper [6]. (B) Schematic diagram showing components and ray paths. Such a device would still be capable of giving useful information today although modern equivalents use a variety of different sources, a diffraction grating in place of the prism, and pixelated detectors.

around 750 BCE. References to lenses or "burning glasses" are found as far back as 424 BCE in "The Clouds" [10], a comic play by Aristophanes which suggests that they were commonplace by then:

> Strepsiades*: You know that lovely, clear stone that one can get at the pharmacies? People can start fires with it.*
>
>   Socrates*: Yes. You mean glass, right?*
>
>   Strepsiades*: That's right, that's the one. Well, what if I went and bought one of them and when I'm in court, and, just as the clerk is about to write up my charges on his wax tablet, I stand back a bit—have him in front of me and the sun behind me—and then… well, couldn't I just make all his writing melt away?*
>
>   **(Translation reproduced by kind permission of George Theodoridis [10])**

Science grows from ideas which are often erroneous but which contain grains of truth, through to partial truths of limited application, and onwards to increasingly sophisticated and complete descriptions of the universe in which past theories are often contained as limiting cases. (It is not obvious that this process has any endpoint and, moreover, it might be compared to a modern scientific education where what one is taught first is almost entirely wrong and has to be unlearned…) After hundreds, if not thousands, of years of experimentation and observation by the Greeks and others, in 50 CE or so, a work believed to be [11] by Heron (or Hero) of Alexandria was published [11, 12]. Entitled $\kappa\alpha\tau o\pi\tau\rho\iota\kappa\alpha$' (*Catoptrica*—On Mirrors, and one of several works with this title), it contained the first grain of true optical explanation—the "principle of least distance." This stated that, when reflected from a mirror, the light contributing to the image has traveled the least possible distance from the object. Although this is not fully correct, from it can be deduced a law of optics: that the angle of incidence is equal to the angle of reflection. Only this geometry corresponds to the least distance of travel for the light involved (Fig. 2).

The law itself had already been established empirically by Euclid around 350 years earlier and published in his *Catoptrica*, a work known to have existed not least because he cites it in a later work, *Optics* [13, 14]. There, he shows how to use the law to measure altitude "when the sun is not shining" (proposition ιθ' pp. 28–30) using

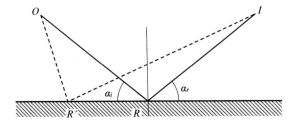

FIG. 2

The path *ORI* corresponds to the angle of incidence α_i being equal to the angle of reflection α_r and is the least distance via the mirror surface from *O* to *I*. For any other path such as *OR'I*, the path length is greater.

similar triangles. (A *Catoptrica* [15] whose attribution to Euclid has been both widely disputed and accepted [16] indeed contains the law of reflection (Ref. [14] pp. 288–289).)

However, it was not until 1900 years after Heron's theory on path length that the underlying reasons for this apparently simple behavior started to be properly understood following the solutions to Paul Dirac's 1932 description of relativistic quantum mechanics [17] (quantum electrodynamics (QED)) by Feynman [18], Schwinger [19], and Tomonaga [20], which we mention in more detail later.

Fig. 2 shows another concept introduced by Ancient Greek thought: that of the ray (ἀκτίς) of light—lines *OR* and *RI* are rays. This concept is still in widespread use and has applications in the visualization of light paths in the computer modeling of optics and in the description of charged particle trajectories in electric and magnetic fields in physics, for example. Of course, rays (in this geometrical sense) are an abstraction—they have no real existence but they were used by the Ancient Greeks as we use them today—scientifically, descriptively, and poetically [21].

The principle of least distance could not explain the phenomenon of refraction. That explanation came nearly 1000 years after Heron, sometime between 1020 and 1030 CE, in Hasan Ibn al-Haytham's seminal *Kitāb al-Manāẓir* (Treasury of Optics) [22,23]. This contains a fundamental modification to Heron's concept and was a step closer to the truth: "Light travels from object to image by the easiest path" [24]. Al-Haytham believed, correctly, that light traveled at a finite and well-defined speed but was unable to measure it. Al-Haytham was strongly influenced by the remarkable work of Ibn Sahl, about whom little is known but who in 921 CE was working at the Persian court [25]. Ibn Sahl was possibly the first to understand what we now call the *refractive index* (see below) and used an ingenious geometrical construction which embodied the concept to manufacture low aberration lenses. The "easiest path" principle was rigorously formulated by Pierre de Fermat (1607–1665) as "the principle of least time" (for example, [26]) and is now known as Fermat's principle [24]. This more sophisticated partial truth contains Heron's principle as far as reflection is concerned for the following reason: Neglecting its interaction with the surface of the mirror, the light travels through a single medium (air, say). Since light effectively travels at constant speed in a particular medium, the path of "least time" corresponds exactly to "least distance" so the two are equivalent. However, for refraction, which occurs when light moves across boundaries between media of different density (e.g., air to glass in a lens or air to water), the effective speed is lower in the denser medium. Fermat's principle almost accounts for both reflection and refraction, and the laws of refraction can be deduced from it as well as those of reflection. However, it turns out that Fermat's principle is not complete as originally stated. A modern form obeyed by all light image forming systems and which contains Fermat's principle in the same way that Fermat's principle contains Heron's can be stated as: "Light contributing most of the intensity to an image has followed optical paths from the source which lie *arbitrarily close* to a stationary path" [27]. (Lenses, mirrors, layers of hot air causing mirages, and lensing by the gravitational fields around stars and galaxies all obey this principle.) What does "arbitrarily close" mean in this context? Suppose you follow the course of a stream down a mountain staying close to the stream, or even wading in it so you essentially move down the same slopes. Then you are arbitrarily close to the stream. And "stationary"? This means that the paths taken by the light are such that a small deviation in the path makes almost no difference to the transit time. The optical paths might be loosely identified with the Ancient Greek concept of a ray but formally, the optical path is a mathematical concept given by the product of distance and refractive index in a single material and added up as the light passes across successive interfaces to get a total optical path.

We now know that the "effective" speed of light is lower in a denser medium because, although the photon (the particle associated with light in modern scientific thought) always travels at $c = 3 \times 10^8 \, \text{ms}^{-1}$, photons trying to plow their way through a dense medium repeatedly interact with electrons. Each interaction takes a very small but finite time, which is added to the transit time along a path so the effective speed is reduced below c. This reduction in speed is what gives rise to the refractive index—the salient property of any optical material. Our current civilization depends on our knowledge of the refractive index and ability to make materials with given indices at will. It underlies everything from fundamental science through the whole of applied optics (cameras, computer displays, and endless other technology) to the communication of trivia via social media (and all other optical communications). Although its effects were, so far as we know, first identified quantitatively by Ibn Sahl [25], rediscovery and mathematical formulation by Willebrord Snell in Leyden in 1621 gives us the key optical law of refraction based on the concept of a stationary optical path: Snell's Law (Fig. 3) and Eq. (1). His law was not published until after his death, when his work was examined by Isaac Vossius and Christiaan Huygens [28,29].

$$\frac{n_1}{n_2} = \frac{\sin \alpha_r}{\sin \alpha_i} = \frac{v_2}{v_1}, \tag{1}$$

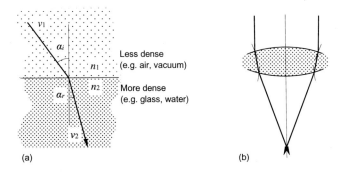

FIG. 3

(A) A ray of light crossing the boundary from a low-density medium to one of a higher density is bent toward a line normal to the surface according to Snell's law. The refractive index n_1 of the low-density medium is less than that n_2 of the high-density medium. Conversely, the effective speed v_1 of the ray in the less dense medium is higher than the speed v_2 in the more dense medium. (B) As a consequence, if the boundary is curved in the correct way, then rays can be brought to a focus and we have a converging lens or, as it was thought of in ancient times, a "burning glass."

where subscripts 1 and 2 refer to the less dense and more dense medium, n is the refractive index of the medium, v is the effective velocity of light in the medium, α_i is the angle of incidence (measured from the normal), and α_r is the angle of *refraction*.

Again, the deceptively simple behavior shown in Fig. 3 was not fully understood (in terms of the behavior of photons) until the advent of QED. Nevertheless, it was refraction (in a prism) which gave us the first spectroscopes such as that of Bunsen and Kirchoff.

Devices such as electron microscopes and mass spectrometers also contain lenses and mirrors among many other components and they have an analogous behavior to their light optical counterparts. However, these *charged particle optics* consist of carefully tailored electrostatic, magnetostatic, and occasionally electrodynamic fields, rather than glass or silvered surfaces (see, for example, Ref. [30,31]). The components have a refractive index which depends on the way the particle energy is changed by the field as well as that energy itself. Nevertheless, charged particles passing through static fields follow trajectories which obey a principle (Pierre de Maupertius, 1744) analogous to Fermat's: "The most probable path of a charged particle through a system of static fields is such that the total work done on the particle is arbitrarily close a stationary value." This is almost exactly what Ibn al-Haytham meant by "path of least effort." It is no coincidence that photons, electrons, ions, and other quantum particles obey similar rules. Underlying their behavior is the conservation of energy and momentum as described globally by Hamilton's principle [32].

2.2 Light, vision, and spectra

In ancient Greek thought there were many differing sets of ideas concerning light and vision. At the extremes (in our view) were: *intromission*, where light from an external source interacts with various media and surfaces on its way to the eye; and *extramission*, where the eye emits rays from an "internal fire" which interact with the surroundings [33] and return. Empedocles (c.494–434 BCE) considered light to be a "streaming substance" emitted by a luminous source with a finite speed [34] which (essentially) reflected from objects and arrived in the eye. This is an intromission theory. Pythagoras (c.570–495 BCE) is said to have taught the contrary: that rays emitted from the eye returned to it [33]. Plato (c.428–347 BCE) apparently combined the two contrasting theories and described how rays from the eye interacted with daylight to produce vision [33] (but this is further discussed in [35]). Aristotle appears to support this view in his *Μετεωρολογικά* [36] (Meteorologica, c.340 BCE) where he describes the rainbow as being the reflection of vision back to the Sun, but in his essay on the senses and the soul *Περὶ Ψυχῆς* (Peri Psyches, c.350 BCE) [33,37] he states:

> *That is why in the case of reflection it is better, instead of saying that the sight issues from the eye and is reflected, to say that the air, so long as it remains uninterrupted, is affected by shape and color [37].*

His background argument, that color is something imposed on a pervasive transparent medium, looks deceptively like an early version of the "luminiferous ether" theory.

Extramission theory has proven ridiculously hard to kill off even though it was comprehensively debunked [23] by al-Haytham in *Kitāb al-Manāzir* around 1000 years ago. Indeed, a surprisingly large number of modern humans, even with college educations, still persist in the erroneous belief that vision is caused by extramission (see for example Ref. [38]).

The first indications of a theory involving a corpuscular nature for light are perhaps to be found among the Ancient Greek Atomists [39]. An particle theory of light is summarized in Lucretius' poem De Rerum Natura (On the nature of things) written sometime between 99 and 55 BCE. For example, [40]:

> *In the first place, we have constant means of observing how swift in their motion are those bodies which are light, and which consist of minute particles. Of which kind is the Sun's light and his heat; for this reason, that they are compounded of minute primary atoms, which are, as it were, struck out…*

Here, radiant heat is included in the same class as light as would have been evident to the Greeks from the action of "burning glasses." As is well-known, elsewhere in the same work, Lucretius discusses aspects of the Atomists' theory of matter.

Over the next 2000 years or so ideas about light as geometrical rays, traveling waves, a stream of particles, or a disturbance in some medium were developed, forgotten, redeveloped in different cultures, hotly disputed, and passionately defended. The idea that light was a disturbance in a medium called the luminiferous ether dominated more recent European thought perhaps because of analogies with water and sound waves which do travel through a medium. However, the ether was made redundant by Maxwell's theory of electromagnetism in 1865 [41], and its existence was put in serious doubt by the Michaelson-Morley experiment in 1887 [42]. In 1905, Einstein's special theory of relativity [43] destroyed the notion completely.

Newton is often described as a proponent of a particle theory of light; for example, in his *Optiks* [44] he mentions particles of light a few times and at the end poses the question [44]a:

> *Are not the Rays of Light very small Bodies emitted from shining substances?*

However, he also suggests that the rays were shaken off shining bodies by vibrations in the bodies, and that reflection from and transmission through polished glass plates was related to "Fits of easy reflexion and easy transmission" as a function of the plate's thickness [44]b (implying a wave-like behavior). On responding to Robert Hook's criticism [45] of his work, he does not seem to have been particularly dogmatic about particle theory [46] and at the end of *Optiks* both compares the behavior of light to water [44]c and positions his criticism of wave theory as unfinished work [44]d, so it is unfortunate that "his" corpuscular view was seen as competing with the wave theories of Christiaan Huygens [47], Hooke [48], and others during the 17th century. In fact, a wave theory would have been able to explain all of Newton's own observations whereas a corpuscular theory could not (at that time). Nevertheless, Newton's profound skill in fabricating optical components and superb experiments on the behavior of the Sun's light when refracted through prisms, especially when apertured and collimated using a lens, and his explanation of the rainbow [44]e are the foundation of modern spectroscopy. Of course, Newton's own eyes were the detectors in his spectroscopes, observing the dispersion of sunlight on his wall and on sheets of paper.

Newton's spectra were produced using the phenomenon of refraction of light in a glass prism. Modern scientific instruments contain many examples of prism-like action [30,31]—for example, the magnetic sector in a mass spectrometer deflects a beam of ions in a vacuum through angles which depend on their momenta; this is essentially what a prism does for light. High-mass ions are deflected less by a magnetic field than low-mass ones with the same kinetic energy providing one method for separating the masses to obtain a mass spectrum. However, for visible light, high-energy (blue) photons are deflected *more* at an air glass interface than low-energy (red) photons because blue photons interact more strongly with the electrons in a solid than do red ones so their change in effective speed and, therefore, refractive index is greater.

So much for reflection and refraction. However, especially where electromagnetic radiation, electrons, and neutrons are concerned, another phenomenon known as diffraction becomes important, often as the centerpiece of the analytical technique itself.

In 1665, while Newton was still a student, a remarkable work by the Italian Jesuit priest Francisco Maria Grimaldi (1618–1663) was published, entitled *Physico-Mathesis De Lumine, Coloribus et Iride* (Physics and Mathematics of Light, Colors and Rainbows) [49]. Newton was clearly aware of this work and it undoubtedly guided some of the discoveries in his own *Optiks* and indeed later work such as that of Thomas Young (1773–1829). Grimaldi discovered the phenomenon of diffraction and christened it with the name we now use taken from the Latin *diffringo*—to shatter. (For Newton, the phenomenon was "inflection.") At the beginning of his work, Grimaldi writes:

> *Lumen propagatur seu diffunditur non sòlùm Directè, Refractè, ac Reflexè, sed etiam alio quodam Quarto modo, DIFFRACTE [sic].*
>
> *(Light is propagated or diffused not only directly, and by refraction, and reflection, but also by a fourth means: diffraction.) (M.G.D.)*

It is impossible to overstate the importance of Grimaldi's discovery not only to many of the analytical techniques in this book but also in wider science and technology. Diffraction effects first established the wave theory of light, then later the wave theories in quantum physics (from observations based on diffraction of X-rays, electrons, and neutrons, for example). Diffraction lies at the heart of imaging systems whatever radiation they are based on and helps to monochromate (filter out a narrow band of wavelengths) or analyze much of the radiation used in analytical techniques or anywhere else. Finally, having been instrumental in the establishment of pure wave theories, diffraction helped to destroy them (insofar as their being a correct underlying model for radiation) and see them replaced with QED in the middle of the 20th century.

Grimaldi begins his first experimental description thus (see also Fig. 4):

> *Aperto in fenestra foraminulo perquam parvo AB, inroducatur per illud in cubiculum, alioqui valdè obscurum, lumen Solis Cælo serenissimo, cuius diffusio erit per conum, vel quasi conu ACDB visibilem si aër fuerit refertus atomis puluereis, vel si in eo excitetur aliquis fumus.*
>
> *(In the window I uncovered an extremely small hole AB to admit the most fair light of the heavenly Sun to a very dark room. The light will be distributed in a cone, or rather the nearly conical form ACDB, and could be made visible if the air were to be filled with dust motes or if someone were to infuse it with smoke.) (M.G.D.)*

Grimaldi then describes the diffraction fringes he observed on a screen around the shadow of a tiny rectangle of opaque material *FE* placed in the cone of light.

In a letter to Robert Boyle, written in February 1678 [50], Newton proposed that:

> *…there is diffused through all places an ethereal substance capable of contraction and dilation, strongly elastic, and in a word, much like air in all respects but far more subtil. 2. I suppose this ether pervades all gross bodies but yet so as to stand rarer in their pores than in free spaces … and this I suppose (with others) to be the cause why light incident on those bodies is refracted toward the perpendicular; … 3. I suppose the rarer ether within bodies, and the denser without them to grow gradually into one another at some little distance from the superficies of the body … And this may be the cause why light in Grimaldo's experiment, passing by the edge of a knife or other opaque body is turned aside, and, as it were, refracted [sic] and by that refraction makes several colours.*

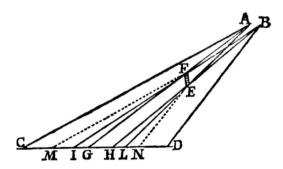

FIG. 4

Grimaldi's own diagram of his first diffraction experiment. Note that his text makes it clear that AB is a single aperture (not two as some commentators have thought) and the rays shown originate from the edges at A and B. CD is "a white coated board or a white sheet of paper stretched to cover the area of the cone ACBD." This was placed on the floor of his cell.

Here, Newton uses an ether-based hypothesis to explain both refraction at a surface and Grimaldi's diffraction. Although this luminiferous ether later became (albeit temporarily) a vehicle for the support of light waves, Newton saw it as influencing a stream of particles; the changes in density of the ether were bending the particle beam through interaction with a faster system of vibrations.

In 1803, Thomas Young described and quantified Grimaldi's experiment among others in his Bakerian Lecture at the Royal Society [51] in which he acknowledges "the ingenious and accurate Grimaldi." (Not to be confused with his "two slit" experiment—see below—although that clearly derives from Grimaldi's original.) He combined both Newton's and Grimaldi's measurements with his own and showed that the results were consistent with light being a wave whose color was determined by wavelengths which he established. In 1807 he published the experiment which now bears his name—Young's two slit experiment—first for water waves and then for light [52,53]. This experiment, which we will describe in Section 2.3, appears so convincing as far as the wave nature of light is concerned that, along with equivalents such as the elegant two-mirror experiment of Augustine Fresnel (1788 1827) (see p. 55 of [54]) and Fresnel's amazingly complete optical experiments and mathematical theory (still the basis of most optical modeling today) [54–56], it overturned the corpuscular theory of light against the strong opposition of the European scientific establishment.

A vital consequence of the discovery of diffraction was the invention of the diffraction grating—the device which, rather than the prism, lies at the heart of modern optical spectroscopes. The fact that a set of closely spaced, parallel, opaque fibers or wires (and later ruled lines on glass and other media too) can be made to disperse light more powerfully and controllably than a prism was perhaps first recorded in 1673 by James Gregory [57] in a letter to the mathematician John Collins. He observed the spectrum dispersed by a feather from a pinhole light source and suggested that Collins might ask Newton for the explanation. The earliest quantitative investigation [58] was by the American astronomer David Rittenhouse [59] in 1786, but it was Fraunhofer who independently discovered the properties of wire gratings in the early 1800s and used them in his spectroscope [5].

How beguiling were, and are, the analogies between sound and water waves on the one hand and light on the other. Of course, they are still used today, especially in undergraduate and high school teaching. All three exhibit diffraction and interference effects, for example. However, with sound and water the nature of the wave is obvious—*longitudinal* pressure waves in the case of sound, and an interfacial effect where the interface between two media of different density (e.g., water and air) is displaced vertically to produce a *transverse* wave in the case of water. In a longitudinal pressure wave, the displacement of atoms or molecules which leads to alternate compression and rarefaction of the medium is in the direction of travel. In a transverse wave the displacement is orthogonal to the direction of travel. But what was the nature of the wave in the case of light? This was established by James Clerk Maxwell between around 1855 and 1865 [41]. He derived a set of equations which described as a single phenomenon both the behavior observed outside wires carrying variable currents—which we now recognize as the emission of radio

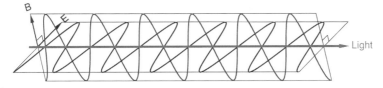

FIG. 5

The production of a plane polarized beam of light according to Maxwell. An oscillating electric field creates an oscillating magnetic field orthogonal to itself (or, equally, an oscillating magnetic field similarly creates an electric field). Light propagates in the direction of the arrow as a result.

waves—and the emission of light (Fig. 5). The wave behavior consisted of an oscillating electric field orthogonal to an oscillating magnetic field with the wave traveling in a direction orthogonal to both—i.e., a transverse electromagnetic wave. (It should perhaps be mentioned that if we create an oscillating electric field, for example, by moving electrons first one way, then the other along a wire (radio), or by forcing electronic transitions in atoms (light), an oscillating magnetic field orthogonal to the electric field is created as part of the same phenomenon. The two are inseparable, as is the coincident emission of electromagnetic radiation.)

At the end of the 19th century then, it seemed fairly certain that light was a wave whereas the electron, discovered by Joseph John Thomson in 1896 [60] was a particle. This simple picture unraveled quickly. In 1905, Albert Einstein published his work on the photoelectric effect [61]. He showed that classical wave theory could not explain the emission of electrons by a surface irradiated with light. On the contrary, the phenomenon was consistent with light arriving as discrete quanta (little packets of energy or the particle we now call a photon) each with an energy specific to the wavelength. So light was apparently a wave or a particle, depending on how you looked for it. Then, in 1927, George Paget Thomson (son of J. J. Thomson) and, independently, C. Davisson and L. H. Germer demonstrated the phenomenon of electron diffraction [62,63]. So the electron was apparently a particle or a wave, depending on how you looked for it. Indeed the same is true for all subatomic particles (at least for those which can be observed directly such as neutrons, protons, and so on) and this gave rise to the concept of *wave-particle duality*.

For both electrons and light, if your experiment looked for a wave, a wave you would see. If it searched for particles, then there they were. To see some resolution of this paradox, and for other reasons too, we need to look in more depth at Young's two slit experiment.

2.3 The two slit experiment

The two slit experiment, usually named for Thomas Young, is one of the most crucial 19th-century experiments underlying both optics and, indeed, the whole of modern physics. It gives an accurate starting point for understanding many of the techniques described in this book because it demonstrates diffraction and interference from the simplest multisource case—just two sources. It is therefore a prelude to understanding neutron, X-ray, and electron diffraction where the multiple source is an ensemble of many atoms.

On November 24, 1803, Thomas Young delivered a lecture entitled *Experiments and Calculations Relative to Physical Optics* [51] as part of the Bakerian lecture series at the Royal Society. Therein he replicated and explained Grimaldi's experiment, as we have already mentioned, using a wave description of light and linking the concepts of color and wavelength. Then, in 1807, he published a series of lectures that he had delivered at the Royal Institution, covering many aspects of physics, geography, and astronomy. It is here, in Lecture XXIII, *On the Theory of Hydraulics* [52]a, that we first find a description of the "two slit" experiment, described for waves on water. In Fig. 6A from Plate XX of his lectures he describes how circular ripples from two stones dropped simultaneously into a pool at A and B add together to form regions where overlapping wave crests next to overlapping wave troughs double the height of the waves, whereas where troughs and crests overlap, the wave cancels out. This phenomenon is known as *interference*. In the same lecture he also describes how parallel waves meeting a wall with a small aperture create concentric circular waves radiating outward on the other side (if the aperture is small enough)—see Fig. 6B. This is Grimaldi's *diffraction* but for water waves. If the wall has two apertures close together, a system of intersecting circular waves results, and interferes (almost) exactly as for the circular ripples. Later on, in Lecture XXXIX [52]b, *On the Nature of Light and Colors*, Young advances the wave theory of light and describes the optical version of his two slit experiment. Light from a single source (initially filtered so as to present one color) meets an opaque screen with two small parallel slits made close together. The slits

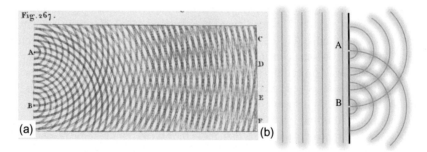

FIG. 6

(A) Thomas Young's diagram showing the interference of water waves originating as rings from points A and B [53]. (B) These rings might arise (for example) from a system of parallel sea waves striking a harbor wall containing two narrow apertures, A and B. These will act as secondary sources of circular waves. Where the wave crests intersect, the crest will be enhanced. Where the troughs intersect, the trough will be deepened. Where a crest meets a trough, the sea will be flat.

can be shown individually to act for light as the holes in the harbor wall do for water: They apparently behave as sources of cylindrical waves, circular in cross section as in Fig. 6B. Unlike the case for water waves, however, a screen is needed to scatter light from the intensity pattern in the space beyond the slits into the eye so that it can be seen. Fig. 7A shows Young's visualization of the resulting two slit diffraction pattern caused by interference due to different ray paths through the two slits.

Young is saying that Newton's belief that light consisted of a stream of particles is wrong. He is proposing, with proof, something as remarkable and revolutionary then as, say, the concepts of the Higgs boson or dark energy are now—that light is a wave. His experiment can be analyzed using rays as in Fig. 7B. Part of a wave crest originating in the left-hand slit and moving along path a to a point P on the screen a distance r from the plane of symmetry will add to a wave crest from the right-hand slit moving along path b if they arrive simultaneously. Following the crests, subsequent troughs will also arrive simultaneously and produce a "deeper" trough so the amplitude of the wave measured at P will be doubled and a bright band will be observed. Conversely, if a trough traveling along a arrives at P simultaneously with a crest moving along b (or vice versa), the two will cancel out and the recorded amplitude will be zero—a dark band will be seen. Whether one of these two extremes actually occurs at P, or whether some intermediate amplitude is observed, depends on the path difference (the difference in length) between paths a and b. This in turn depends on r. Using only Pythagoras' theorem we can write:

$$a = \sqrt{d^2 + \left(r - \frac{s}{2}\right)^2}, \quad b = \sqrt{d^2 + \left(r + \frac{s}{2}\right)^2}.$$

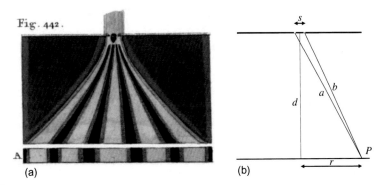

(a)   (b)

FIG. 7

(A) Young's fig. 442 from Plate XXX of his Royal Institution Lectures [53]. A parallel beam of light enters from the top through two parallel slits and impinges on a screen A in a series of alternating light and dark fringes—the two slit diffraction pattern. If the screen is first placed close to the slits and then moved gradually away, the fringes separate along curved paths as shown in the upper part of the figure. (B) A ray interpretation: A screen is placed a distance d from the slits and we consider what happens at a point a distance r from the center where arrive two rays a and b, one from each slit.

Then the path difference $\delta = b\text{-}a$ is given by:

$$\delta = d \left[\sqrt{1 + \left(\frac{r + \frac{s}{2}}{d} \right)^2} - \sqrt{1 + \left(\frac{r - \frac{s}{2}}{d} \right)^2} \right],$$

where we have taken a factor of d out of the square root expressions. The presence of the square roots explains why the separation of the bands follows a curved path with increasing distance to the screen d. However, in Young's diagram (Fig. 6A) we can see that the separation tends to become proportional to distance for large d. Indeed, if $d \gg r + s/2$, the expression for δ can be greatly simplified using the binomial approximation for the square root terms.

We find:

$$\delta \cong d \left[1 + \frac{1}{2} \left(\frac{r + \frac{s}{2}}{d} \right)^2 - 1 - \frac{1}{2} \left(\frac{r - \frac{s}{2}}{d} \right)^2 \right]$$

$$= \frac{rs}{d}$$

For disturbances from the two slits to produce a bright band centered on P, this path difference must be a whole number n of wavelengths λ (equally we can say that the disturbances from the two slits arrive in phase). Conversely, for a dark region, the path difference must be $n + 1/2$ wavelengths (the disturbances arrive in antiphase). So for a maximum intensity:

$$r = \frac{n \lambda d}{s}, \tag{2a}$$

and for a minimum:

$$r = \frac{\left(n + \frac{1}{2} \right) \lambda d}{s}. \tag{2b}$$

Using similar expressions, Young was able to calculate the wavelengths corresponding to different colors by measuring the positions of bright and dark bands in the diffraction pattern for their monochromatic light.

Independently, the same ideas, strongly rooted in the work of Huygens [47], had occurred to Augustin Fresnel (1788–1827). Professionally, Fresnel was a civil engineer whose optical research was initially done in his spare time. He was driven by a profound religious belief that it was his duty to reveal the miraculous aspects of nature created by God. In September 1815, at the age of 27, he had fallen foul of Napoleon's police and had been suspended from his work (see notes on p. 5 of [55]). With time on his hands, he started a correspondence with François Arago at the French Academy of Sciences to whom he expressed his first ideas. In reply, Arago sent Fresnel a reading list including Grimaldi, Newton, Jordan, Brougham, and Young. Fresnel's response, during was what to be his final 12 years of a life cut short by tuberculosis, was a remarkable and complete wave theory of optics with numerous novel devices to match. His work encompassed every optical phenomenon then known including polarization, where he invented the terminology we use, and Fresnel more than anyone else overturned the rather dysfunctional corpuscular theories of Newton (often using Newton's data amplified by his own). Fresnel's devices and his optical formalism underpin all of modern applied optics. Much of the optical equipment he designed is of great practical importance and still used today. One example is the Fresnel lens which he designed to intensify lighthouse beams from candles and oil lamps, today used in combination with LEDs [64].

Fresnel's 1822 version of Young's experiment used two slightly angled plane mirrors to reflect the light from a single source as if it came from two (identical) sources behind the mirrors [54]. It would have been able to produce a more intense diffraction pattern because it was able to use more light from the source.

We have already seen how electrons and other particles can have wave-like properties depending on the way they are observed. Young's experiment was replicated for electrons by Claus Jönsson in 1960 [65] and by Anton Zeilinger and coworkers using cold neutrons in 1988 [66] demonstrating in a simple way a wave nature in these particles, which had long been known from other diffraction and quantum phenomena. Demonstrations with atoms and even very large molecules have followed in recent years, showing the quantum nature of such particles.

However, the simple classical wave picture of diffraction phenomena in light beams (and by extension, other particle fluxes) had already started to unravel around the beginning of the 20th century. At that time, there was an evolving idea that the "indivisible units" or energy packets in a beam somehow acted together to produce wave phenomena and that if a beam were sufficiently attenuated, diffraction phenomena would be modified. Around 1900, the detection of individual particles was not possible because the necessary devices had not yet been invented. Nevertheless, in 1909, measurements by Taylor [67] showed that even at the limits of detectability, using photographic exposures of several months, diffraction phenomena remained unchanged—except of course in intensity. Clearly, the pattern did not depend on the particle flux.

In the period 1930–34, the first devices which were (in principle) capable of detecting single photons and single electrons were invented—the photomultiplier and electron multiplier [68,69]. By the beginning of the 21st century, imaging single particle detectors (i.e., pixelated detectors) had become common and are now ubiquitous in spectroscopy and diffraction measurements and can even record video from discrete particle arrivals. Young's experiment has been repeated many times at particle fluxes so low that there is probably only one particle traveling through the experiment at a time (in the case of light, using equipment which is accessible to high schools and undergraduate laboratories) [70–73]. Yet the diffraction pattern remains unchanged. So, in general:

- In any experiment producing a diffraction pattern (i.e., apparently demonstrating wave nature) the beam intensity can be reduced so that only one particle is in transit through the instrument at a time—e.g., to a few individual particles a second—but the diffraction pattern remains; it just takes longer to acquire. This is remarkable; the disturbance in any detector due to a single particle (photon, electron, neutron, etc.) has settled a billion times sooner than the next particle is likely due—there is no interaction between particles.

- Any experiment which attempts to localize the particle trajectory on its way from the source to the detector destroys the diffraction pattern. For example, you are not allowed to know which of the two slits in Young's experiment the particle went through. Indeed, the question has no meaning.

The phenomenon which underlies all spectroscopy, diffraction, and tomography measurements dealt with in this book, and, indeed, the whole science of imaging—*interference*—is somehow encoded into the fundamental properties of each particle in a beam. A single photon (or electron, or neutron, etc.) transiting the two slit paths has a high probability of arriving at the detector at the position of a diffraction maximum and a low probability of arriving in a minimum. This behavior is quite unlike that of sound or water waves. Somehow it seems as if the particle "knows where it should arrive." The explanation of this counterintuitive physical behavior came in 1948 and onwards from R. P. Feynman, J. Schwinger, and S. Tomonaga in the theory of quantum electrodynamics (QED) [18–20]. Here's what Feynman says about waves and particles in his public lectures on QED in 1983 [74]:

> *I want to emphasize that light comes in this form—particles. It is very important to know that light behaves like particles, especially for those of you who have gone to school, where you were probably told something about it behaving like waves. I'm telling you the way it* does *behave—like particles.*

Feynman's statement can be made about electrons, free protons and neutrons, many other subatomic particles, and larger entities such as atoms and molecules. A beam formed from any of these is a stream (or *flux*) of particles. But these are not the particles of popular imagining—individual chunks of something with a clearly defined position at some time or other. One photon (for example) can be regarded as going from the source to the point of detection by all possible routes *at once*. It does not even have to go in a straight line; it can meander back toward the source and so on. Each of the (infinite) possible paths has two key quantities: a probability that the photon will go that way and a transit time. The latter determines the difference between the photon's phase at the start and the phase it would have at the detector if it had taken that particular path. If some high-probability routes are closed (for example, in an attempt to determine the route "actually" taken), the diffraction pattern will be modified or destroyed. To determine the total probability that the photon will be detected somewhere, all the possible phase differences weighted by the path probabilities have to be added together (a mathematical process known as *path integration*). To see how this might be done and to get a general feel for other implications of QED in quite simple optics see Feynman's "QED, The Strange Theory of Light and Matter" [74].

Feynman describes why, in reflection, the angle of incidence equals the angle of reflection for a plane mirror: Photon paths far away from path ORI in Fig. 2 destructively interfere at the reflected image so that only paths arbitrarily close to ORI contribute to the intensity. Remember that this is true, even if only one photon per minute bounces off the mirror; photons *do not* produce any of these effects by interfering with each other. (In fact, photon-photon interactions are so incredibly rare that only a handful has ever been observed in the whole of modern physics.) It is the combination of possible paths for each individual photon which creates the interference. (The same is true for the other particles dealt with in this book when they interfere.) Suppose some of the paths are removed by, say, dividing the mirror up into reflecting and nonreflecting strips. Then, the angle of incidence no longer needs to equal the angle of reflection. If white light is incident on the structure, the constituent colors will be reflected at different angles to make a spectrum—the resulting device is a form of diffraction grating called a reflection grating and is common in optical spectroscopy.

In a strange historical parallel, Leonardo da Vinci (who drew sketches and made notes suggesting that he may have discovered interference of water waves [75,76]) described a similar experiment in around 1509—he imagined a mirror divided up into thin strips so that the "passage of individual rays" could be investigated [75]. It is not certain that he attempted the experiment. Had he done so, and, had the strips been sufficiently narrow, the reflecting properties of the mirror would have been completely changed in a way which might have advanced applied optics by several hundred years.

It is rare for a scientific model to invalidate previous models completely. Rather it encompasses them as limiting cases and takes human knowledge to new destinations. The remarkable thing about QED is that the equations which arise from the path integrals explain why photons, and all the other particles, can exhibit the observed wavelike behavior while actually being quantum particles. So, for example, electron, X-ray, and neutron diffraction, all covered in this book, are still almost exclusively described using the same wave theories as they were upon their discovery. However, some phenomena, such as the scattering of X-rays from very tightly bound electrons, need a fuller QED approach [77].

QED itself became the scaffolding for the modern theory of the atomic nucleus and the interactions of quarks—quantum chromodynamics (QCD) [74]. The physical implications of Grimaldi's *diffringo* and the Young-Fresnel two source experiment are still being explored at the forefront of optics, directly feeding into quantum computing, novel optoelectronic devices, and, without doubt, future analytical instrumentation. The history of optics, and with it those of spectroscopy, diffraction, and tomography, is far from over.

3 The eye as a spectroscope

A rainbow is a familiar example of a spectrum, although it is a complicated one. The colors and structure differ considerably from optical spectra created in the laboratory even from a narrow pencil of the Sun's rays defined by a hole or slit and passing through a prism. For example, double reflection within raindrops can produce a second bow concentric with the first and interference effects due to the size of the water droplets can create so-called supernumerary bows. Examples of these phenomena are shown in Fig. 8.

Light from the Sun enters individual raindrops at a critical distance from the observer, so that the angle between the rays from the Sun entering the drop and the rays leaving it to enter the eye is about 42° [78]. Such conditions exist along an arc with the Sun behind the observer as first described by Aristotle (see, for example Ref. [36] p. 251 et seq.] (although his explanation of the colors was quite wrong). The basic explanation of the arc and the colors came in the 14th century from Kamāl al-Dīn al-Fārisī and Theodoric of Freiberg [22]: For a primary bow, at each drop, light is first refracted at the air-water interface, then reflected from the back of the drop, and then refracted again on its passage back into the air and thence into the eye. The secondary (fainter) bow seen outside the primary bow is created when two reflections occur in each drop. The mechanism behind supernumerary bows (Fig. 8b) is still under discussion today [79]. The different colors present in the Sun's approximately white light refract at slightly different angles and so come from raindrops at different altitudes and are separated in the image formed on the observer's retina. The rainbow, like other beauty, exists in the eye of the beholder. The distribution of wavelengths across the bow is continuous, but human vision groups them into bands.

Human color vision originates in specialized cells in the retina of the eye known as cones [80]. There are three types of cone each sensitive to a different but

FIG. 8

On the left is a typical double bow photographed against typical Cornish rain clouds. It's morning and the Sun is low in the east, so the bow is high in the west. Note the reversal of the colors in the secondary bow. The space between the bows is perceptibly darker as compared to the sky low on the left and right (Alexander's dark band). On the right is a primary bow with supernumerary bows beneath. Three *pinkish-violet* bands are discernible. The rain in this case is extremely fine—almost mist. (Unenhanced photos, MGD.)

overlapping range of colors (bandwidth). Because normal human color vision depends on three color ranges or bands, it is called trichromatic [81]. The relative strength with which light of a particular color (i.e., wavelength) stimulates cone cells sensitive to different bandwidths is processed by the brain's visual cortex into perceived color. So, even though there are only three color ranges, the eye can discern around 10 million colors. (It is believed that a small percentage of humans (possibly all of them female) has tetrachromatic vision and could, in principle, discern around 100 million colors [82].) However, even in the eye of a single observer, the relationship between the actual color of an object (i.e., the wavelength or mixture thereof) and the perceived color can be contextual. For example, perceived color can be modified or even created by the surrounding colors (Fig. 9).

This is quite different from the result in an optical spectroscope receiving the same photons where different colors will be dispersed across the detector and distinguished according to the resolution of the device, not their surroundings. Many spectroscopic detectors can count single photons which the eye cannot do. At its limit, human vision requires around 10 photons arriving within a few tens of milliseconds to stimulate a receptor.

Nevertheless, the naked eye forms a starting point for the application of spectroscopy in heritage science. For example, the "Blue Wool" standard of the International Organization for Standardization was introduced in conservation in the 1950s and has been extensively studied in that context since [4]. A standard photosensitive set of color swatches is exposed to the ambient conditions adjacent to a displayed textile or painting with (say) the left-hand half of the swatches covered. After some time, visual comparison of the exposed and unexposed halves gives a qualitative measure of the degree of fading and an indication of the suitability or otherwise of the display environment, not necessarily limited to the intensity and spectral distribution of the lighting. A more modern, quantitative, and sensitive development of this concept is the LightCheck range of sensors [3] which use a light-sensitive emulsion whose color changes progressively on exposure. The color of the sensor after some degree of exposure to the same ambient as an artifact can be compared visually

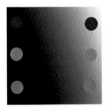

FIG. 9

A simple example of contextual color. The *blue dots* are identical as are the *pink dots* although, to many people, the dots on the right will appear darker. The *gray dots* are also identical and a single shade of *gray*, although the one on the right may appear to be graded in shade.

to a calibrated color chart to estimate the exposure to light. This direct use of the spectroscopy built into the human vision system is simple to use and far less expensive than photometric measurements with instruments.

4 Radiation beams

The techniques described in this book all depend on the interaction of a beam of radiation with a sample. These beams are streams of quantum particles, although describing them as if they were waves is convenient under some circumstances (it can make a mathematical analysis of what is going on much easier). In the context of this book, we need to describe just two different types of beam:

- Electromagnetic beams—we have already referred to these generically as "light" or photons in this introduction. For our purposes light runs from γ-rays through X-rays, ultraviolet, visible, infrared, to terahertz radiation in order of increasing wavelength. In their original meanings, the terms γ-ray and X-ray referred not to different regions in the electromagnetic spectrum, but to the source of the light. The atomic nucleus was the source of γ-rays, whereas X-rays were produced from the atomic electron cloud. However, short wavelength light is also produced by the violent acceleration or deceleration of charged particles (bremsstrahlung) and the annihilation of matter with antimatter (producing γ-energies). Rather than invent a lot of different rays depending on the source, the assignment of γ- or X- now comes more or less down to energy range but with some overlap.
- Matter beams—these are beams of electrons, protons or heavier ions, or neutrons. Unlike light, these transport mass and sometimes electrical charge from their source to the sample.

The two kinds of beam exhibit a range of very similar interactions with matter, and also a range of different behaviors, any of which might be the basis for an analytical technique. For example, they can be coherently scattered and observed as a diffraction pattern, or they can be absorbed in the sample and excite the emission of other types of radiation. Heavy, high-energy matter beams are also scattered more mechanically (analogously to billiard balls) loosing characteristic amount of energy before they leave the material. Different kinds of interaction will take place at the same time, although the results from any particular one will only be observed if the right detectors are in place. The various interactions give different information on sample composition and structure and this is often complementary. However, an experiment to do (say) diffraction will usually ignore the absorption which is occurring, and vice versa. Or, it may be that the experimental conditions (especially the wavelength associated with the radiation) favor one kind of interaction over the others. Overall, both kinds of beam exhibit both wave-like and particle-like properties, and the detection of either may be the basis of an analytical technique.

4.1 **Beam energy and momentum**

If we are using radiation beams to probe heritage materials (or any other materials for that matter), how do we start to understand what the likely interactions will be and, most importantly, whether the beam will damage the materials? Clearly, we need to know something about both the beam and material properties.

Starting with the beam, important beam parameters to consider are the energy (usually given as if it were the kinetic energy of a single particle in the beam), the power (total kinetic energy delivered per second), and the densities of these—energy and power delivered per unit area or absorbed per unit volume. The beam energy or beam power absorbed per unit volume of sample material become especially important when discussing radiation damage caused by the beam. We will discuss whether energy density or power density is important in the section on sample damage.

The SI unit of energy is the joule (J) and this is how energy is measured on the macroscopic scale. For example, the energy required to boil the water for a cup of coffee is about 90 kJ. The energy expended in picking up the cup to take a sip is around 2.5 J (accounting only for the work done on the weight of the cup and contents). The joule is an inconveniently large quantity for describing the quantum phenomena underlying our techniques and so the electron volt (eV) is more often used. One reason for this is both simple and practical. Any singly charged subatomic particle such as the electron or the proton will acquire a kinetic energy of 1 eV when falling through the electric field created by a potential difference of 1 V in a vacuum. The same is true of any singly charged ion. Therefore, the energy of a charged particle is easily related to the amount of acceleration it receives from the technology producing the beam. For example, in an electron microscope using a 20,000 V (20 kV) power supply for acceleration, the electrons in the beam will each have an energy of 20 keV. One joule is equivalent to approximately 6.25×10^{18} electron volts (i.e., 1 J = a little over 6 million million million eV).

Declaring the beam energy in eV is useful from other perspectives: The energies of atomic bonds between atoms (i.e., the energy required to break the bond) lie typically in the range 1.5–12 eV so a radiation beam with an energy greater than this has the potential to break bonds *if* it interacts in a particular way. *If* that happens the sample chemistry will be altered *unless* the bond reforms after breaking. Likewise, the displacement energy of an atom in a solid (the energy required to move an atom completely off its natural site) is rather larger, typically 20–50 eV, so a radiation beam with an energy greater than this has the potential to relocate atoms in a solid *if* the kinetic energy in the beam can be coupled into atomic motion. *If* that happens the sample chemistry and structure will be changed. This is potentially the case where the incoming particle mass is of the same order as the mass of the atoms. So, protons and other ions are more likely to cause atomic displacements than electrons or photons (at least at the energies normally encountered in our techniques, and unless the electron or photon beam deposits so much power density that it heats the sample). The ionization energies of electrons inside atoms (the energy it takes to remove the electron from the atom) vary from something below 10 eV to many keV depending on the atomic number of the atom (the number of electrons it has) and which of its electrons we mean.

For example, to split the valence electron from the proton in a (free) hydrogen atom takes a minimum of 13.6 eV, for the element cesium it takes 3.89 eV (the lowest value in the periodic table), whereas to eject the most tightly bound electrons in a uranium atom takes over 115 keV. Across the stable elements in the periodic table, a range of binding energies between these values is found. So, a radiation beam with an energy greater than that of particular atomic binding energy can ionize that level (and those with lower binding energies). Such ionization will typically decay in a time somewhere between a few tens of nanoseconds for a binding energy in the eV range to a femtosecond or less in the keV range, resulting in the release of electrons or photons (the detection of which might form the basis of our analytical technique) and coincidental bond breaking—not just between the deionizing atom and its neighbors, but further afield if the radiation emitted is above some threshold.

For particles which have no charge and therefore cannot be accelerated by an electric field, one can distinguish between massless particles such as photons whose energy can be calculated from their wavelength or frequency and massive particles such as neutrons where the (kinetic) energy can be calculated from classical or relativistic formulae depending on their mass and velocity or from their wavelength. In fact, the energy of any particle beam is just another way of giving its wavelength.

Particles can also be characterized in terms of their momentum p which, for a particle of matter, is given by:

$$p = mv, \tag{3}$$

where m is the particle mass and v is its speed. For particles like electrons in an electron microscope and the protons and other particles used in techniques such as RBS and PIXE, the value of m needs to be increased above the particle mass when at rest m_0 to account for relativistic effects:

$$m = \frac{m_0}{\sqrt{1 - \frac{v^2}{c^2}}}, \tag{4}$$

where c is the speed of light in vacuo ($\cong 3 \times 10^8$ ms^{-1}). (Actually 2.9979×10^8 ms^{-1} but 3×10^8 ms^{-1} is accurate enough for most purposes.) For example, for $v = 3 \times 10^7$ ms^{-1}, m is 0.5% larger than m_0 and rises steeply as v increases above this value. Electrons achieve this speed at 2.6 keV and protons or neutrons at 4.7 MeV (because they are nearly 2000 times heavier).

4.2 Wavelength and frequency

The wavelength λ of a system of waves is the distance between the wave crests (or troughs or any equivalent distance; see Fig. 10). The frequency f or sometimes ν (Greek nu) is the number of whole waves per second measured at a point fixed relative to the observer. These two quantities are simply related by the velocity c_w at which the wave is traveling:

$$f = \frac{c_w}{\lambda} \tag{5}$$

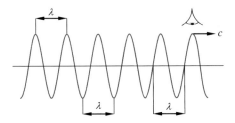

FIG. 10

Equivalent representations of wavelength λ. If the wave propagates at speed c then c/λ waves pass the observer each second, defining the frequency ν.

(Some readers may be familiar with the concept of a standing wave such as occurs on a vibrating string with fixed ends, or in certain kinds of instrumentation such as an interferometer. The wave crests are stationary. However, such a standing wave is formed by the interference between two identical traveling waves moving in opposite directions, and it is the speed of those which is relevant in that case.)

Wavelength and frequency are therefore just reciprocal ways of measuring the same underlying property of a system, and, as the wavelength goes up, the frequency goes down and vice versa. How does this relate to the concept of a beam of quantum particles? Wavelength is also a property of a moving particle as we now explain.

In 1923, Arthur Holly Compton discovered that X-rays could loose a fraction of their energy to electrons in a solid, resulting in an increase in the X-ray wavelength—a phenomenon now known as Compton scattering [83]. His measurements demonstrated for the first time that photons have the property of *momentum* and that it is related to their wavelength through Planck's constant h:

$$p\lambda = h. \tag{6}$$

The photon has no mass so that the fact that it had momentum was (and still is) astonishing, but very useful. Apart from many scientific uses, practical applications include light sails for spacecraft [84] and highly efficient and compact laser-cooled refrigeration systems [85] which might, one day, find heritage applications.

The following year, Louis de Broglie [86] suggested that this relationship should apply to both matter and light beams, with the result that a wavelength could be associated just as well with a beam of electrons, protons, neutrons, or other material particles as it could with light. This is no random association; if you were to do Young's two slit experiment for electrons, and calculate the electron wavelength from the diffraction pattern (e.g., by rearranging Eq. 2a), it is the de Broglie wavelength that you would get. So, for both light and matter beams, the wavelength is inversely proportional to the particle momentum.

For a light beam the wavelength λ is also related to the photon energy E by:

$$\lambda = \frac{hc}{E}, \tag{7}$$

where c is the speed of light. To get λ in nanometers when E is in eV, we can write:

$$\lambda/\text{nm} = \frac{1.2398 \times 10^3}{E}. \tag{8}$$

For a matter beam the corresponding expression is:

$$\lambda = \frac{hc^2}{E_k v}\left(1 - \sqrt{1 - \frac{v^2}{c^2}}\right), \tag{9}$$

where E_k is the kinetic energy of the particle. This expression is more complicated because it has to contain a correction for relativistic effects. For speeds below 10% c it reduces to:

$$\lambda \cong \frac{h}{\sqrt{2mE_k}}. \tag{10}$$

Again, to get λ in nm when E_k is in eV, the appropriate constants are:

$$\lambda/\text{nm} = \frac{1.1705 \times 10^{-15}}{\sqrt{mE_k}}. \tag{11}$$

Eq. (5) can be used with the appropriate value of c_w (e.g., c in the case of light) to get the frequency from the wavelength. So frequency and wavelength are directly related to the underlying particle properties of momentum and/or kinetic energy for all particles.

4.3 Particles in a beam

The techniques described in this book all use an incident particle beam to obtain the information from the sample. The beam may be focused (converging to a small spot) as in a scanning electron microscope (SEM), quasiparallel as from some neutron sources, or diverging as in many kinds of laboratory X-ray diffractometer. In the most familiar example, the beam might be white light, or it might be a monochromatic beam of any one of electrons, helium ions, neutrons, photons, protons, or other particles. White light is a mixture of all the visible energies, so the terminology *white beam* is often used to describe any kind of radiation beam that contains a wide continuum. Although monochromatic means "only one color" (more generally "only one energy"), it is impossible to produce such a beam in reality for both practical and fundamental reasons. All particle beams have a finite *bandwidth*, usually defined by the physical processes used to produce the beam and very often associated with the temperature of the source. In Fig. 11, we show a Lorentzian line shape which approximates that of many monochromatic light sources. The full width at half maximum (FWHM) is used to characterize the bandwidth $\Delta\lambda$. Depending on the beam content and its source, there are many other line shape possibilities too, and a complication is that different line shapes and techniques use different definitions for the bandwidth. Where the quality of the output data improves as the bandwidth of the beam decreases (which is often the case), the complexity, physical size, and cost of a corresponding analytical instrument will generally increase.

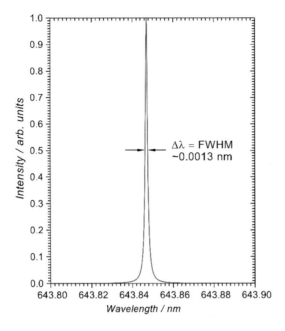

FIG. 11

One monochromatic line at 643.8 nm from the spectrum of a spectral calibration source. The line has a bandwidth $\Delta\lambda$ of 0.0013 nm defined as the full width at half maximum (FWHM) of its Lorentzian profile.

Monochromatic beams typically have a ratio of the energy spread ΔE to the energy E in the range $10^{-2} > \Delta E/E > 10^{-6}$ depending on the nature of the beam, although some laser linewidths can be hundreds of times narrower still.

4.4 More beam parameters

As we described in previous sections, the energy of a beam is usually given as the energy of a single particle in the beam, and the particle wavelength, frequency, and momentum can all be calculated from the energy. An often-used alternative to frequency as defined above is the angular frequency ω (see Fig. 12) where:

$$\omega = 2\pi f, \tag{12}$$

and π is the familiar constant 3.14159…

Another parameter which is often used in infrared, Raman, and other spectroscopies is the wavenumber κ where:

$$\kappa = \frac{1}{\lambda}. \tag{13}$$

Whereas f is the number of wavelengths per second, κ is the number per meter.

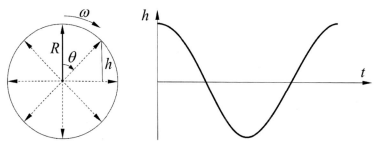

FIG. 12

Imagine an arrow of length R rotating at a constant angular velocity ω (radians per second) so that its tip moves in a circle. After a time t, the angle of rotation θ is given by $\theta = \omega t$ so that if the length $h = R\cos\omega t$ is plotted for one complete rotation (2π radians), h traces out one cycle of the (co)sinusoidal wave as shown. If there are (say) 10 rotations per second then $f = 10$ Hz and $\omega = 2\pi f$.

There are many other physical parameters used to characterize beams and their geometry. Many of these are more important to instrument designers than to users. Unfortunately, due to historical usage and a lack of early or even consistent standardization, the terminology in common use differs from one analytical technique to another even for the same quantity. So, when an analyst or instrument expert quotes a parameter to a client, it is vital that the client makes sure exactly what their informant means by the terminology used. This becomes particularly important for parameters that are essential to understanding the data, those which might need to be reproduced to get comparable data from different analytical runs, or to ensure that an analysis is made within the nondestructive envelope of the technique. We list some of these quantities here, where possible using definitions for the IUPAC gold book [87] or ISO sources [88] but mentioning other meanings encountered in common usage. Where we describe multiple definitions we will use Roman numbering, thus (i), (ii), (iii)… If, later on in this chapter, it becomes necessary to use a particular definition we will indicate which possibility we mean like this, dose(ii):

- *Incident angle*: The angle at which the beam hits the surface. This quantity needs some care because some techniques measure the angle of incidence between the beam and the sample surface whereas others quote the angle between the beam and the surface normal. This makes a big difference to the calculation of quantities such as those given in Eqs. (14) and (16) below. These are given for angle of incidence measured between the beam and the surface. If the other definition is used the cosine of the angle must be used, not the sine.
- *Scan, raster*: In some instruments (e.g., the SEM) a small beam is scanned over a larger analyzed area which for a normally incident beam will usually be a square. For a charged particle beam, this is done using electric or magnetic fields to deflect the beam. For light, it can be done using mirrors or by moving the

sample instead. In modern instruments the scan or raster is done in discrete steps, which results in the beam dwelling on a spot or pixel on the sample for a tightly controlled time. Data can be collected from each pixel and related to the position on the sample to form an image.

- *Microprobe image*: An image collected by a scanned microbeam as described above.
- *Microscope or full field image*: An image collected as in a simple optical microscope by the eye—the whole field is illuminated and the detector collects the whole image in parallel.
- *Footprint*: The shape filled by the beam on the sample surface, usually circular, elliptical, square, rectangular, or trapezoidal (Fig. 13). If the beam is normally incident and not scanned, the footprint is the same as the cross section of the beam. Otherwise the area of the footprint A will be given by:

$$A = \frac{A_{Beam}}{\sin \varphi}, \tag{14}$$

if the beam is not scanned, where A_{Beam} is the cross-sectional area of the beam and φ is the angle of incidence measured between the beam and the surface. If the beam is scanned, the area of the footprint depends on the geometry of the scan. In some instruments the footprint may be larger than the sample, especially at small angles of incidence, in which case it is important to mount the sample on a surface whose spectrum or diffraction pattern does not interfere with the data from the sample.

- *Current*: For a charged particle beam (e.g., electrons, protons) the amount of charge transported per second by the beam. Typically measured in μA or nA for analytical beams.

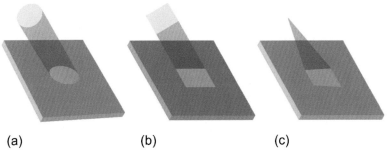

(a)   (b)   (c)

FIG. 13

Examples of beam footprint on a sample surface at nonnormal incidence: (A) A beam of circular cross section has an elliptical footprint which will become very eccentric at small angles to the surface. For example, at 1° to the surface a 1 mm diameter beam will form an ellipse 1 mm × 60 mm. (B) Similarly, a square section beam has a rectangular footprint unless it is (C) diverging or converging, in which case the footprint is trapezoidal. A small beam scanned in a square raster but not normally incident also has a trapezoidal footprint and, consequently, the dose or image magnification will vary across the scanned area.

- *Flux*: (i) The number of particles per unit area per second flowing through a plane orthogonal to the beam. (ii) The energy flow per unit area per second through a plane orthogonal to the beam. May be given in particles $s^{-1}m^{-2}$, $Js^{-1}m^{-2}$, Wm^{-2}, but other units of energy and length (e.g., eV, μm, μW) may be substituted. (iii) The total number of particles per second in the beam in which case the units will be particles s^{-1}, Js^{-1}, W, etc. For example, taking this definition, the flux F of singly charged particles in a $1\,\mu A$ beam is:

$$F = \frac{I}{q_e} = \frac{10^{-6}}{1.602...\times 10^{-19}} \cong 6.24 \times 10^{12} \text{ particles s}^{-1}, \tag{15}$$

where q_e is the elementary charge. Flux also has several other meanings.
- *Intensity*: (i) Loosely means the same thing as flux[(i) (ii)] but may, more specifically, refer to flux[(ii)] above. (ii) May also refer to the height of a peak in a spectrum or diffraction pattern.
- *Range*: Some estimator of the distance a beam particle travels in the sample [89], often measured from the surface along its original direction of travel, but may also refer to the distance measured normal to the surface (this quantity can also be known as the projected range). Not all particles stopping in a material travel the same distance, so this will be a mean value calculated according to a set of rules relevant to the type of particle. If the range R is measured along the original direction of travel, then the projected range R_p is given by:

$$R_p = R \sin\varphi. \tag{16}$$

- *Attenuation length*: This is a range-like parameter especially relevant to particles like X-rays [90]. It is the distance into the sample, measured normal to the sample surface, over which the beam flux decreases by a factor $1/e$ where e is the base of natural logarithms and equals 2.71828... It is a transcendental number like π. It arises naturally from an absorption law of the form:

$$I = I_0 e^{-\frac{z}{\Lambda}}, \tag{17}$$

where I is the flux or another intensity parameter at depth z, I_0 is the same parameter measured at the surface, and Λ is the attenuation length.
- *Interaction volume*: (i) The mean volume of sample which interacts with a single particle in the beam. Depending on the particle and its energy, this may be anything from a teardrop shape extending in from the surface where incident particles experience subsurface scattering in the material to a small cylindrical volume surrounding the particle trajectory if they pass straight through. (ii) The total volume over which the beam interacts with the sample. If R_p or Λ is known (substitute for L in Eq. 17) then we use a rule of thumb to get an interaction volume V_{int} for the purpose of calculating power density:

$$V_{int} \approx 2LA, \tag{18}$$

where A is the area of the footprint.

- *Dose*: (i) Total number of beam particles per unit sample volume or (ii) total beam energy input per unit sample volume in a measurement or (iii) the total energy absorbed per unit mass. (iv) Obsolete term for areic dose (see below). Units may be (i) particles m^{-3}, (ii) $J m^{-3}$, or (iii) Gy ($J kg^{-1}$). Other units of energy and length might be used. Dose defined as in (i–iii) will vary with depth in a manner which is dependent of the type of beam, and across the irradiated area if the flux$^{(i)}$ varies across the beam. For example, if the beam is absorbed according to a law like Eq. (17) then the dose will be a maximum at the surface and decay exponentially with depth. If there is a projected or most probable range for the particles, then the dose as particles per unit volume will initially increase with depth, then decrease when $z > R_p$. Especially where a large fraction of the incident particles are scattered out of the sample, or pass through it, a more relevant quantity when estimating the likelihood of any damage will be retained dose.
- *Retained dose*: The dose according to definitions (i–iii) actually retained in the sample assuming some of the beam is scattered back out or passes through.
- *Areic dose*: Usually the total number of particles per unit sample surface area delivered during a measurement. Typically measured in particles cm^{-2}.
- *Fluence*: For beams which are parallel, or nearly so, the fluence is synonymous with the areic dose.
- *Power density*: This is an important quantity when studying damage in materials and making recommendations for safe limits. If the dose$^{(i)(ii)}$ is delivered in a short time, then the power density will be higher than if the same dose is delivered more slowly. This becomes especially important if the interaction between the beam and the sample produces heat. A low power input to the analytical volume will allow heat to dissipate harmlessly, whereas a high power input may not. Beam heating effects can be a problem when using high-flux microfocus beams, especially of electrons or X-rays. Units are Wm^{-3} with cm, mm, and μm often used in place of m.
- *Beam profile, current density distribution, beam shape*: The beam flux$^{(i)(ii)}$ usually varies across the beam, being highest in the center and falling off at the edges. Beam profile etc. are terms which describe the quantitative behavior of the flux$^{(i)(ii)}$ or current distribution as a function of distance from the beam axis. Beams are often assumed to be Gaussian (Fig. 14A) but exactly what functional form they follow depends entirely on the beam source characteristics and the subsequent beam optics. Fig. 14B shows a flat-topped or trapezoidal beam (not to be confused with a trapezoidal footprint). Such a beam produces much more uniform illumination of the sample than a quasi-Gaussian beam and can be produced by illuminating an aperture with just the central region of a broad beam and imaging the aperture onto the sample. Even if a beam (or the sample) is scanned it is important to know something about the beam shape, because scanning of either the sample or the beam will be done digitally. The beam will dwell on a pixel for a controlled period of time before moving either rapidly or (hopefully) in a blanked state to the next. If the pixel spacing $d \leq \sigma$ where σ is the standard deviation of the Gaussian, then the beam dose is distributed very

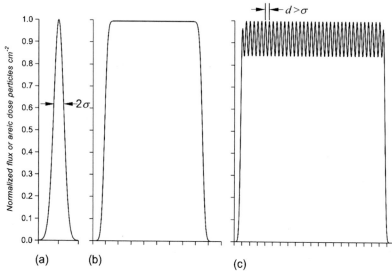

FIG. 14

Examples of beam profiles/shapes: (A) Gaussian or quasi-Gaussian profile characterized by a width parameter such as the standard deviation σ ($\sigma \approx$ half the beam diameter at 0.61 of the maximum intensity). (B) A flat-topped or trapezoidal beam profile, or the areic dose on the sample produced by scanning a beam like (A) with pixel spacing $d \le \sigma$. (C) The effect of scanning a beam in a digital raster with $d > \sigma$ is to produce regions of high dose[i–iii] in the sample and consequently higher power/energy density. In the example $d = 2.5\sigma$.

uniformly as again shown in Fig. 14B. Fig. 14C shows the effect of $d > \sigma$ on the flux in the scanned area or dose on the sample. The maxima are regions of high energy or power input which might have implications for sample damage.

4.5 Information depth

Most of the techniques described in this book give information typical of the top few micrometers of the sample. However, if that is insufficient, a proton beam of more than 60 MeV will provide an analysis over a depth of several millimeters and may even do so under conditions where the protons pass completely through the sample, leaving very little trace of their passing (Chapter 11). Thermal or cold neutrons are even more penetrating as are X-rays in the 100 keV energy range. See Chapters 5 and 8 for more information. Alternatively, if the sample may be suitably prepared, the analysis of a cross section normal to the surface can give an indefinitely large depth range.

The information depth is partly determined by either the range of the beam in the material or the range of the escaping particles used by the technique. The escape depth is the distance measured normal to the surface over which all but a fraction 1/e of the secondary or scattered radiation which is sought by the technique will escape. The depth of penetration (defined for light, but loosely applicable to other types of beam) is the depth over which the incident beam intensity decreases to 1/e of its value at the surface (see also attenuation length in Section 4.4). The smaller of these two parameters will strongly influence, but not necessarily determine, the information depth of the technique. Except for the case of coherent scattering (as for electrons, X-rays, or neutrons observed in diffraction) the energies and types of the incoming and escaping particles will be different and so, therefore, will the penetration and escape depths. So when, for example, electrons (less penetrating) liberate X-rays (more penetrating) which can also release further X-rays as they move through the material, the information depth will exceed the electron range [91].

These characteristic depths depend on the energy of the particles involved and also the strengths of their interactions with the target material. Visible light has a hugely varying range in materials (from microns to meters depending on transparency), as do the ultraviolet and infrared ends of the spectrum (Chapters 2, 4, and 7). An electron moving in matter plows through a fog of other electrons, both free electrons in a conductor and electrons bound to atoms. It interacts very strongly with both (Chapter 3). An X-ray (Chapters 5, 6, and 9) also interacts primarily with the electrons, but far less strongly than an incident or escaping electron. In most analytical instrumentation operating at energies <20 keV, penetration and escape considerations limit an analysis using electrons or X-rays or both to within a few microns of the surface. A neutron, on the other hand, interacts predominantly with the atomic nuclei in the material. Since the nuclei occupy <1/1,000,000,000,000th of the total sample volume, matter looks like mostly empty space to neutrons, so they are very penetrating indeed and meV (milli electron volt) neutrons will penetrate several centimeters of material (Chapters 5 and 8). Protons, helium ions, and alpha particles interact with electrons because of their charge, but at the typical energies at which they are used have far too much momentum to be more than a little slowed down by the electron "fog." They also interact with the atomic nuclei, because of both coulomb repulsion and nuclear reactions. The range of MeV protons will be tens of microns, typically, but their information depth from elastic scattering or X-ray excitation is somewhat less (Chapter 10). Higher-energy protons are not only more penetrating but can excite characteristic γ-rays from the atomic nuclei in the material, providing spectroscopy deeper into the material than with X-rays (Chapter 11).

5 Destructive, nondestructive, invasive, and noninvasive techniques

5.1 Destructive and nondestructive

While there exist some specialized adaptations of analytical techniques for heritage applications, most analyses will be undertaken using standard instrumentation. In either case, the underlying method will have been developed out of a set of tools for investigating fundamental aspects of physics. Even before its adoption in heritage science, a technique's range of applications will have been vast, and usually in areas where damage to the sample was inconsequential so long as the data were accurate. So, it is unfortunate that the term "nondestructive" ever entered the lexicon of heritage analytical science, especially when applied to radiation-in, radiation-out analytical techniques.

In its original meaning, at a time in the 1960s–80s when techniques, applications, and instruments were evolving very rapidly, "nondestructive" as applied to techniques using energetic radiation (i.e., particles with energies above the bond strength) meant that the information carrier, whatever it was, had left the sample before, or unaffected by, any changes that the bombardment wrought in the sample (see Fig. 15).

The badge "nondestructive" became attached to many techniques and was written into many text books and instrument manuals and was later taken at face value—"does no damage," even by some of a technique's practitioners. This was despite the fact that it is well-known that bombardment with radiation with energies above a few eV can break bonds, initiate chemical reactions, or even cause foreign atoms to accumulate in the sample if ions are used as a probe. Even the very low-energy neutrons used in neutron diffraction and tomography can, in unfavorable circumstances, be absorbed if the sample contains certain atomic nuclei and make a sample radioactive enough to require safe storage for a while (Chapters 5 and 8). Intense or repeated irradiation with light, especially at the blue to UV end of the spectrum, can cause pigments to fade. Indeed, deliberately induced fading of an area too small to see with the naked eye, so-called *microfading*, has become a textile conservation tool [92]. A further danger is that many kinds of input radiation—electrons, X-rays, and infrared light among them—can cause localized sample heating if the power density (watts per unit area) is too high. All of these pitfalls will be avoided by a skilled analyst who will, if necessary, conduct tests on similar materials to determine safe levels of operation.

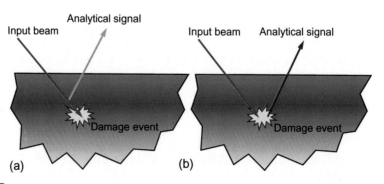

FIG. 15

Original definitions of destructive and nondestructive for a radiation-in, radiation-out technique. (A) Analytical signal leaves the sample before or spatially decoupled from any damage event—technique was badged as nondestructive. (B) Analytical signal is caused by the damage event or originates from a region where damage has accumulated—technique was badged as destructive.

Arguments over terminology also mask an important difference between destructive and nondestructive techniques as defined above: In a nondestructive technique, damage will likely accumulate in the sample as the input radiation dose increases so that the analytical precision has its upper limit defined by the maximum dose which can be allowed before the damage starts to affect the results or becomes detectable in the sample, whichever is the lower. In the case of an acknowledged destructive technique (like mass spectrometry, for example), the analytical precision is limited by the amount of sample which can be consumed per datum. As we describe in more detail below, the latter circumstances often result in a much more sensitive analysis, which makes *microdestructive* techniques a possibility (Chapter 7).

Of course, it is very often true that a nondestructive technique does no detectable damage to the sample, and is nondestructive in the way that curatorial staff would hope. This is because the input radiation energy, dose, and/or input power density are below some threshold. However, it is important to understand that a threshold exists, so that a technique which is nondamaging when applied using standard laboratory instrumentation might become damaging when much higher radiation doses and power densities become available, for example, in a focused microbeam system or at an infrastructural facility such as a synchrotron. Moreover, some samples may absorb the incident radiation more efficiently than others and consequently be more susceptible to damage. If there is any doubt as to whether a technique is sufficiently nondestructive in the context in which it is to be used then it should be tested on noncritical samples first. Such samples may include artifact fragments, material from badly damaged or very common artifacts, and simulants. In the case of very valuable and rare artifacts, especially paintings and textiles, it may be necessary to wait a few months after irradiation of a substitute at the analytical dose to make sure that no long-term reactions are initiated by the radiation. In the last 20 years, the potential for damage and the cost–benefit consequences have started to be well-documented for heritage and other applications [1,93,94].

It is also true that many analyses of heritage materials are carried out on fragments of material in any case—paint flakes, dust, chips, and splinters, for example. In such cases, the need to get certain information from the sample (for example, to inform the conservation of a complete but degrading artifact) may override any consideration of sample damage. It is rare that sufficient information can be gathered using one technique alone, so the order in which techniques are used needs to be carefully thought out. The least destructive techniques should be used first.

5.2 Microdestructive techniques

A microdestructive technique consumes or damages (or both if damage is left behind) a microvolume of the sample where a microvolume is no more than a few tens of μm^3 ($1\,\mu m^3 = 1\,fL$). As a rule, destructive analytical techniques have a very high yield (i.e., amount of signal emitted per input beam particle) and may be thousands of times more sensitive than nondestructive methods. This is especially true of techniques based on mass spectrometry (which is inherently background free and where just a few ions need to be counted to give a detectable signal at a given mass) and laser ablation techniques (Chapter 7). So, if a microvolume is removed from the material by laser ablation or sputtering due to ion bombardment, a sensitive analysis can be obtained. However, this will be a very localized analysis which may need to be repeated in different parts of the sample to obtain an adequate measure of the sample composition. It is also worth remembering that electron microscopes and other micro- or nano-beam instruments can handle, analyze, and image very small samples indeed. Dust, paint flakes, splinters, chips, debris from restoration, and so on may be analyzed to reveal a wealth of information on technology, provenance, and use as well as deterioration.

5.3 Noninvasive analysis

Most analytical instruments, especially those for laboratory use, have limitations on the size and shape of the sample which they can accommodate. Moreover, there may be a requirement to put the sample into a harsh environment such as a vacuum, which will (at least) result in loss of water vapor from the surface and potential damage from dehydration. To use such an instrument a piece must be removed from any artifact too large to fit. In a noninvasive analysis the instrument can accommodate the whole sample in a natural environment even if only a small volume is to be analyzed. For this to be possible, both the primary radiation and the signal it generates must be able to travel through the air (or at least an air pressure above 610 Pa below which dehydration occurs).

Noninvasive instruments can be divided into two categories: those which are portable and small enough to take to the artwork and perform an analysis in situ, and infrastructural facilities where the instrumentation is too big to move. (An exception to this division is the environmental scanning electron microscope (ESEM) which is a laboratory SEM with a large sample chamber which can house artifacts a few centimeters on a side in a gaseous environment up to around 1/40 of an atmosphere which can be high enough to avoid dehydration [95].)

Instruments in the former category include handheld X-ray fluorescence (XRF) spectrometers which can provide an elemental analysis averaged over the top few microns of the sample surface, and many kinds of spectrometers using visible light and nearby wavelengths. In fact, almost every technique described in this book is available in a portable format, although in some cases (e.g., neutron tomography using a portable MeV source [96]) radiation safety and sample damage considerations would probably exclude their use in a museum.

Infrastructural instrumentation includes synchrotrons (usually, but not exclusively, for X-ray generation), particle accelerators (e.g., for MeV and higher energy protons and helium ions), and various neutron sources such as nuclear reactors and spallation sources. Typically, the advantages that infrastructural facilities have over both portable and laboratory instrumentation are much higher beam fluxes (faster and more sensitive analysis), greatly improved beam quality (e.g., narrower bandwidth at high flux for higher resolution), smaller spot sizes (high-sensitivity analysis of small areas and microvolumes, imaging), large sophisticated detectors (fast parallel acquisition), and beams such as cold or thermal neutrons which cannot be produced on small-scale equipment.

5.4 How to approach a truly nondestructive analysis

In this section we will assume that nondestructive means what a conservator or curator would reasonably expect: "does no detectable damage and has no after effects." A first, but not definitive, test of such a technique would be that a repeat analysis gave exactly the same result to within experimental error. This keeps the analysis within the envelope defined by Fig. 15A. A more robust test would be that of experience: foreknowledge that analysis of similar materials using the same technique and similar beam dose, analyzed area, and delivery rate had done no measurable harm. Here are some suggestions for keeping samples safer:

- Set the instrument up on an unimportant sample or a reference material. If necessary, provide a suitable sample of similar material and elemental composition to a more critical sample. Some samples are more suitable for setting up an analysis than others and the instrument operator or scientist responsible should be able to advise.
- If it is feasible and if the instrument operator is inexperienced, carry out your own dose- or power-related damage tests using simulants or noncritical samples. This is especially important on an infrastructural instrument. We often combine this with setting up when starting a synchrotron XRD run.
- Radiation beams such as synchrotron X-rays and protons which can travel significant distances through the air often generate ozone and nitrogen oxides along the beam path. These chemical species are very reactive and can attack the sample, causing changes in its surface chemistry. An example is given in Fig. 16. The simple solution is to protect the analyzed area with a helium jet.
- Always use the minimum dose and power density consistent with obtaining statistically useful data. If a microfocus beam is not required, then use a broad beam. For example, we have found that metal patinas suffer significant modification during microfocus X-ray bombardment [97], so we always try to analyze and image such materials using full field rather than scanned microfocus techniques.

Except where the technique demands it, avoid intense bombardment with strongly absorbed wavelengths or energies which might heat the sample. For example, red laser light will strongly absorb on a green pigment (it is not reflected; if it were, then the pigment would not be green!).

If each input particle creates damage in the sample, then the final damage level may be dose (particles absorbed per unit volume) dependent. However, if the energy deposited by the input beam thermalizes (turns to heat) in the sample then the power density may become the determining parameter of damage. Assuming the former for a moment, if we define the beam flux[iii] Φ as the number of particles per second in

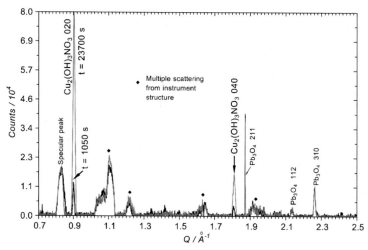

FIG. 16

Growth of copper hydroxy nitrate (gerhardite) due to interaction between the surface of a disc of copper reference material, X-ray generated nitrogen oxides and ozone from the air, and natural humidity, observed while setting up a high-sensitivity X-ray diffraction experiment on the XMaS beamline at the ESRF. The reflections at $Q = 0.9$ and 1.75 grow by a factor of nearly 8 over an interval of around 6.25 h. All other reflections in the range $0 < Q < 6.5$ remained stable. The gerhardite 020 peak is more than 3000 times smaller than the Cu 111 reflection so the effect is not large, but best avoided with a helium jet.

the beam and use Eq. (18) as the estimate of interaction volume then the average dose[i] D' will be:

$$D' \approx \frac{\Phi t}{V_{int}}, \tag{19}$$

where t is the time for which the beam was input to the position on the sample at that flux. The dose[ii] in terms of energy input D'' is:

$$D'' = 1.602 \times 10^{-19} D' E, \tag{20}$$

where E is the beam energy in eV and the factor of 1.602×10^{-19} converts the parameter to joules.

If the power is the critical parameter (and it can be very difficult to detect heating in an area irradiated by a microbeam except by some monitoring of chemical changes in the sample [97]) then an estimate of the power density P in Wm^{-3} is given by:

$$P = 1.602 \times 10^{-19} \frac{\Phi E}{V_{int}}. \tag{21}$$

If the beam does not have a flat-topped profile or is moved in large steps (see Fig. 14), then these equations will underestimate the energy and power densities and a more complicated approach using the beam profile will be required.

There is always pressure on designers to make beams ever smaller and more intense for both laboratory and infrastructural instrumentation. This improves the lateral resolution of the instrument and the sample throughput. This is not always a positive development for heritage analysis. More care with instrument parameters may be required in future as techniques continue to develop.

References

[1] A. Zucchiatti, F. Agulló-Lopez, Potential consequences of ion beam analysis on objects from our cultural heritage: an appraisal, Nucl. Instrum. Methods Phys. Res. B 278 (2012) 106–114.

[2] E.J. Schofield, R. Sarangi, A. Mehta, A.M. Jones, A. Smith, J.F.W. Mosselmans, A.V. Chadwick, Strontium carbonate nanoparticles for the surface treatment of problematic sulfur and iron in waterlogged archaeological wood, J. Cult. Herit. 18 (2006) 306–312.

[3] M. Bacci, C. Cucci, A.-L. Dupont, B. Lavédrine, C. Loisel, S. Gerlach, H. Roemich, G. Martin, LightCheck, new disposable indicators for monitoring lighting conditions in museums, in: Preventive Conservation, vol. II, ICOM Committee for Conservation, 2005, pp. 569–572.

[4] M. Bacci, C. Cucci, A.A. Mencaglia, A.G. Mignani, S. Porcinai, Calibration and use of photosensitive materials for light monitoring in museums: blue wool standard 1 as a case study, Stud. Conserv. 49 (2004) 85–98.

[5] J.V. Fraunhofer, Neue Modifikation des Lichtes durch gegenseitige Einwirkung und Beugung der Strahlen, und Gesetze derselben, in: E. Lommel (Ed.), Joseph von Fraunhofer's Gesammelte Schriften, Verlag der Königlich Akademie, München, 1888, pp. 51–112.

[6] G. Kirchhoff, R. Bunsen, Chemische Analyse durch Spectralbeobachtungen, Ann. Phys. 186 (1860) 161–189 (Originally Ann. Phys. Chem. 110 (1860) 11-39).

[7] J.M. Enoch, History of mirrors dating back 8000 years, Optom. Vis. Sci. 83 (2006) 775–781.

[8] https://www.britishmuseum.org/collection/object/W_-90959 (accessed 20.02.2021).

[9] J.E. Curtis, J.E. Reade, Art and empire: treasures from Assyria in the British Museum, British Museum, London, 1995.

[10] Aristophanes, Clouds, 423-417 BCE, Translated by G. Theodoridis, 2007 https://bacchicstage.wordpress.com/aristophanes/clouds/ (accessed 14.02.2021).

[11] A. Jones, Pseudo-Ptolemy De Speculis, SCIAMVS 2 (2001) 145–186.

[12] H. Alexandrini, in: L. Nix, W. Schmidt (Eds.), Opera Quae Supersunt Omnia Vol II, B. G. Teubner, Leipzig, 1900.

[13] H.E. Burton, The optics of euclid, J. Opt. Soc. Am. 35 (1945) 357–372.

[14] Euclidis, Optica, in: I.L. Heiberg (Ed.), Euclidis, Opera Omnia, vol. VII, B. G. Teubner, Leipzig, 1895.

[15] Euclidis, Catoptrica, in: I.L. Heiberg (Ed.), Euclidis, Opera Omnia, vol. VII, B. G. Teubner, Leipzig, 1895.

[16] A.M. Smith, Ptolemy and the foundations of ancient mathematical optics: a source based guided study, Trans. Am. Philos. Soc. 89 (Part 3) (1999) 1–172.

[17] P.A.M. Dirac, Relativistic quantum mechanics, Proc. Roy. Soc. A 136 (1932) 453–464.

[18] R.P. Feynman, Space-time approach to non-relativistic quantum mechanics, Rev. Mod. Phys. 20 (1948) 367–403.

[19] J. Schwinger, Quantum electrodynamics. I. A covariant formulation, Phys. Rev. 74 (1948) 1439–1461.

[20] S. Tomonaga, On infinite field reactions in quantum field theory, Phys. Rev. 74 (1948) 224–225.

[21] H.G. Liddell, R. Scott, A Greek-English Lexicon, eighth Ed., American Book Company, New York, 1882.

[22] J. Al-Khalili, Advances in optics in the medieval Islamic world, Contemp. Phys. 56 (2015) 109–122.

[23] J. D. Smith, The remarkable Ibn al-Haytham, Math. Gaz. 76 (1992) 189-198. JSTOR, www.jstor.org/stable/3620392, (accessed 16.02.2021).

[24] R. Rashed, Fermat et le principe du moindre temps, Comptes Rendus Mecanique 347 (2019) 357–364.

[25] R. Rashed, A pioneer in anaclastics: Ibn Sahl on burning mirrors and lenses. Isis 81 (1990) 464–491. JSTOR, www.jstor.org/stable/233423, (accessed 16.02.2021).

[26] P. Fermat, Letter to C. De La Chambre 1st Jan 1662 (CXII), in P. Tannery, C. Henry (Eds) Oeuvres De Fermat Vol II, Gauthier-Villars et Fils, 1894, 457–463. https://archive.org/details/oeuvresdefermat942ferm/, (accessed 16.02.2021).

[27] R. P. Feynman, R. B. Leighton, M. Sands, The Feynman Lectures on Physics, Addison-Wesley, 1963, (Chapter 26).

[28] I. Vossius, De lucis natura et proprietate, Apud Ludovicum and Danielem Elzevirios, Amstelodami (Amsterdam), 1662, 42. https://archive.org/details/delucisnaturaetp00voss/, (accessed 16.02.2021).

[29] C. Huygens, Dioptrique, in: D.J. Korteweg (Ed.), C. Huygens, Oevres Complètes, vol. XIII, Martinus Nijhoff, Den Haag 1916. https://www.dbnl.org/tekst/huyg003oeuv13_01/huyg003oeuv13_01_0015.php, (accessed 16.02.2021).

[30] A. Septier (Ed.), Focussing of Charged Particles Vols.I and II, Academic Press, New York, 1967.

[31] P. Grivet, in: A. Septier (Ed.), Electron Optics Parts I and 2, 2nd English Edition, P. W. Hawkes (Trans.), Pergamon Press, Oxford, 1972.

[32] W.R. Hamilton, On a general method in dynamics, Phil. Trans. Roy. Soc. (1834) 247–308, and Second essay on a general method in dynamics, Phil. Trans. Roy. Soc. (1835) 95-144, David R Wilkins (Ed.), https://www.emis.de/classics/Hamilton/, (accessed 17.2.2021).

[33] J.I. Beare, The Ancient Greek Philosophy of Vision, in Greek Theories of Elementary Cognition from Alcmaeon to Aristotle, Clarendon Press, Oxford, 1906, pp. 9–92.

[34] S. Sambursky, Philoponus' interpretation of Aristotle's theory of light, Osiris 13 (1958) 114–116.

[35] K. Ierodiakonou, Theophrastus on Plato's theory of vision, Rhizomata 7 (2019) 249–268.

[36] Aristotle, Meteorologica book III, in: Meteorologica, Translated by H. D.P. Lee, Harvard University Press, Cambridge, Massachusetts, 1952, (Chapter IV).

[37] Aristotle, On the Soul, 350 BCE, Translated by J. A. Smith, http://classics.mit.edu/Aristotle/soul.html, (accessed 17.2.2021).

[38] G.A. Winer, A.W. Rader, J.E. Cottrell, Testing different interpretations for the mistaken belief that rays exit the eyes during vision, J. Psychol. 137 (2003) 243–261.

[39] S. Berryman, Ancient atomism (2016), in: Stanford Encyclopedia of Philosophy, https://plato.stanford.edu/entries/atomism-ancient/, (accessed 18.02.2021).

[40] Lucretius, On the Nature of Things, A Philosophical Poem in Six Books (c. 50 BCE), Translated by the Rev. J. S. Watson, George Bell and Sons, London, 1880, p. 153.

[41] J.C. Maxwell, A dynamical theory of the electromagnetic field, Phil. Trans. Roy. Soc. 155 (1865) 459–512.

[42] A.A. Michelson, E. Morley, Am. J. Sci 34 (1887) 333–345.

[43] A. Einstein, Zur Elektrodynamik bewegter Körper, Ann. Phys. 17 (1905) 891–921.

[44] I. Newton, Optiks—A Treatise of the Reflections, Refractions, Inflexions and Colors fourth ed., William Innys, London, 1730, https://www.gutenberg.org/ebooks/33504 (accessed 18.02.2021) (a) Book 3 Query, 29; (b) Book 2, Propositions XIII and XIV for example; (c) Book 3, Query 17;(d) Book 3 Query 28; (e) Book 1, Proposition IX.

[45] R. Hook, Considerations of Mr. Hook upon Mr. Newton's Discourse of Light and Colours, The Royal Society, 1672, https://makingscience.royalsociety.org/s/rs/items/RBO_4_45 (accessed 23.02.2019).

[46] I. Newton, Mr. Isaac Newtons Answer to some Considerations upon his Doctrine of Light and Colors", Phil. Trans. Roy. Soc. 7 (1672) 5084-5103, https://royalsocietypublishing.org/doi/10.1098/rstl.1672.0051 (accessed 23.02.2019).

[47] C. Huygens, Treatise on Light (1678), Translated by S. P. Thompson, University of Chicago Press, 1912, https://www.gutenberg.org/ebooks/14725 (accessed 20.02.2019).

[48] R. Hook, Micrographia, The Royal Society, London, 1665, (Observation, IX, X) https://www.gutenberg.org/ebooks/15491, (accessed 19.02.2021).

[49] F. M. Grimaldo (Grimaldi), Physico-Mathesis de Lumine, Coloribus, et Iride, Hieronymi Bernia, Bologna, 1665, https://books.google.co.uk/books?id=_T9jRcV6qOMC&printsec=frontcover&source=gbs_book_other_versions_r&redir_esc=y#v=onepage&q&f=false, (accessed 15.02.2019).

[50] I. Newton, Letter from Newton to Boyle, in Correspondance of Scientific Men. vol. II, Oxford University Press, 1841, (Letter CCLXXV).

[51] T. Young, The Bakerian lecture, experiments and calculations relative to physical optics, Phil. Trans. Roy. Soc 94 (1804) 1–16, https://royalsocietypublishing.org/doi/10.1098/rstl.1804.0001, (accessed 15.02.2019).

[52] T. Young, A Course of Lectures on Natural Philosophy and the Mechanical Arts, vol. I Taylor and Walton, 1845, (Lectures (a) XXIII and (b) XXXIX).

[53] T. Young, A Course of Lectures on Natural Philosophy and the Mechanical Arts Vol. II—Plates, Taylor and Walton, London, 1845.

[54] A. Fresnel, in: É. Verdet, L. Fresnel (Eds.), Oeuvres Complètes, H de Senarmont, vol. II, Imprimerie Impériale, Paris, 1866.

[55] A. Fresnel, in: É. Verdet, L. Fresnel (Eds.), Oeuvres Complètes, H de Senarmont, vol. I, Imprimerie Impériale, Paris, 1868.

[56] A. Fresnel, in: É. Verdet, L. Fresnel (Eds.), Oeuvres Complètes, H de Senarmont, vol. III, Imprimerie Impériale, Paris, 1870.

[57] J. Gregory, Letter from Gregory to Collins, in Correspondance of Scientific Men Vol II., Oxford University Press, 1841, (Letter CCXIII).

[58] I.D. Bagbaya, On the history of the diffraction grating, Sov. Phys. Usp. 15 (1973) 660–661.

[59] F. Hopkinson, D. Rittenhouse, An optical problem proposed by Mr Hopkinson and solved by Mr Rittenhouse, Trans. Am. Phil. Soc. 2 (1786) 201–206, https://www.jstor.org/stable/1005186 (accessed 13.02.2021).

[60] J. J. Thomson, Cathode rays, Phil. Mag. 44 (1897) 293-316, https://doi.org/10.1080/14786449708621070, (accessed 01.1.2020).

[61] A. Einstein, Über einen die Erzeugung und Verwandlung des Lichtes betreffenden heuristischen Gesichtspunkt, Ann. Phys. 17 (1905) 132–148.

[62] G.P. Thomson, A. Reid, Diffraction of cathode rays by a thin film, Nature 119 (1927) 890.

[63] C.J. Davisson, L.H. Germer, Reflection of electrons by a crystal of nickel, PNAS 14 (1928) 317–322.

[64] IALA, Guideline 1049, The use of modern light sources in traditional lighthouse optics, Edition 2, International Association of Marine Aids to Navigation and Lighthouse Authorities, 2007.

[65] C. Jönsson, Elektroneninterferenzen an mehreren künstlich hergestellten Feinspalten, Z. Phys. 161 (1964) 454–475.

[66] A. Zeilinger, R. Gähler, C.G. Shull, W. Treimer, W. Mampe, Rev. Mod. Phys. 60 (1988) 1067–1073.

[67] G.I. Taylor, Interference fringes with feeble light, Proc. Camb. Phil. Soc. 15 (1909) 114–115.

[68] H. Iams, B. Salzberg, The secondary emission phototube, Proc. Inst. Radio Eng. 23 (1935) 55–64.

[69] B.K. Lubsamdorzhiev, On the history of photomultiplier tube invention, Nucl. Instrum. Methods Phys. Res. A 567 (2006) 236–238.

[70] A. Tonomura, J. Endo, T. Matsuda, T. Kawasaki, H. Ezawa, Demonstration of single-electron buildup of an interference pattern, Am. J. Phys. 57 (1989) 117–120.

[71] S. Frabboni, G.C. Gazzadi, G. Pozzi, Nanofabrication and the realization of Feynman's two-slit experiment, Appl. Phys. Lett. 93 (2008) 073108:1–3, https://doi.org/10.1063/1.2962987.

[72] W. Rueckner, J. Peidle, Young's double slit experiment with single photons and quantum eraser, Am. J. Phys 81 (2013) 951–958.

[73] R.S. Aspden, M.J. Padget, G.C. Spalding, Video recording true single-photon double-slit interference, Am. J. Phys 84 (2016) 671–677.

[74] R.P. Feynman, QED—The Strange Theory of Light and Matter, Princeton University Press, Princeton, 2014.

[75] M. Kemp, Leonardo Da Vinci—Experience, Experiment and Design, V&A Publications, London, 2011.

[76] E. MacMurdy, The Notebooks of Leonardo Da Vinci Vol. II.

[77] M.N. Piancastelli, K. Jankala, L. Journel, T. Gejo, Y. Kohmura, M. Huttula, M. Simon, M. Oura, Phys. Rev. A 95 (2017) 061402-1–016402-6, https://doi.org/10.1103/PhysRevA.95.061402.

[78] H. B, The theory of the rainbow, Nature 59 (1899) 616–618, https://www.nature.com/articles/059616b0 (accessed 03.03.2019) (In the M.S. and on the wesite, the author's name is just given as H. B., no further information is available.

[79] P. Laven, Supernumerary arcs of raibows: Young's theory of interference, Appl. Opt. 56 (2017) G104–G112, https://doi.org/10.1364/AO.56.00G104 (accessed 20.02.2021).

[80] C.A. Villee, E.P. Solomon, C.E. Martin, D.W. Martin, L.R. Berg, P.W. Davis, Biology, second ed. Saunders College Publishing, Philadelphia, 1989 (Chapter 48).

[81] B.B. Lee, The evolution of concepts of color vision, Neurociencias 4 (2008) 209–224.

[82] K.A. Jameson, S.M. Highnote, L.M. Wasserman, Richer color experience in observers with multiple photopigment opsin genes, Psychon Bull Rev. 8 (2001) 244–261.

[83] A.H. Compton, A quantum theory of the scattering of X-rays by light elements, Phys. Rev. 21 (1923) 483–502.

[84] B. Fu, E. Sperber, F. Eke, Solar sail technology—a state of the art review, Prog. Aerosp. Sci. 86 (2016) 1–19.

[85] M. P. Hehlen, M. Sheik-Bahrae, R. I. Epstein, Solid state optical refrigeration, in: J-C. G. Bunzli, V. K. Pecharsky (Eds.), Handbook on the Chemistry and Physics of Rare Earths, vol. 45, Elsevier, 2014 (Chapter 265).

[86] L. De Broglie, A tentative theory of light quanta, Phil. Mag. 47 (1924) 446–458. https://doi.org/10.1080/14786442408634378.

[87] https://goldbook.iupac.org/ (accessed 20.02.2021).

[88] https://www.iso.org/standard/41628.html, (accessed 20.02.2021).

[89] https://www.nist.gov/pml/stopping-power-range-tables-electrons-protons-and-helium-ions, (accessed 21.02.2021).

[90] https://henke.lbl.gov/optical_constants/atten2.html, (accessed 21.02.2021).

[91] J. I. Goldstein, D. E. Newbury, J. R. Michael, N. W. M. Ritchie, J. H. J. Scott, D. C. Joy, Scanning Electron Microscopy and X-Ray Microanalysis, fourth ed. Springer, 2018, https://doi.org/10.1007/978-1-4939-6676-9.

[92] J. Druzik, C. Pesme, Comparison of Five Microfading Tester (MFT) Designs, RATS Postprints 2 (2010) 14–29.

[93] O. Enguita, T. Calderón, M.T. Fernández-Jiménez, P. Beneitez, A. Millan, G. García, Damage induced by proton irradiation in carbonate based natural painting pigments, Nucl. Instrum. Methods Phys. Res. B 219–220 (2004) 53–56.

[94] N.P. Barradas, D. Benzeggouta, M. Doebeli, C. Jeynes, A. Vantomme, I. Vickeridge, in: D. Benzeggouta, I. Vickridge (Eds.), Handbook on Best Practice for Minimising Beam Induced Damage During IBA V1.0, Spirit, 2011.

[95] E. Doehne, ESEM applications: from cultural heritage conservation to nano-behaviour, Michrochim. Acta 155 (2006) 45–50, https://doi.org/10.1007/s00604-006-0505-1.

[96] P. Andersson, E. Andersson-Sunden, H Sjöstrand, S. Jacobsson-Svärd, Neutron tomography of axially symmetric objects using 14 MeV neutrons from a portable neutron generator, Rev. Sci. Instrum. 85 (2014) 085109-1–085109-11, https://doi.org/10.1063/1.4890662.

[97] A. Adriaens, P. Quinn, S. Nikitenko, M.G. Dowsett, Real time observation of X-ray-induced surface modification using simultaneous XANES and XEOL-XANES, Anal. Chem. 85 (2013) 9556–9563, https://doi.org/10.1021/ac401646q.

Raman and infrared spectroscopy in conservation and restoration

Anastasia Rousaki[a] and Peter Vandenabeele[a,b]

[a]*Raman Spectroscopy Research Group, Department of Chemistry, Ghent University, Ghent, Belgium* [b]*Archaeometry Research Group, Department of Archaeology, Ghent University, Ghent, Belgium*

Abbreviations

ATR	attenuated total reflectance
CCD	charge-coupled device
DRIFT	diffuse reflectance infrared Fourier transform spectroscopy
FORS	fiber-optic reflectance spectroscopy
FT	Fourier transform
FTIR	Fourier transform infrared spectroscopy
I_p	polymerization index
IR	infrared
LIBS	laser-induced breakdown spectroscopy
Micro-SORS	microspatially offset Raman spectroscopy
PB15	copper phthalocyanine
SERRS	surface enhanced resonance Raman scattering
SERS	surface enhanced Raman scattering
SORS	spatially offset Raman spectroscopy
TERS	tip enhanced Raman spectroscopy
UV	ultra-violet
Vis-NIR-FORS	visible near-infrared fiber optics reflectance spectrometry
XRF	X-ray fluorescence

Spectroscopy, Diffraction and Tomography in Art and Heritage Science. https://doi.org/10.1016/B978-0-12-818860-6.00004-0

1 Raman and infrared spectroscopy in conservation and restoration

Raman and infrared (IR) spectroscopy, also called vibrational spectroscopies, are two of the most frequently used molecular techniques applied to cultural heritage objects including archeological artifacts and pieces of art. Their dynamics, abilities and potentials are extensively described in the scientific literature [1,2].

This chapter covers the working principle of both techniques with emphasis on their advances and approaches, especially in the case of works of arts. Special attention is given to the analysis conducted outside the laboratory and directly onto the artwork. This is underlined especially in the case of immovable large-scale objects or fragile artifacts.

Before addressing more thoroughly the techniques involved in the next sections of this chapter, we would like to discuss the terminology we use in this chapter. The techniques can be, at first, either micro-destructive or nondestructive. The micro-destructive character of the analysis refers to the consumption of the sample during the study. Then, a technique can be invasive or noninvasive in reference to the extraction of micro-samples [3–5]. Generally, in the case of both Raman and infrared spectroscopy, benchtop techniques are usually considered invasive, as typically sampling is involved; however, they are nondestructive [3–5]. In some cases, small artifacts can be placed directly in the beam for investigation. Here other terms should be considered, those including direct and in situ analysis [3–5]. Direct analysis is conducted straight on the artifact while in situ (or on-site) analysis is the research performed on the field and in the natural environment or exposition place of the work of art [3–5]. Instrumental advances have shaped both techniques and made it possible to be applied outside the laboratory, directly and on the field, noninvasively and nondestructively with the use of mobile instrumentation.

2 Introduction to vibrational spectroscopy

Spectroscopy is a term used to describe the interaction of electromagnetic radiation with matter [6,7]. It can include a wide range of interactions, and matter can be addressed either at an atomic, molecular and nucleic level. In this chapter, interactions on the molecular level and more specifically those concerning vibrational spectroscopy, infrared and Raman spectroscopy, are described [2,6–11].

At ambient temperatures (i.e., at temperatures above absolute zero or $0\,K$) molecules vibrate. When a specific amount of energy is added to these molecules, e.g., by heating them or by irradiating them with electromagnetic radiation of a specific wavelength, they start vibrating at higher frequencies: they are excited to an excited vibrational state [2,8–11]. According to the quantum theory, molecules can only vibrate at specific frequencies, corresponding to specific energy levels [2,8–11]. Therefore the vibrational energy (E_v) of a molecule is proportional to the square root of the strength of the chemical bond (κ), and is inversely proportional to the square root of the reduced mass of the molecule (μ), with Planck's constant (h) divided by 2π being the proportionality constant (Eq. 1):

$$E_v = \frac{h}{2\pi} \cdot \sqrt{\frac{k}{\mu}} \tag{1}$$

The reduced mass (μ) of a bi-atomic molecule is given by:

$$\frac{1}{\mu} = \frac{1}{m_1} + \frac{1}{m_2} \tag{2}$$

where m_1 and m_2 being the masses of the constituting atoms. As a consequence, heavy molecules vibrate at lower energies and the stronger the bond, the higher the vibrational energy [2,8–11]. Moreover, molecular vibrations can occur differently as chemical bonds can be stretched or bended (deformation), and κ is also dependent on the type of vibration that is considered [2,8–11]. Vibrational spectroscopic techniques use electromagnetic radiation to study vibrational energy levels of the molecules (Fig. 1).

As the energy of these molecular vibrations depends on the chemical bond and the reduced mass of the molecule, these energy levels are different for all types of molecules. By measuring the energy that is needed during the transitions between vibrational energy levels, it is possible to identify the molecules.

In the case of infrared spectroscopy, the molecule is irradiated with polychromatic light and it absorbs one or more specific wavelengths as it undergoes a transition from the vibrational ground state to the first excited vibrational state (Fig. 1a). The absorbed energy corresponds exactly with the transition energy between these

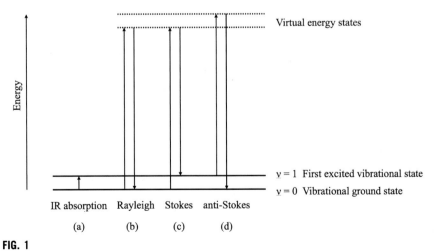

FIG. 1

An energy diagram illustrating the most basic transitions in infrared and Raman spectroscopies ($\nu = 0$: vibrational ground state; $\nu = 1$: first excited vibrational state).

states, and is proportional to the change in molecular vibrational frequency [1,7]. The energy (E) of the absorbed radiation is proportional to the frequency (ν) of the infrared radiation, or inversely proportional to the wavelength (λ). In vibrational spectroscopy, this energy is typically expressed as wavenumber (ω), with cm^{-1} as units. The relationship is given in Eq. (3), with h being the Planck's constant and c the speed of light in vacuum.

$$E = h\nu = h\frac{c}{\lambda} = hc\omega \tag{3}$$

It is essential that the molecules can only absorb infrared radiation if there is a change in dipole moment (polarization) during the vibration [7,11].

Raman spectroscopy is another spectroscopic technique to study the molecular vibrations. In this case, the molecule is irradiated with monochromatic light (i.e., electromagnetic radiation with a specific energy). Typically, lasers are used in this case. If a molecule is irradiated with this monochromatic light, it might also absorb the energy of the electromagnetic radiation and is excited to a so-called virtual state, but as this energy does not correspond with the energy difference between the two energy levels of the molecule, the energy is immediately released, and the light is scattered, while the molecule returns to the ground state (Fig. 1b). This elastically scattered electromagnetic radiation has exactly the same energy and wavelength as the laser and is called the Rayleigh scattering [2,8–11]. However, a small fraction of the molecules does not relax to the ground state, but to the first excited vibrational state (Fig. 1c). In this case, the scattered radiation has a lower energy and longer wavelength than the laser, and the energy difference corresponds to the energy difference between the ground state and the first vibrational excited state. As a consequence, by measuring the energy difference between the incident radiation and this so-called Stokes scattering, it is possible to study the molecular vibrations [2,8–11]. It is also possible that the molecule was already in a vibrational excited state, and after the absorption-scattering process it relaxes till the vibrational ground state (Fig. 1d). In that case, the scattered radiation has a higher energy than the irradiating laser. This phenomenon is called the anti-Stokes radiation. Both Stokes and anti-Stokes forms of inelastic scattering are the forms of Raman scattering, but as the Stokes scattering is typically more intense than the anti-Stokes scattering, usually only the Stokes part is studied [2,8–11]. It has to be mentioned that the occurrence of Raman scattering is far less favorable than the elastic scattering, and therefore, during Raman experiments, the Rayleigh line has to be filtered [2,8–11]. Considering that the Raman mechanism goes via a forbidden virtual state, this effect is relatively weak, and the scattered radiation can easily be overwhelmed by other light sources, such as ambient light or fluorescence caused by the electronic transitions in the sample. However, if the energy of the excitation laser happens to correspond to an allowed transition (e.g., to an excited electronic state), the Raman effect is much more likely to occur, and the corresponding Raman band is much more intense [11]. This highly favorable case is called the resonance Raman effect, and can be harvested in some studies to achieve higher sensitivities. Contrary to the infrared spectroscopy where a change in polarization occurs, a vibration is only Raman active, if there is a change in the polarizability during the vibration [11]. Considering this remarkable difference between the two related techniques, it can be proven that in the centrosymmetric molecules, vibrations that are Raman active cannot be observed by the infrared spectroscopy, and vice versa [11]. A deep insight of the physical phenomena that occur when the light interacts with the matter on a molecular level can be found elsewhere [2,6–11].

3 Raman spectroscopy

Although the theoretical background for the inelastic scattering of the electromagnetic radiation was described in 1923 by A. Smekal (as mentioned by Krishnan and Shankar in 1981) [12], it was not before 1928 that C.V. Raman and K.S. Krishman had experimentally verified the theory [2,8–11]. For this, C.V. Raman was awarded the Nobel prize in physics in 1930. The introduction of lasers and the technological advances of the following years helped the technique slowly but steadily grow. Half a century after the experiments of C.V. Raman and K.S. Krishman, Rosasco et al. [13] and Delhaye and Dhamelincourt [14] advanced the technique by coupling a microprobe or microscope with a Raman spectrometer. In the 1980s, the technique was first applied to study the art [15,16].

Nowadays, a variety of Raman spectrometers, both benchtop and mobile ones, are commercially available. To understand their advantages and disadvantages, each type will be discussed separately, followed by some common applications in the literature on the physicochemical analysis of cultural heritage objects.

3.1 Laboratory Raman analysis

Laboratory analysis on fragments or small artifacts can be carried out with the benchtop Raman spectrometers. These laboratory systems are distinguished for their stability, outstanding performance and high-quality results. Nowadays, Raman spectrometers dedicated to laboratory analysis are used with high-magnification microscopes, enabling the analysis to be conducted with a high spatial resolution [4,5,17,18]. The identification of different phases vibrating in close wavenumber proximity can be accomplished as many Raman systems can achieve a spectral resolution down to $1\,cm^{-1}$ [4,5,17,18]. In general terms, Raman spectroscopy is a nondestructive technique as long as the power of the laser is sufficiently controlled in order to avoid damage or thermal degradation on the sample and/or artifact [3–5]. The technique can characterize both inorganic and organic components alone or in mixtures, with a minimal or even with no sample preparation [3–5]. Sample pretreatment is performed when the stratigraphy of consequent layers is to be analyzed. One approach is to embed the samples in resin mixtures, so that one sees the cross-section of the sample, and after polishing and positioning it under the Raman microscope for investigation.

In general, Raman spectroscopy is one of the most favorable techniques in the field of archaeometry [3–5], with many review articles underlining its importance, performance and versatility both in the laboratory and outside the laboratory [3–5,17–21].

3.1.1 Applications

The first examples of the application of Raman spectroscopy to cultural heritage objects were focused on the examination of illuminated mediaeval manuscripts [15,16,22,23]. Since then, Raman spectroscopy, and more specifically micro-Raman spectroscopy, has been applied to a large variety of artworks, with glass [24–26], ceramics [27–29], rock art paintings [30–37], different murals [38,39], gems and gemstones [40,41], being some of them. The technique has proven to be a powerful tool for the attribution of the molecular vibrations to the correct substance by the creation of meticulous online tools/libraries and spectral databases [42–44] and article databases [45–48]. Moreover, specific studies have been performed on certain materials including corrosion products [49,50]. Some examples come from the research on the red iron oxide, commonly known as hematite (α-Fe_2O_3) [37,51,52]. Hematite and goethite (α-FeOOH can transform under elevated temperatures. Haematite can be obtained from goethite [52,53]. The differentiation of the processes used to obtain a specific color can arise questions especially in the case of rock art painting research. Firing can incorporate distortion in the crystal structure of hematite that is reflected in the Raman spectrum [37,52]. The main problem is that lowering of the crystal structure of the red oxide can also arise from many mechanisms other than thermal treatment, e.g., biodegradation, etc. [52], or disordered effect [51,52] and crystal morphology [51] thus making the identification of processes though the Raman spectroscopy is rather dubious. Furthermore, oxides and oxyhydroxides containing, for example, manganese (Mg) or iron have relatively low Raman scattering capabilities and are laser-sensitive materials [54,55].

Raman spectroscopy is an ideal approach to study crystalline and amorphous inorganic materials but also including those that can be found in polymorphs. Titanium dioxide (TiO_2) can unambiguously be identified and distinguished in its three polymorphs: rutile, anatase, brookite, although the latter is generally not used in artworks [56]. Calcium carbonate ($CaCO_3$), another material found in the works of art both as an intentional choice of the artist or postdepositional product, can be characterized by the Raman spectroscopy and is possible to differentiate between calcite, aragonite and vaterite [57–60].

Micro-Raman spectroscopy is an ideal tool for the analysis of glass and glazes of art origin [25,61–63]. The technique is not only able to characterize the glassy matrix in depth, but also can specify the existence of possible pigments used, fluxes, opacifiers, etc. because of its true confocality on a micrometer scale. Typically, in a Raman spectrum of a glass, the Si—O bending and Si—O stretching modes/massifs are located at around $500\,cm^{-1}$ and $1000\,cm^{-1}$ [63,64]. The ratio of their areas, A_{500}/A_{1000}, known as the polymerization index (I_p), is linked to a glass family group and firing temperature [63,64]. The pioneering work for acquiring the I_p is described by Colomban [64].

Synthetic organic pigments used in modern art can also be studied, including polymorphs, e.g., the copper phthalocyanine (PB15) family [65]. The solid characterization of synthetic organic pigments is an analytical research topic that includes many challenges, but Raman spectroscopy is able to produce reliable results when applied to these 20th and 21st century materials [47,66–68] found in, for example, graffiti [69]. Towards the identification of synthetic organic pigments, a dedicated online library can be found online [42]. Besides the synthetic organic pigments, natural organic products, varnishes and binding media can be explored with micro-Raman spectroscopy [70,71].

3.1.2 Selection of an appropriate laser

Although different applications have been presented for micro-Raman spectroscopy, the selection of lasers was not underlined until now. Indeed, one of the most powerful characteristics of the technique is the ability to use powerful monochromatic lasers. This variation of lasers, or the choice of the correct laser upon case, includes not only the physical phenomena (e.g., resonance Raman, suppression of fluorescence) behind the selection, but also follows instrumental developments (e.g., the availability of multi-channel detectors that are sensitive in the infrared region triggered the introduction of dispersive spectrometers with 1064 Nd:YAG lasers).

Initially, C.V. Raman used filtered sunlight for his experiments [11]. Nowadays, lasers used in Raman spectrometers can vary from the ultra-violet (UV) to the visible to the IR region of the electromagnetic spectrum. UV lasers are typically of little use in cultural heritage studies. Due to their high-power density, they are keen to damage the samples or artifacts. It is important to select, where possible, the most appropriate laser for the study on hand. Different phenomena have to be considered. If the frequency of the laser (short wavelength) is higher, the energy of the radiation will be higher. From the laws of physics, it is known that the intensity of the Raman effect is proportional to the 4th power of the laser frequency. Hence, the Raman signal will be more intense if the wavelength is shorter. Moreover, when considering the sensitivity, one should also take the detector sensitivity into account. Typically, charge-coupled device (CCD) detectors are much more sensitive in the visible region of the spectrum compared to the infrared region.

On the other hand, by selecting an appropriate laser, some forms of interferences can be avoided, or the signal can be enhanced. One serious interference in the Raman spectroscopy is the occurrence of fluorescence. This phenomenon occurs when the molecule absorbs sufficient energy to be excited to an electronic state. If a short-wavelength laser is selected (i.e., high energy radiation), it is much more likely that the molecule is excited to a higher electronic state, and fluorescence may occur. Typically, the fluorescence effect is much more likely than the Raman effect, and it can easily overwhelm the Raman spectrum. Therefore selecting a long-wavelength laser (e.g., in the IR region) might be advantageous, despite the lower Raman scattering possibilities and the lower sensitivities of the detectors.

However, fluorescence avoidance and evaluation of the sensitivity of the spectrometer are not the only parameters when selecting an appropriate laser. Sometimes it is possible to select a laser that makes the virtual state exactly to correspond with an allowed electronic state. Despite this can cause fluorescence, the Raman signal is seriously enhanced in this case also as the transition passes through an allowed state. When selecting an appropriate laser, it is possible to take advantage of this resonance Raman effect to detect molecules with higher sensitivity. This effect is rather beneficial when studying certain pigments such as lazurite $[(Na,Ca)_8(AlSiO_4)_6(SO_4,Cl, S)_2]$, indigo, etc. [72,73], but also other compounds like carotenoids [74,75], a pigment in plants, a resulting product of bacteria, lichens, etc.

3.2 Direct and on-site Raman spectroscopy

One of the major advantages of Raman spectroscopy is that reliable results can be achieved, also when the technique is applied in situ. It is undeniable that laboratory Raman systems are very stable spectrometers, some being a true confocal apparatus, providing analysis on a micrometer scale. The drawback of such benchtop systems is that the analysis should be carried out on samples (or artifacts) that fit in the spectrometer chamber, thus jeopardizing the size of the work of art. Instrumental advances brought the Raman technique on the field and outside the laboratory enhancing the analytical experience according to the ethics of conservation and preservation of cultural heritage.

Mobile Raman systems with long fiber optic cables or fixed optical heads are nowadays commercially available. Generally, these are typically not coupled to a microscope when measuring in situ. Thus focusing on the surface is not achieved through microscope objectives, but by evaluating the signal-to-noise ratio in any given focal position. The latter is more straightforward when using positioning equipment such as tripods and articulating arms. When it is not possible to carry such accessories onto the field, dedicated caps can be slid over the probe head lens, ensuring a stable focal length but also blocking the environmental signal [76]. Although mobile Raman spectrometers are sometimes suffering wavenumber instability that can be corrected with thorough calibration of the system [76,77], have larger measuring spots compared to benchtop systems and may be subjected to environmental interferences, they are recommended when noninvasive and nondestructive analyses should be carried out.

Before studying the mobility and performance of Raman systems in depth, the nomenclature based on their portability should be addressed. A distinction between mobile instruments can be made. In general, these can be described as transportable, mobile, handheld and palm size [3,19]. The different categories are synthesized based on the weight of the total packaging, the autonomous operation, etc. In art analysis, mobile instruments with long fiber optics are preferable compared to their handheld counterparts, as the latter use fixed optical heads (difficult positioning) and often use preset measuring conditions.

Although benchtop Raman spectrometers are extensively tested and their performance is established, mobile Raman spectrometers appeared in the art analysis literature approximately two decades ago. Since the first efforts in 2001, when one was to measure directly onto the artifact using an altered benchtop Raman system [78], to 2004 when the first mobile Raman spectrometer for art analysis with an in-house built hard- and software was constructed [79], to today where advanced mobile systems are commercially available, measuring directly and on site met new milestones progressing the noninvasive experience.

Nowadays, *state-of-the-art* mobile Raman systems using different hardware and software characteristics can be found widely available in the market. According to the materials analyzed, one can opt for a dual or mono laser systems with long fiber optics or fixed optical heads, equipped with silicon-based or solid state physics detectors. Besides, the dispersive 1064 nm excitation Raman systems (i.e., those equipped with a 1064 nm excitation Nd:YAG laser) are able to be brought onto the field for direct analysis, while the long laser wavelength counts for fluorescence suppression [80]. Moreover, a new handheld Raman system emerged onto the market changing the way how dual laser systems were traditionally operated. Indeed, the Bravo system manufactured by Bruker uses the sequentially shifted excitation technology as a tool to correct for fluorescence emission [80]. The basis for this approach is based on the fact that the absolute Raman band position (expressed in nm) shifts when a different excitation wavelength is used, while broad features such as fluorescence are little influenced by small shifts in the laser wavelength. Thus the output of such an instrument is a reconstructed spectrum, free from artifacts and fluorescence emission, projecting the spectral data in a large spectral range including the fingerprinting and C—H stretching area. Indeed, several studies have demonstrated the use of Bravo on cultural heritage objects [80–82] with concerns pointed to the elevated fixed laser power.

The abilities and advantages of mobile Raman instrumentation are underlined by numerous studies carried out directly on the artefact, giving the results not only regarding the identification of the materials used, but also on their preservation state.

3.2.1 Applications

Noninvasive Raman analysis is carried out, among others, on illuminated manuscripts [83], wall paintings [84,85], glass [86,87], gems and gemstones [88] and mosaics [80,89]. Impressive work has been done in the case of measuring directly on rock art paintings [76,90–92]. Usually, the aim of such studies is twofold: to identify and characterize the materials of the general rock art paintings and substrata and to conclude on the preservation state of these magnificent prehistoric art performances. Some typical main chromophores that can be found on the pigmented rock art paintings are hematite (α-Fe_2O_3), ferric oxy-hydroxides, green earths, manganese oxides, etc. Degradation processes and excess weathering are visualized on the rock art surfaces and can be recorded in the form of biodegradation products, including among other Ca-oxalates, other salt formations and carotenoids.

Moreover, a large-scale study involves the Pompeii archeological park, one of the most significant historic sites in the world. Specifically, for the in situ studies, mobile Raman spectroscopy was used successfully for the identification of the chromophores on the Pompeian wall paintings, and also for the research on biodegradation processing affecting the murals as, for example, salt formation or carotenoid identification [93,94].

3.2.2 Comparison of two mobile instruments

A class of materials that is frequently used in conservation practice and that often causes some interference by fluorescence is known under the brand name Paraloid [95–97]. These synthetic polymers were brought to the market originally by the Rohm and Haas company and were frequently used since the 1950s as a consolidation agent, modifier, varnish or binding medium [95–97]. Of the different types of Paraloid, Paraloid B-72 has been frequently used in conservation science, and its properties as a conservation product (e.g., consolidating action and performance durability) have been thoroughly examined [95–97]. They are transparent and soluble in several solvent mixtures, which justify their widespread use.

Raman spectra of six commercially available polymers (Paraloid B44, Paraloid B48N, Paraloid B66, Paraloid B67, Paraloid B72 and Paraloid B82) that are frequently used in conservation were recorded with two different spectrometers. The first instrument is the Bravo handheld spectrometer (Bruker, Ettlingen, Germany), using the sequentially shifted excitation technology and the second instrument is the dispersive i-Raman system (BWTek, Newark, USA) coupled to a 1064 nm excitation Nd:YAG laser.

Fig. 2 shows an overview of the Raman spectra of the six polymers of the Paraloid group acquired with the Bravo system. The spectra demonstrate that for these products the Bravo spectrometer yields good quality Raman spectra, able to differentiate between the different products on hand. All spectra have intense bands centered around c.2900 cm^{-1}, c.1730 cm^{-1} and c.1450 cm^{-1}. The massif around c.2900 cm^{-1} can be associated with the ν(C—H) stretching vibrations, and is relatively intense compared to the spectra of organics as excited with a 785-nm laser. This can be related to the fact that the detector is sensitive in the infrared region contrary to a charge-coupleddevice (CCD) detector where the sensitivity seriously drops at longer wavelengths [5,80,81]. The intense bands around c.1730 cm^{-1} are associated with the ν(C=O) stretching vibration of the conjugated esters, and the intense band around c.1450 cm^{-1} can be assigned to the δ(C—H) deformation [98,99]. Although both Paraloid B66 and B82 are copolymers of methylmetacrylate and buthylmetacrylate, their spectra can clearly be distinguished.

Fig. 3 represents the Raman spectra as recorded with 1064-nm excitation instrument. With this spectrometer, it is not possible to record the ν(C—H) spectral region around 3000 cm^{-1}. However, the low wavenumber end of the spectra seems to be covered better, although the edge of the laser line can also be observed here.

In general, the Raman band positions and relative intensities are similar, although the spectra in Fig. 3 (1064-nm excitation instrument) seem to have a slight, broad

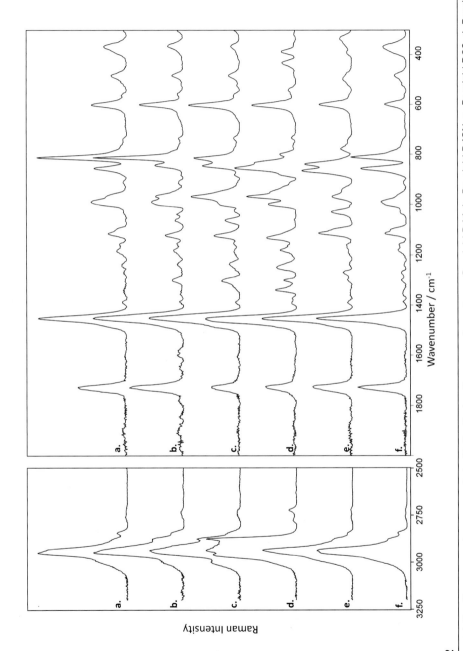

FIG. 2

Representative Raman spectra as recorded with the Bruker Bravo spectrometer. a: Paraloid B44, b: Paraloid B48N, c: Paraloid B66, d: Paraloid B67, e: Paraloid B72, f: Paraloid B82.

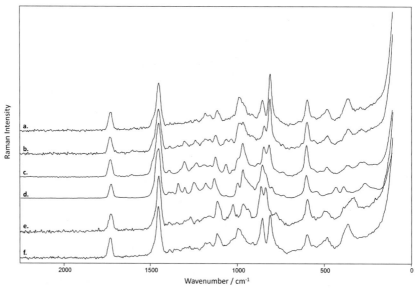

FIG. 3

Representative Raman spectra as recorded with the BWTek spectrometer with 1064 nm excitation wavelength. a: Paraloid B44, b: Paraloid B48N, c: Paraloid B66, d: Paraloid B67, e: Paraloid B72, f: Paraloid B82.

background signal, which is absent in the reconstructed spectra (Fig. 2) obtained with the sequentially shifted excitation algorithm (Bravo system) [80,81]. Nevertheless, also with this spectrometer, it is possible to differentiate between the different polymers.

3.3 Other Raman approaches and techniques

Other approaches, based on Raman spectroscopy, deviated from their 'traditional' use and brought a new perspective onto the scientific field of vibrational spectroscopy. Indeed, Raman spectroscopy did not remain static since the implementation of the lasers and the realization of the first benchtop spectrometers. The selection of the lasers according to the materials used, their scattering capabilities, their resonance effect, and the suppression of fluorescence emission are still some of the main parameters for a traditional spectroscopist to be considered, but other techniques have emerged to deal with several problems coming from the materiality of works of art and also from the need the analysis being as less invasive as possible.

One of the most promising techniques within Raman spectroscopy is surface-enhanced Raman scattering (SERS). The inherent problem of fluorescence emission together with the low-scattering capabilities of certain organic molecules that can be found dispersed in low concentrations in the cultural heritage objects [100,101] is the problems that conventional Raman spectroscopy can only partially solve. Even when changing the laser excitation to suppress fluorescence or benefit from resonance enhancement of the Raman signal, the technique cannot always produce straightforward results if used as it is. To overcome such problems, one can benefit from the considerable enhancement of the Raman signal when the molecules of an analyte are attached to metallic nanoparticles [20,100,101]. The phenomenon of such interaction was first observed in 1974 by Fleischmann et al. [102] and in 1987 Guineau and Guichard [103] applied the technique on cultural heritage objects. Indeed, SERS has been applied successfully on the identification of natural and synthetic organic dyes [100,104,105]. Moreover, some reviews of the technique are available summarizing all the possible intermediate steps of acquiring Raman spectra with strong enhancements [20].

Regarding the use of the approach, SERS is not straightforward when it comes on the extraction of the molecules for analysis and their preparation on the metallic substrates [20,106]. Moreover, someone can benefit from the resonance phenomena when selecting an appropriate laser excitation combined with the already strong enhancement of SERS (surface enhanced resonance Raman scattering [SERRS]) [100], or enhance the Raman signal via a metallic coated tip (tip enhanced Raman spectroscopy [TERS]) [107].

As artworks often consist of layered structures, many questions relate to their stratigraphy. Traditionally, an embedded sample can be studied under the Raman microscope. The need for creating new approaches as less invasive as possible has driven the scientists to create new techniques for characterizing painted layers without jeopardizing the work of art or its unity. The concept of performing stratigraphic analysis in a noninvasive way has been the base for the micro-spatially offset Raman spectroscopy (micro-SORS).

The pioneering idea of spatially separating the laser source and the signal collection area for subsurface analysis was introduced in 2005 by Matousek et al. [108,109]. Following the photon propagation inside the material, the larger the spatial separation the more information from deeper than the surface can be obtained. The transition from SORS, as described by Matousek et al. [108,109], and its adaptation on characterizing painting layers brought the micro-SORS technique into the fields of archaeometry and art analysis [110].

The most straightforward approach of micro-SORS technique is the *defocusing* micro-SORS where commercially available benchtop Raman spectrometers can be used in a nonaltered set-up [111]. The spatial separation in such a configuration is achieved on the focusing z-axis. The more defocused from the zero or imaged position (or in focus position) the more information is acquired from the sublayers under a turbid one [112]. *Full* micro-SORS benefits from the spatial separation of the laser from the information area on the x-axis both in the cases of internal and external beam delivery [113]. *Fiber-optics* micro-SORS is proposed as an alternative approach by using 2-μm thick glass fibers for delivering the laser and collecting the Raman signal [114]. The spatial separation is achieved again on the x-axis, and is proposed as a solid basis for miniaturization of the technique with possible application on in situ studies.

The micro-SORS technique has been applied in numerous studies concerning the cultural heritage research [110,115,116] with portable configurations being developed for direct and on-field analyses [117,118].

Chemical imaging of the surfaces of works of art is a tool of great importance when it comes on visualizing the technique of the artist. The spatial distribution of the pigment molecules relates the chemical information with the area under study. The benchtop Raman spectrometers for laboratory use are equipped with stable XYZ stages, allowing the free movement of the microscopes on three axes, and thus a mapping of an area can be retrieved. Postprocessing of the large data collection (the mappings are usually single-point measurements in a grid, a line, etc.) reveals the imaging fingerprint of the compounds used. The high quality of the Raman mappings and the consequent molecular images secured as benchtop spectrometers are true confocal, stable systems using advantageous spatial and spectral resolutions. The main restriction of such mappings is that these can only be performed on samples visualizing the surface, cross-sections performing stratigraphic analysis or objects small enough to fit the system chamber. The transition from micro-Raman to macro-Raman mappings is not only an upscaling of the size of the chemical images created, but a true upscaling of the hardware and software used. An attempt to create chemical images noninvasively from the 19th century porcelain cards was described in 2016 by Lauwers et al. [119]. Although other techniques, such as X-ray fluorescence (XRF), are already adapted to mobile instrumentations to create mapping set-ups, producing high-quality elemental images [120], macro-Raman mapping systems are still under development.

4 Infrared spectroscopy

A related technique to Raman spectroscopy is Fourier transform infrared spectroscopy (FTIR). This vibrational technique uses infrared radiation to excite the molecules and during vibrations, the molecule undergoes a change in dipole moment [11].

Initially, infrared spectroscopy started to emerge during the 1950s with the first paper published in art analysis in 1966 (as stated by Casadio and Toniolo in 2001) [121]. Fourier transform instruments were only later developed, and this technological advantage gave the new instruments the ability to produce high-quality results with a better spectral resolution and minimum amount of sample [121]. In 2001 Casadio and Toniolo [121] published an extensive review on 40 years of infrared spectroscopy, while in 2010 Prati et al. [122] discussed new advances in FTIR spectroscopy on materials from the works of art. In 2019 Rosi et al. [123] reviewed the recent trends in FTIR spectroscopy. Traditionally, FTIR spectroscopy requires the samples to be crushed and incorporated in salt (KBr)-based pellets for analysis. Nowadays, different techniques, such as attenuated total reflectance (ATR), diffuse reflectance infrared Fourier transform spectroscopy (DRIFT), micro-FTIR spectroscopy, etc. require minimal sample preparation or different sampling methods.

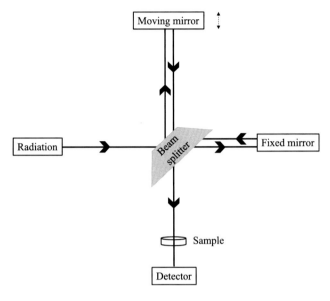

FIG. 4

Simplified representation of the Michelson interferometer.

The key part of an FTIR spectrometer is the Michelson interferometer, which makes the actual difference between the FT and the dispersive systems (Fig. 4). Here the incident radiation is split equally in two parts with one part being reflected and the other one passing through the beam splitter. As the radiation hits the immovable and moving mirrors, it is reflected back to the beam splitter where it recombines. The different distance of the two paths results in constructive or destructive interference. As the moving mirror moves, difference in signal intensities occurs as a function of time. The transition from the function of time to the function of frequency is being achieved by the Fourier transformation.

FTIR studies can be potentially performed in transmission mode or in reflection mode, depending on the nature and the purpose of the analysis [122]. FTIR spectroscopy shares a lot of applications in the field of art analysis as it allows to identify both the inorganic and organic components of the mixtures [124–130].

Attenuated total reflection Fourier transform infrared spectroscopy (ATR-FTIR) is another tool widely used in cultural heritage studies [131–134]. Moreover, it was used as complementary technique in studies related to rock art paintings from La Serena, Spain [135], on San rock art from South Africa [136], and possible compounds from San/Bushman rock art [137]. Usually, it involves an ATR crystal (high refractive index) that is brought in close contact with the sample [131]. Towards the noninvasive analysis of polymers, a portable FTIR using ATR was proven to be successful [138].

The need for noninvasive analysis moved infrared spectroscopy forward, and as it also happened for Raman spectroscopy, new techniques emerged for advancing the direct, on-field or simply noninvasive analysis [139–141]. Diffuse reflectance infrared Fourier transform (DRIFT) spectroscopy is incorporated in cultural heritage studies by the use of mobile systems [142,143]. A novel portable system combining diffuse reflectance and laser-induced breakdown spectroscopy (LIBS) is described by Siozos et al. [144]. Fiber-optic reflectance spectroscopy (FORS) was proven successful in the case of investigating compounds from a sarcophagus [145]. Ultraviolet-visible diffuse reflectance spectrophotometry with optic fibers combined with spectro-fluorimetry was applied on a mediaeval breviary [146]. Finally, portable visible near-infrared fiber optics reflectance spectrometry (Vis-NIR-FORS) is thoroughly discussed and evaluated in the case of restoration product characterization [147].

Although, infrared spectroscopy is a very valuable technique applied on cultural heritage materials, its applications especially with mobile instrumentation are less frequent than those of Raman spectroscopy. Infrared spectroscopy is an ideal technique for the identification of organic and inorganic pigments, including possible weathering products. The chapter in infrared spectroscopy tries to give an overview of some of the advantages of the technique in the field of cultural heritage in the recent years.

5 Conclusions

This chapter underlines the basic fundamentals behind Raman and infrared spectroscopy. Moreover, it tries to describe the two techniques on the basis of their instrumentation and technological advances. Both benchtop and mobile applications are thoroughly described with an orientation on *state-of-the-art* approaches. Indeed, both vibrational techniques are used extensively for the analysis of materials from the works of art, making them the first choice tools in the fields of art conservation and archaeometry.

Acknowledgments

Anastasia Rousaki greatly acknowledges the Research Foundation–Flanders (FWO-Vlaanderen) for her postdoctoral grant (under the project number: 12X1919N).

References

[1] B.H. Stuart, Infrared Spectroscopy: Fundamentals and Applications, John Wiley & Sons Ltd, Chichester, UK, 2004.

[2] R.L. McCreery, Raman Spectroscopy for Chemical Analysis, John Wiley & Sons, New York, USA, 2000.

[3] P. Vandenabeele, H.G.M. Edwards, J. Jehlička, The role of mobile instrumentation in novel applications of Raman spectroscopy: archaeometry, geosciences, and forensics, Chem. Soc. Rev. 43 (2014) 2628–2649.

[4] P. Vandenabeele, H.G.M. Edwards, L. Moens, A decade of Raman spectroscopy in art and archaeology, Chem. Rev. 107 (3) (2007) 675–686.

[5] A. Rousaki, L. Moens, P. Vandenabeele, Archaeological investigations (archaeometry), Phys. Sci. Rev. 3 (9) (2018), https://doi.org/10.1515/psr-2017-0048.

[6] J.L. McHale, Molecular Spectroscopy, Pearson Prentice Hall, Upper Saddle River, NJ, 1999.

[7] C.N. Banwell, Fundamentals of Molecular Spectroscopy, third ed., McGraw-Hill Company Europe, Berkshire, United Kingdom, 1983.

[8] E. Smith, G. Dent, Modern Raman Spectroscopy: A Practical Approach, John Wiley & Sons, Chichester, UK, 2005.

[9] J.R. Ferraro, K. Nakamoto, C.W. Brown, Introductory Raman Spectroscopy, second ed., Academic Press, Amsterdam, 2003.

[10] D.A. Long, Raman Spectroscopy, McGraw-Hill, Maidenhead, UK, 1977.

[11] P. Vandenabeele, Practical Raman Spectroscopy: An Introduction, John Wiley & Sons, Chichester, UK, 2013.

[12] R.S. Krishnan, R.K. Shankar, Raman effect: history of the discovery, J. Raman Spectrosc. 10 (1981) 1–8.

[13] G.J. Rosasco, E. Roedder, J.H. Simmons, Laser-excited Raman spectroscopy for non-destructive partial analysis of individual phases in fluid inclusions in minerals, Science 190 (1975) 557–560.

[14] M. Delhaye, P. Dhamelincourt, Raman microprobe and microscope with laser excitation, J. Raman Spectrosc. 3 (1975) 33–43.

[15] B. Guineau, Microanalysis of painted manuscripts and of colored archeological materials by Raman laser microprobe, J. Forensic Sci. 29 (2) (1984) 471–485.

[16] B. Guineau, C. Coupry, M.T. Gousset, J.P. Forgerit, J. Vezin, Identification de bleu de lapis-lazuli dans six manuscrits à peintures du XIIe siècle provenant de l'abbaye de Corbie, Scr. Theol. 40 (1986) 157–171.

[17] D. Bersani, P.P. Lottici, Applications of Raman spectroscopy to gemology, Anal. Bioanal. Chem. 397 (2010) 2631–2646.

[18] D. Bersani, C. Conti, P. Matousek, F. Pozzi, P. Vandenabeele, Methodological evolutions of Raman spectroscopy in art and archaeology, Anal. Methods 8 (2016) 8395–8409.

[19] P. Vandenabeele, M.K. Donais, Mobile spectroscopic instrumentation in archaeometry research, Appl. Spectrosc. 70 (1) (2016) 27–41.

[20] F. Pozzi, M. Leona, Surface-enhanced Raman spectroscopy in art and archaeology, J. Raman Spectrosc. 47 (2016) 67–77.

[21] P. Colomban, The on-site/remote Raman analysis with mobile instruments: a review of drawbacks and success in cultural heritage studies and other associated fields, J. Raman Spectrosc. 43 (2012) 1529–1535.

[22] R.J.H. Clark, Raman microscopy: application to the identification of pigments on medieval manuscripts, Chem. Soc. Rev. 24 (1995) 187–196.

[23] P. Vandenabeele, B. Wehling, L. Moens, B. Dekeyzer, B. Cardon, A. Von Bohlen, R. Klockenkämper, Pigment investigation of a late-medieval manuscript with total reflection X-ray fluorescence and micro-Raman spectroscopy, Analyst 124 (1999) 169–172.

[24] L.C. Prinsloo, A. Tournié, P. Colomban, A Raman spectroscopic study of glass trade beads excavated at Mapungubwe hill and K2, two archaeological sites in southern Africa, raises questions about the last occupation date of the hill, J. Archaeol. Sci. 38 (12) (2011) 3264–3277.

[25] A. Rousaki, A. Coccato, C. Verhaeghe, B.-O. Clist, K. Bostoen, P. Vandenabeele, L. Moens, Combined spectroscopic analysis of beads from the tombs of Kindoki, lower Congo Province (Democratic Republic of the Congo), Appl. Spectrosc. 70 (1) (2016) 76–93.

[26] P. Colomban, H.D. Schreiber, Raman signature modification induced by copper nano-particles in silicate glass, J. Raman Spectrosc. 36 (2005) 884–890.

[27] R.J.H. Clark, M.L. Curri, The identification by Raman microscopy and X-ray diffraction of iron-oxide pigments and of the red pigments found on Italian pottery fragments, J. Mol. Struct. 440 (1–3) (1998) 105–111.

[28] D. Parras, P. Vandenabeele, A. Sánchez, M. Montejo, L. Moens, N. Ramos, Micro-Raman spectroscopy of decorated pottery from the Iberian archaeological site of Puente Tablas (Jaén, Spain, 7th–4th century B.C.), J. Raman Spectrosc. 41 (1) (2010) 68–73.

[29] L. Medeghini, P.P. Lottici, C. De Vito, S. Mignardi, D. Bersani, Micro-Raman spectroscopy and ancient ceramics: applications and problems, J. Raman Spectrosc. 45 (2014) 1244–1250.

[30] M. Tascon, N. Mastrangelo, L. Gheco, M. Gastaldi, M. Quesada, F. Marte, Micro-spectroscopic analysis of pigments and carbonization layers on prehispanic rock art at the Oyola's caves, Argentina, using a stratigraphic approach, Microchem. J. 129 (2016) 297–304.

[31] H. Morillas, M. Maguregui, J. Bastante, G. Huallparimachi, I. Marcaida, C. García-Florentino, F. Astete, J.M. Madariaga, Characterization of the Inkaterra rock shelter paintings exposed to tropical climate (Machupicchu, Peru), Microchem. J. 137 (2018) 422–428.

[32] H. Gomes, P. Rosina, H. Parviz, T. Solomon, C. Vaccaro, Identification of pigments used in rock art paintings in Gode Roriso-Ethiopia using micro-Raman spectroscopy, J. Archaeol. Sci. 40 (11) (2013) 4073–4082.

[33] A. Hernanz, J.F. Ruiz-López, J.M. Gavira-Vallejo, S. Martin, E. Gavrilenko, Raman microscopy of prehistoric rock paintings from the Hoz de Vicente, Minglanilla, Cuenca, Spain, J. Raman Spectrosc. 41 (2010) 1394–1399.

[34] A. Hernanz, J. Chang, M. Iriarte, J.M. Gavira-Vallejo, R. de Balbín-Behrmann, P. -Bueno-Ramírez, A. Maroto-Valiente, Raman microscopy of hand stencils rock art from the Yabrai mountain, inner Mongolia autonomous region, China, Appl. Phys. A: Mater. 122 (2016) 699.

[35] L.C. Prinsloo, W. Barnard, I. Meiklejohn, K. Hall, The first Raman spectroscopic study of San rock art in the Ukhahlamba Drakensberg Park, South Africa, J. Raman Spectrosc. 39 (2008) 646–654.

[36] M. Iriarte, A. Hernanz, J.F. Ruiz-Lopez, Santiago Martín, μ-Raman spectroscopy of prehistoric paintings from the Abrigo Remacha rock shelter (Villaseca, Segovia, Spain), J. Raman Spectrosc. 44 (2013) 1557–1562.

[37] A. Rousaki, C. Bellelli, M. Carballido Calatayud, V. Aldazabal, G. Custo, L. Moens, P. Vandenabeelee, C. Vázquez, Micro-Raman analysis of pigments from hunter–gatherer archaeological sites of North Patagonia (Argentina), J. Raman Spectrosc. 46 (2015) 1016–1024.

[38] P. Vandenabeele, S. Bodé, A. Alonso, L. Moens, Raman spectroscopic analysis of the Maya wall paintings in Ek'Balam, Mexico, Spectrochim. Acta A 61 (10) (2005) 2349–2356.

[39] M. Maguregui, U. Knuutinen, J. Trebolazabala, H. Morillas, K. Castro, I. Martinez-Arkarazo, J.M. Madariaga, Use of in situ and confocal Raman spectroscopy to study the nature and distribution of carotenoids in brown patinas from a deteriorated wall painting in Marcus Lucretius House (Pompeii), Anal. Bioanal. Chem. 402 (2012) 1529–1539.

[40] G. Barone, D. Bersani, P.P. Lottici, P. Mazzoleni, S. Raneri, U. Longobardo, Red gemstone characterization by micro-Raman spectroscopy: the case of rubies and their imitations, J. Raman Spectrosc. 47 (2016) 1534–1539.

[41] D. Bersani, G. Azzi, E. Lambruschi, G. Barone, P. Mazzoleni, S. Raneri, U. Longobardo, P.P. Lottici, Characterization of emeralds by micro-Raman spectroscopy, J. Raman Spectrosc. 45 (2014) 1293–1300.

[42] https://soprano.kikirpa.be/. (Accessed 22 June 2020).

[43] https://rruff.info/. (Accessed 22 June 2020).

[44] http://www.irug.org/. (Accessed 22 June 2020).

[45] K. Castro, M. Pérez-Alonso, M.D. Rodríguez-Laso, L.A. Fernández, J.M. Madariaga, On-line FT-Raman and dispersive Raman spectra database of artists' materials (e-VISART database), Anal. Bioanal. Chem. 382 (2005) 248–258.

[46] L. Burgio, R.J.H. Clark, Library of FT-Raman spectra of pigments, minerals, pigment media and varnishes, and supplement to existing library of Raman spectra of pigments with visible excitation, Spectrochim. Acta A 57 (7) (2001) 1491–1521.

[47] N.C. Scherrer, Z. Stefan, D. Francoise, F. Annette, K. Renate, Synthetic organic pigments of the 20th and 21st century relevant to artist's paints: Raman spectra reference collection, Spectrochim. Acta A 73 (3) (2009) 505–524.

[48] A. Coccato, J. Jehlička, L. Moens, P. Vandenabeele, Raman spectroscopy for the investigation of carbon-based black pigments, J. Raman Spectrosc. 46 (2015) 1003–1015.

[49] I. Costantini, P.P. Lottici, D. Bersani, D. Pontiroli, A. Casoli, K. Castro, J.M. Madariaga, Darkening of lead- and iron-based pigments on late Gothic Italian wall paintings: energy dispersive X-ray fluorescence, μ-Raman, and powder X-ray diffraction analyses for diagnosis: presence of β-PbO_2 (plattnerite) and α-PbO_2 (scrutinyite), J. Raman Spectrosc. 51 (2020) 680–692.

[50] G.D. Smith, R.J.H. Clark, The role of H2S in pigment blackening, J. Cult. Herit. 3 (2) (2002) 101–105.

[51] D. Bersani, P.P. Lottici, A. Montenero, Micro-Raman investigation of Iron oxide films and powders produced by sol–gel syntheses, J. Raman Spectrosc. 30 (1999) 355–360.

[52] D.L.A. de Faria, F.N. Lopes, Heated goethite and natural hematite: can Raman spectroscopy be used to differentiate them? Vib. Spectrosc. 45 (2007) 117–121.

[53] H.G.M. Edwards, E.M. Newton, J. Russ, Raman spectroscopic analysis of pigments and substrata in prehistoric rock art, J. Mol. Struct. 550–551 (2000) 245–256.

[54] D.L.A. De Faria, S. Venâncio Silva, M.T. de Oliveira, Raman microspectroscopy of some iron oxides and oxyhydroxides, J. Raman Spectrosc. 28 (1997) 873–878.

[55] F. Buciuman, F. Patcas, R. Craciun, D.R.T. Zahn, Vibrational spectroscopy of bulk and supported manganese oxides, Phys. Chem. Chem. Phys. 1 (1999) 185–190.

[56] H.G.M. Edwards, N.F. Nik Hassan, P.S. Middleton, Anatase a pigment in ancient artwork or a modern usurper? Anal. Bioanal. Chem. 384 (2006) 1356–1365.

[57] S. Martinez-Ramirez, S. Sanchez-Cortes, J.V. Garcia-Ramos, C. Domingo, C. Fortes, M.T. Blanco-Varela, Micro-Raman spectroscopy applied to depth profiles of carbonates formed in lime mortar, Cem. Concr. Res. 33 (2003) 2063–2068.

[58] P.C. Gutiérrez-Neira, F. Agulló-Rueda, A. Climent-Font, C. Garrido, Raman spectroscopy analysis of pigments on Diego Velázquez paintings, Vib. Spectrosc. 69 (2013) 13–20.

[59] H.G.M. Edwards, S.E. Jorge Villar, Jan Jehlicka, Tasnim Munshi, FT–Raman spectroscopic study of calcium-rich and magnesium-rich carbonate minerals, Spectrochim. Acta A 61 (10) (2005) 2273–2280.

[60] L.C. Prinsloo, Rock hyraces: a cause of San rock art deterioration? J. Raman Spectrosc. 38 (2007) 496–503.

[61] M.B.D.T. da Costa, A.M. Arruda, L. Dias, R. Barbosa, J. Mirão, P. Vandenabeele, The combined use of Raman and micro-X-ray diffraction analysis in the study of archaeological glass beads, J. Raman Spectrosc. 50 (2) (2019) 250–261.

[62] A. Coccato, M. Costa, A. Rousaki, B.-.O. Clist, K. Karklins, K. Bostoen, A. Manhita, A. Cardoso, C. Barrocas Dias, A. Candeias, L. Moens, J. Mirão, P. Vandenabeele, Micro-Raman spectroscopy and complementary techniques (hXRF, VP-SEM-EDS, μ-FTIR and Py-GC/MS) applied to the study of beads from the Kongo Kingdom (Democratic Republic of the Congo), J. Raman Spectrosc. 48 (2016) 1468–1478.

[63] P. Colomban, A. Tournie, L. Bellot-Gurlet, Raman identification of glassy silicates used in ceramics, glass and jewellery: a tentative differentiation guide, J. Raman Spectrosc. 37 (2006) 841–852.

[64] P. Colomban, Polymerization degree and Raman identification of ancient glasses used for jewelry, ceramic enamels and mosaics, J. Non-Cryst. Solids 323 (2003) 180–187.

[65] C. Defeyt, J. Van Pevenage, L. Moens, D. Strivay, P. Vandenabeele, Micro-Raman spectroscopy and chemometrical analysis for the distinction of copper phthalocyanine polymorphs in paint layers, Spectrochim. Acta A 115 (2013) 636–640.

[66] P. Vandenabeele, L. Moens, H.G.M. Edwards, R. Dams, Raman spectroscopic database of azo pigments and application to modern art studies, J. Raman Spectrosc. 31 (2000) 509–517.

[67] W. Fremout, S. Saverwyns, Identification of synthetic organic pigments: the role of a comprehensive digital Raman spectral library, J. Raman Spectrosc. 43 (2012) 1536–1544.

[68] F. Schulte, K.-W. Brzezinka, K. Lutzenberger, H. Stege, U. Panne, J. Raman, Spectroscopy 39 (2008) 1455–1463.

[69] C. Cucci, G. Bartolozzi, M. De Vita, V. Marchiafava, M. Picollo, F. Casadio, The colors of Keith Haring: a spectroscopic study on the materials of the mural painting tuttomondo and on reference contemporary outdoor paints, Appl. Spectrosc. 70 (1) (2016) 186–196.

[70] P. Vandenabeele, B. Wehling, L. Moens, H. Edwards, M. De Reu, G. Van Hooydonk, Analysis with micro-Raman spectroscopy of natural organic binding media and varnishes used in art, Anal. Chim. Acta 407 (2000) 261–274.

[71] C. Daher, C. Paris, A.-.S. Le Hô, L. Bellot-Gurlet, J.-.P. Échard, A joint use of Raman and infrared spectroscopies for the identification of natural organic media used in ancient varnishes, J. Raman Spectrosc. 41 (2010) 1494–1499.

[72] M.C. Caggiani, P. Acquafredda, P. Colomban, A. Mangone, The source of blue colour of archaeological glass and glazes: the Raman spectroscopy/SEM-EDS answers, J. Raman Spectrosc. 45 (2014) 1251–1259.

[73] N.D. Bernardino, V.R.L. Constantino, D.L.A. de Faria, Probing the indigo molecule in Maya blue simulants with resonance Raman spectroscopy, J. Phys. Chem. C 122 (21) (2018) 11505–11515.

[74] L. Bergamonti, D. Bersani, D. Csermely, P.P. Lottici, The nature of the pigments in corals and pearls: a contribution from Raman spectroscopy, Spectrosc. Lett. 44 (2011) 453–458.

[75] J. Jehlička, H.G.M. Edwards, A. Oren, Raman spectroscopy of microbial pigments, Appl. Environ. Microbiol. 80 (11) (2014) 3286–3295.

[76] A. Rousaki, C. Vázquez, V. Aldazábal, C. Bellelli, M. Carballido Calatayud, A. Hajduk, E. Vargas, O. Palacios, P. Vandenabeele, L. Moens, The first use of portable Raman instrumentation for the in situ study of prehistoric rock paintings in Patagonian sites, J. Raman Spectrosc. 48 (2017) 1459–1467.

[77] D. Hutsebaut, P. Vandenabeele, L. Moens, Evaluation of an accurate calibration and spectral standardization procedure for Raman spectroscopy, Analyst 130 (2005) 1204–1214.

[78] P. Vandenabeele, F. Verpoort, L. Moens, Non-destructive analysis of paintings using Fourier transform Raman spectroscopy with fibre optics, J. Raman Spectrosc. 32 (2001) 263–269.

[79] P. Vandenabeele, T.L. Weis, E.R. Grant, L.J. Moens, A new instrument adapted to in situ Raman analysis of objects of art, Anal. Bioanal. Chem. 379 (2004) 137–142.

[80] A. Rousaki, M. Costa, D. Saelens, S. Lycke, A. Sánchez, J. Tuñón, B. Ceprián, P. Amate, M. Montejo, J. Mirão, P. Vandenabeele, A comparative mobile Raman study for the on field analysis of the Mosaico de los Amores of the Cástulo Archaeological Site (Linares, Spain), J. Raman Spectrosc. (2019) 1–11.

[81] C. Conti, A. Botteon, M. Bertasa, C. Colombo, M. Realini, D. Sali, Portable sequentially shifted excitation Raman spectroscopy as an innovative tool for in situ chemical interrogation of painted surfaces, Analyst 141 (15) (2016) 4599–4607.

[82] F. Pozzi, E. Basso, A. Rizzo, A. Cesaratto, T.J. Tague Jr., Evaluation and optimization of the potential of a handheld Raman spectrometer: in situ, noninvasive materials characterization in artworks, J. Raman Spectrosc. 50 (2019) 861–872.

[83] D. Lauwers, V. Cattersel, L. Vandamme, A. Van Eester, K. De Langhe, L. Moens, P. Vandenabeele, Pigment identification of an illuminated mediaeval manuscript De Civitate Dei by means of a portable Raman equipment, J. Raman Spectrosc. 45 (2014) 1266–1271.

[84] G. Barone, D. Bersani, A. Coccato, D. Lauwers, P. Mazzoleni, S. Raneri, P. Vandenabeele, D. Manzini, G. Agostino, N.F. Neri, Nondestructive Raman investigation on wall paintings at Sala Vaccarini in Catania (Sicily), Appl. Phys. A: Mater. 122 (2016) 838.

[85] J.M. Madariaga, M. Maguregui, S. Fdez-Ortiz de Vallejuelo, U. Knuutinen, K. Castro, I. Martinez-Arkarazo, A. Giakoumaki, A. Pitarch, In situ analysis with portable Raman and ED-XRF spectrometers for the diagnosis of the formation of efflorescence on walls and wall paintings of the Insula IX 3 (Pompeii, Italy), J. Raman Spectrosc. 45 (2014) 1059–1067.

[86] P. Ricciardi, P. Colomban, A. Tournié, V. Milande, Nondestructive on-site identification of ancient glasses: genuine artefacts, embellished pieces or forgeries? J. Raman Spectrosc. 40 (2009) 604–617.

[87] P. Colomban, A. Tournié, On-site Raman identification and dating of ancient/modern stained glasses at the Sainte-Chapelle, Paris, J. Cult. Herit. 8 (3) (2007) 242–256.

[88] Z. Petrová, J. Jehlička, T. Čapoun, R. Hanus, T. Trojek, V. Goliáš, Gemstones and noble metals adorning the sceptre of the Faculty of Science of Charles University in Prague: integrated analysis by Raman and XRF handheld instruments, J. Raman Spectrosc. 43 (2012) 1275–1280.

[89] I. Marcaida, M. Maguregui, H. Morillas, N. Prieto-Taboada, M. Veneranda, S.F.-.O. de Vallejuelo, A. Martellone, B. De Nigris, M. Osanna, J.M. Madariaga, In situ non-invasive multianalytical methodology to characterize mosaic tesserae from the House of Gilded Cupids, Pompeii, Herit. Sci. 7 (2019) 3.

[90] A. Rousaki, E. Vargas, C. Vázquez, V. Aldazábal, C. Bellelli, M. Carballido Calatayud, A. Hajduk, O. Palacios, L. Moens, P. Vandenabeele, On-field Raman spectroscopy of Patagonian prehistoric rock art: pigments, alteration products and substrata, TrAC Trends Analyt. Chem. 105 (2018) 338–351.

[91] A. Tournié, L.C. Prinsloo, C. Paris, P. Colomban, B. Smith, The first in situ Raman spectroscopic study of San rock art in South Africa: procedures and preliminary results, J. Raman Spectrosc. 42 (2011) 399–406.

[92] S. Lahlil, M. Lebon, L. Beck, H. Rousselière, C. Vignaud, I. Reiche, M. Menu, P. Paillet, F. Plassard, The first in situ micro-Raman spectroscopic analysis of prehistoric cave art of Rouffignac St.-Cernin, France, J. Raman Spectrosc. 43 (2012) 1637–1643.

[93] M. Maguregui, U. Knuutinen, I. Martínez-Arkarazo, A. Giakoumaki, K. Castro, J.M. Madariaga, Field Raman analysis to diagnose the conservation state of excavated walls and wall paintings in the archaeological site of Pompeii (Italy), J. Raman Spectrosc. 43 (2012) 1747–1753.

[94] N. Prieto-Taboada, S. Fdez-Ortiz de Vallejuelo, M. Veneranda, I. Marcaida, H.R. Morillas, M. Maguregui, K. Castro, E. De Carolis, M. Osanna, J.M. Madariaga, Study of the soluble salts formation in a recently restored house of Pompeii by in-situ Raman spectroscopy, Sci. Rep.-UK 8 (2018) 1613.

[95] C.V. Horie, Materials for Conservation: Organic Consolidants, Adhesives and Coatings, Architectural Press, Imprint of Butterworth-Heinemann, Oxford, 1987.

[96] R.J. Gettens, G.L. Stout, Painting Materials: A Short Encyclopedia, Dover Publications, Inc., New York, 1966.

[97] J.S. Mills, R. White, The Organic Chemistry of Museum Objects, Butterworth-Heinemann, Oxford, England, Boston, 1994.

[98] D. Lin-Vien, N. Colthup, W. Fateley, J. Grasselli, The Handbook of Infrared and Raman Characteristic Frequencies of Organic Molecules, Academic Press, London, 1991.

[99] P. Hendra, J. Agbenyega, The Raman Spectra of Polymers, Wiley, New York, 1994.

[100] M. Leona, Microanalysis of organic pigments and glazes in polychrome works of art by surface-enhanced resonance Raman scattering, Proc. Natl. Acad. Sci. U. S. A. 106 (35) (2009) 14757–14762.

[101] M. Leona, J. Stenger, E. Ferloni, Application of surface-enhanced Raman scattering techniques to the ultrasensitive identification of natural dyes in works of art, J. Raman Spectrosc. 37 (2006) 981–992.

[102] M. Fleischmann, P.J. Hendra, A.J. McQuillan, Raman spectra of pyridine adsorbed at a silver electrode, Chem. Phys. Lett. 26 (1974) 163–166.

[103] B. Guineau, V. Guichard, Identification of natural organic colorants by resonance Raman microspectroscopy and by surface-enhanced Raman effect (SERS) (Identification des colorants organiques naturels par microspectrometrie Raman de resonance et

par effet Raman exalte de surface (SERS)), in: K. Grimstad, J. Hill (Eds.), ICOM Committee for Conservation: 8th Triennial Meeting, Sidney, Australia, 6–11 September, Vol. 2, The Getty Conservation Institute, Marina del Rey, 1987, pp. 659–666.

[104] A. Cesaratto, M. Leona, F. Pozzi, Recent advances on the analysis of polychrome works of art: SERS of synthetic colorants and their mixtures with natural dyes, Front. Chem. 7 (2019) 105.

[105] F. Casadio, M. Leona, J.R. Lombardi, R. Van Duyne, Identification of organic colorants in fibers, paints, and glazes by surface enhanced Raman spectroscopy, Acc. Chem. Res. 43 (6) (2010) 782–791.

[106] E. Fazio, F. Neri, A. Valenti, P.M. Ossi, S. Trusso, R.C. Ponterio, Raman spectroscopy of organic dyes adsorbed on pulsed laser deposited silver thin films, Appl. Surf. Sci. 278 (2013) 259–264.

[107] D. Kurouski, S. Zaleski, F. Casadio, R.P. Van Duyne, N.C. Shah, Tip-enhanced Raman spectroscopy (TERS) for in situ identification of indigo and iron gall ink on paper, J. Am. Chem. Soc. 136 (24) (2014) 8677–8684.

[108] P. Matousek, I.P. Clark, E.R.C. Draper, M.D. Morris, A.E. Goodship, N. Everall, M. Towrie, W.F. Finney, A.W. Parker, Subsurface probing in diffusely scattering media using spatially offset Raman spectroscopy, Appl. Spectrosc. 59 (2005) 393–400.

[109] P. Matousek, M.D. Morris, N. Everall, I.P. Clark, M. Towrie, E. Draper, A. Goodship, A.W. Parker, Numerical simulations of subsurface probing in diffusely scattering media using spatially offset Raman spectroscopy, Appl. Spectrosc. 59 (12) (2005) 1485–1492.

[110] C. Conti, M. Realini, C. Colombo, P. Matousek, Subsurface analysis of painted sculptures and plasters using micrometre-scale spatially offset Raman spectroscopy (micro-SORS), J. Raman Spectrosc. 46 (2015) 476–482.

[111] C. Conti, M. Realini, C. Colombo, K. Sowoidnich, N.K. Afseth, M. Bertasa, A. Botteon, P. Matousek, Noninvasive analysis of thin turbid layers using microscale spatially offset Raman spectroscopy, Anal. Chem. 87 (2015) 5810–5815.

[112] P. Matousek, C. Conti, M. Realini, C. Colombo, Micro-scale spatially offset Raman spectroscopy for non-invasive subsurface analysis of turbid materials, Analyst 141 (2016) 731–739.

[113] C. Conti, M. Realini, C. Colombo, P. Matousek, Comparison of key modalities of micro-scale spatially offset Raman spectroscopy, Analyst 140 (2015) 8127–8133.

[114] P. Vandenabeele, C. Conti, A. Rousaki, L. Moens, M. Realini, P. Matousek, Development of a fiber-optics microspatially offset Raman spectroscopy sensor for probing layered materials, Anal. Chem. 89 (17) (2017) 9218–9223.

[115] A. Botteon, C. Colombo, M. Realini, S. Bracci, D. Magrini, P. Matousek, C. Conti, Exploring street art paintings by microspatially offset Raman spectroscopy, J. Raman Spectrosc. 49 (2018) 1652–1659.

[116] A. Rousaki, A. Botteon, C. Colombo, C. Conti, P. Matousek, L. Moens, P. Vandenabeele, Development of defocusing micro-SORS mapping: a study of a 19th century porcelain card, Anal. Methods 9 (45) (2017) 6435–6442.

[117] M. Realini, C. Conti, A. Botteon, C. Colombo, P. Matousek, Development of a full micro-scale spatially offset Raman spectroscopy prototype as a portable analytical tool, Analyst 142 (2017) 351–355.

[118] M. Realini, A. Botteon, C. Conti, C. Colombo, P. Matousek, Development of portable defocusing micro-scale spatially offset Raman spectroscopy, Analyst 141 (2016) 3012–3019.

[119] D. Lauwers, P. Brondeel, L. Moens, P. Vandenabeele, In situ Raman mapping of art objects, Philos. T. Roy. Soc. A 374 (2016), https://doi.org/10.1098/rsta.2016.0039, 20160039.

[120] G. Van Der Snickt, H. Dubois, J. Sanyova, S. Legrand, A. Coudray, C. Glaude, M. Postec, P. Van Espen, K. Janssens, Large-area elemental imaging reveals Van Eyck's original paint layers on the Ghent altarpiece (1432), rescoping its conservation treatment, Angew. Chem. Ger. Edit. 129 (2017) 4875–4879.

[121] F. Casadio, L. Toniolo, The analysis of polychrome works of art: 40 years of infrared spectroscopic investigations, J. Cult. Herit. 2 (2001) 71–78.

[122] S. Prati, E. Joseph, G. Sciutto, R. Mazzeo, New advances in the application of FTIR microscopy and spectroscopy for the characterization of artistic materials, Acc. Chem. Res. 43 (6) (2010) 792–801.

[123] F. Rosi, L. Cartechini, D. Sali, C. Miliani, Recent trends in the application of Fourier Transform Infrared (FT-IR) spectroscopy in Heritage Science: from micro-to non-invasive FT-IR, Phys. Sci. Rev. 4 (11) (2019), https://doi.org/10.1515/psr-2018-0006.

[124] M. Albini, S. Ridolfi, C. Giuliani, M. Pascucci, M.P. Staccioli, C. Riccucci, Multi-spectroscopic approach for the non-invasive characterization of paintings on metal surfaces, Front. Chem. 8 (2020), https://doi.org/10.3389/fchem.2020.00289.

[125] L. Damjanović, M. Gajić-Kvaščev, J. Đurđevića, V. Andrić, M. Marić-Stojanović, T. Lazić, S. Nikolić, The characterization of canvas painting by the Serbian artist Milo Milunović using X-ray fluorescence, micro-Raman and FTIR spectroscopy, Radiat. Phys. Chem. 115 (2015) 135–142.

[126] G. Germinario, I.D. van der Werf, L. Sabbatini, Chemical characterisation of spray paints by a multi-analytical (Py/GC–MS, FTIR, μ-Raman) approach, Microchem. J. 124 (2016) 929–939.

[127] A. Sarmiento, M. Pérez-Alonso, M. Olivares, K. Castro, I. Martínez-Arkarazo, L.A. Fernández, J.M. Madariaga, Classification and identification of organic binding media in artworks by means of Fourier transform infrared spectroscopy and principal component analysis, Anal. Bioanal. Chem. 399 (2011) 3601–3611.

[128] M. Veneranda, J. Aramendia, L. Bellot-Gurlet, P. Colomban, K. Castro, J.M. Madariaga, FTIR spectroscopic semi-quantification of iron phases: a new method to evaluate the protection ability index (PAI) of archaeological artefacts corrosion systems, Corros. Sci. 133 (2018) 68–77.

[129] D. Bikiaris, S. Daniilia, S. Sotiropoulou, O. Katsimbiri, E. Pavlidou, A.P. Moutsatsou, Y. Chryssoulakis, Ochre-differentiation through micro-Raman and micro-FTIR spectroscopies: application on wall paintings at Meteora and Mount Athos, Greece, Spectrochim. Acta A 6 (1999) 3–18.

[130] V. Ganitis, E. Pavlidou, F. Zorba, K.M. Paraskevopoulos, D. Bikiaris, A post-Byzantine icon of St Nicholas painted on a leather support. Microanalysis and characterisation of technique, J. Cult. Herit. 5 (2004) 349–360.

[131] A. Rizzo, Progress in the application of ATR-FTIR microscopy to the study of multi-layered cross-sections from works of art, Anal. Bioanal. Chem. 392 (2008) 47–55.

[132] E.A. Willneff, B.A. Ormsby, J.S. Stevens, C. Jaye, D.A. Fischer, S.L.M. Schroeder, Conservation of artists' acrylic emulsion paints: XPS, NEXAFS and ATR-FTIR studies of wet cleaning methods, Surf. Interface Anal. 46 (2014) 776–780.

[133] M. Bertasa, E. Possenti, A. Botteon, C. Conti, A. Sansonetti, R. Fontana, J. Striova, D. Sali, Close to the diffraction limit in high resolution ATR FTIR mapping: demonstration on micrometric multi-layered art systems, Analyst 142 (2017) 4801–4811.

[134] I.M. Cortea, R. Cristache, I. Sandu, Characterization of historical violin varnishes using ATR-FTIR spectroscopy, Rom. Rep. Phys. 68 (2) (2016) 615–622.

[135] P. Rosina, H. Collado, S. Garcês, H. Gomes, N. Eftekhari, M. Nicoli, C. Vaccaro, Benquerencia (La Serena - Spain) rock art: an integrated spectroscopy analysis with FTIR and Raman, Heliyon 5 (2019), https://doi.org/10.1016/j.heliyon.2019.e02561.

[136] A. Bonneau, D.G. Pearce, A.M. Pollard, A multi-technique characterization and provenance study of the pigments used in San rock art, South Africa, J. Archaeol. Sci. 39 (2012) 287–294.

[137] L.C. Prinsloo, A. Tournié, P. Colomban, C. Paris, S.T. Bassett, In search of the optimum Raman/IR signatures of potential ingredients used in San/Bushman rock art paint, J. Archaeol. Sci. 40 (2013) 2981–2990.

[138] J. Bell, P. Nel, B. Stuart, Non-invasive identification of polymers in cultural heritage collections: evaluation, optimisation and application of portable FTIR (ATR and external reflectance) spectroscopy to three-dimensional polymer-based objects, Herit. Sci. 7 (2019) 95.

[139] D. Saviello, L. Toniolo, S. Goidanich, F. Casadio, Non-invasive identification of plastic materials in museum collections with portable FTIR reflectance spectroscopy: reference database and practical applications, Microchem. J. 124 (2016) 868–877.

[140] F. Rosi, C. Miliani, C. Clementi, K. Kahrim, F. Presciutti, M. Vagnini, V. Manuali, A. Daveri, L. Cartechini, B.G. Brunetti, A. Sgamellotti, An integrated spectroscopic approach for the non-invasive study of modern art materials and techniques, Appl. Phys. A Mater. Sci. Process. 100 (2010) 613–624.

[141] C. Invernizzi, T. Rovetta, M. Licchelli, M. Malagodi, Mid and near-infrared reflection spectral database of natural organic materials in the cultural heritage field, Int. J. Anal. Chem. 2018 (2018), https://doi.org/10.1155/2018/7823248, 7823248.

[142] J.M. Madariaga, M. Maguregui, K. Castro, U. Knuutinen, I. Martínez-Arkarazo, Portable Raman, DRIFTS, and XRF analysis to diagnose the conservation state of two wall painting panels from Pompeii deposited in the Naples National Archaeological Museum (Italy), Appl. Spectrosc. 70 (1) (2016) 137–146.

[143] M. Manfredi, E. Barberis, M. Aceto, E. Marengo, Non-invasive characterization of colorants by portable diffuse reflectance infrared Fourier transform (DRIFT) spectroscopy and chemometrics, Spectrochim. Acta A 181 (2017) 171–179.

[144] P. Siozos, A. Philippidis, D. Anglos, Portable laser-induced breakdown spectroscopy/diffuse reflectance hybrid spectrometer for analysis of inorganic pigments, Spectrochim. Acta B 137 (2017) 93–100.

[145] I. Roberta, B. Susanna, C. Emma, M. Barbara, An integrated multimethodological approach for characterizing the materials and pigments on a sarcophagus in St. Mark, Marcellian and Damasus catacombs, Appl. Phys. A 121 (2015) 1235–1242.

[146] M. Aceto, A. Agostino, G. Fenoglio, V. Capra, E. Demaria, P. Cancian, Characterisation of the different hands in the composition of a 14th century breviary by means of portable XRF analysis and complementary techniques, X-Ray Spectrom. 46 (2017) 259–270.

[147] N. Odisio, M. Calabrese, A. Idone, N. Seris, L. Appolonia, J.M. Christille, Portable Vis-NIR-FORS instrumentation for restoration products detection: statistical techniques and clustering, Eur. Phys. J. Plus 134 (2019) 67, https://doi.org/10.1140/epjp/i2019-12469-5.

Spectroscopy and diffraction using the electron microscope

3

Philippe Sciau[a] and Marie Godet[b]

[a]CEMES, CNRS, Toulouse University, Toulouse, France [b]IPANEMA, CNRS USR 3461, MCC, UVSQ, Paris-Saclay University, Gif-sur-Yvette, France

1 Basic principles and main outlines

Electron microscopy is an analytical technique, whose story begins in the early 20th century, a few decades after the discovery of electron and its properties [1,2]. The way was opened by Louis de Broglie and Hans Busch's work which established the wave aspect of electrons and laid the foundation of electron optics respectively. The resolution limit of microscopes depends on the wavelength of the primary beam used. The smaller the wavelength, the better the resolution and the higher the magnification. The wavelengths of electron beams are normally much smaller ($\ll 1$ nm) than the wavelength (380–740 nm) of light sources used in optical microscopes. Thus the idea to design an electron microscope was quickly considered to increase the magnification by several orders of magnitude. Only a few years after Louis de Broglie and Hans Busch's work, the first transmission electron microscope was developed in Germany allowing images to be obtained with a resolution of a few tens of nanometers.

The other advantage of electrons is related to their energy range, which is similar to that of X-rays. Electron spectroscopy provides information concerning the elemental composition, the valence and the speciation of elements. The small wavelength of electrons makes it possible to perform diffraction experiments on crystals and determine their crystallographic characteristics.

Based on photonic or optical microscopes, the transmission electron microscope (TEM) was the first developed as previously mentioned. The photon source (white light) is replaced by the electron gun and the optical lenses by magnetic lenses. In the first microscopes, the enlarged image was projected on a phosphorescent screen or on photographic plates. Charge-coupled device (CCD) sensors now replace them. Over time, various spectrometers were integrated in the microscopes to analyze the X-ray emission or the energy lost by the electrons after passing through the sample. The improvement of electron lenses with the correction of asymmetrical beam distortions and then the development of spherical and chromatic aberration

Spectroscopy, Diffraction and Tomography in Art and Heritage Science. https://doi.org/10.1016/B978-0-12-818860-6.00001-5

correctors allow us nowadays to reach the atomic resolution. Thus a wide variety of nanometer- and atomic-resolution information can be obtained, including not only the atomic positions but also the nature of the atoms and how they are bonded to each other. Indeed, TEM is an essential tool for nanoscience in both biological and materials fields, and also in that of cultural heritage materials, as we will show in Section 4.

A few years after the achievement of the TEM prototype, the basics of scanning electron microscopy (SEM) were laid, i.e., the use of an electron beam scanner to produce an image, and the first scanning electron microscope was then designed by Manfred von Ardenne in Berlin in 1937. Unlike in TEM, the images are not formed by the electron lenses, which are only used to focus the electron beam on the sample surface in order to have the smallest footprint. The images result directly from the interactions between electrons and atoms in the sample. They are rebuilt from the recording of an emission signal (secondary electrons, backscattering electrons, X-ray fluorescence, cathodoluminescence, specimen current, diffracted electrons, etc.) step-by-step during the scanning of the sample by the electron beam. Various signals can be used to obtain images depending on the detectors available in the microscope. However, since it is not possible to equip a microscope with all the different detector types, most of them have standard equipment including at least a secondary electron detector. Other detectors are added to suit the intended application. As the propagation of electrons is very limited in air, conventional electron microscopes operate under vacuum and require the deposition of a conductive layer on the surface of nonconducting samples. In addition, samples producing a significant amount of vapor, such as wet biological specimens or oil-bearing rock, must be either dried or cryogenically frozen before examination. However, since the 1980s, environmental scanning electron microscopy (ESEM) has been developed thanks to differential pumping between the SEM column and the specimen chamber [3,4]. While the SEM column is under a high vacuum (as in conventional SEM), the sample and detectors are in a higher-pressure gaseous environment. A series of small apertures allows the beam to pass from the high vacuum in the electron gun to the suitable pressure at the level of the sample. Insulating materials can likewise be investigated without carbon or metallic coating deposition. Gas molecules surrounding the sample allow the removal of the surface charges induced by the electron beam. The charge neutralization does not require a high pressure. Pressures of a few Pa are enough for that and some microscopes, without being true ESEM, were developed to operate with specimen chamber under partial vacuum. One has to note that the presence of gaseous environment around the sample degrades the performances of microscopes. Operating under low pressures can be a good compromise to study insulating samples for which the deposition of a conductive layer can be critical. It is often difficult to take off the coating and its presence can alter some valuable samples. There are several examples of investigation of cultural heritage materials in the literature, taking advantage of the benefits of ESEM. For instance, the gaseous environment of the sample in an ESEM allows one to study the carbonation of hydroxides, which is an essential point in the consolidation and strengthening of decayed carbonate stones [5]. In a review paper focused on the analytical methods applicable to the study of ancient gilded art-objects from the European cultural heritage, the authors showed some examples of ESEM investigations [6].

2 Electron/matter interactions

The different analytical techniques implemented in electron microscopy are based on electron/matter interactions, which describe the effect of an electron beam on the atoms. The atoms are too small to be observed directly, but the disturbances they generate on electron beams are measurable and can be used to obtain indirect information about their chemical nature and spatial organization, as we shall show. Detailed descriptions of the different types of interactions can be found in scientific books or on online websites such as that of Swiss Federal Institute of Technology in Zurich (ETH Zurich) dedicated to electron microscopy (Properties of Electrons, their Interactions with Matter and Applications in Electron Microscopy, by Frank Krumeich. https://www.microscopy.ethz.ch/downloads/Interactions.pdf). In this chapter we will focus only on the outlines needed to understand the images and the spectroscopies presented.

Schematically, three cases can be considered in the electron/atom interaction: (i) the electron does not interact with the atom, (ii) its trajectory is modified by the atom or (iii) it is absorbed by the atom (Fig. 1). Since the electron is a charged particle, it strongly interacts with atoms. Thus recording electrons that have gone through matter without being deviated or absorbed is only possible for very thin samples.

Two cases must be considered for the modification of a trajectory according to whether or not there is energy transfer between electrons and atoms. If electrons lose no energy—elastic scattering case—the coherence of associated electron waves is preserved and interference phenomena can occur among the scattered waves emitted by each atom. For instance, when the atoms are organized in a 3D periodic lattice

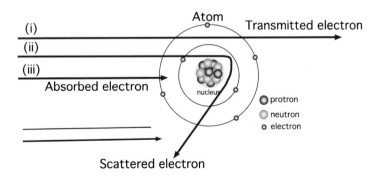

FIG. 1

Schematic representation of electron/atom interactions.

(i.e. a crystal), the scattered waves are in phase only in particular spatial directions. For the other directions, the resultant intensity of the scattered wave is zero. This phenomenon is called diffraction and consequently, from the study of the spatial distribution of scattered intensity, it is possible to determine the atomic organization, i.e., to solve the crystal structure. Even when the atoms are not in a lattice, the study of elastic scattering can give interesting information concerning their spatial distribution. As soon as the atoms are not randomly distributed, the spatial distribution of elastic scattering intensity presents local variations, which can be used to obtain information about the atomic organization.

Scattering can be either forward (small angles, mainly due to electron-electron interaction) or backward (high angles, mainly due to electron-nucleus interaction). The electrons that have been backscattered (called backscattered electrons) are linked to the atomic number Z of the atoms encountered. The bigger the nucleus (high Z) the more electrons are scattered. In other words, the backscattered intensity increases with the atomic number. Therefore the images obtained by recording the backscattered electrons are sensitive to chemical composition, they are called Z-contrast images. Furthermore, it is not necessary to limit the measurements to electrons that have kept their energy. Electrons, which have lost a small amount of energy, can also be used in the imaging process. Backscattered electron (BSE) imaging is widely used in SEM, as we will see in the next section.

When an accelerated electron of several kV is absorbed by an atom, the latter is ionized, i.e., one of its electrons is ejected. The ejected electrons are called secondary electrons and can be collected to build images of the sample surface. Their energies (<50 eV) are lower than those of the backscattered electrons and only the secondary electrons coming from the surface or the near-surface regions of the sample can escape and be caught by the detector. They are more sensitive to the topography of the sample surface as shown in Fig. 2. Indeed, the number of secondary electrons that can leave the sample surface is higher at edges than in flat areas, leading then to an increase of brightness at edges.

After ionization, the atoms do not remain in the excited state. There is a relaxation step where the atoms come back to ground state. An electron of a more external shell

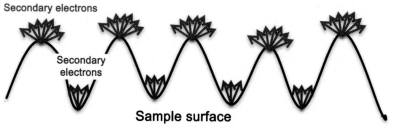

FIG. 2

Sensitivity of secondary electrons to the surface topography. Significantly more electrons can leave the surface at edges (*red color*) than in flats (*blue color*).

in the atom replaces the ejected electron and the energy difference between the two configurations is released. If the ejected electron comes from a core shell, the energy can be released through the emission of X-ray photons or Auger electrons. Since the core electrons do not participate in chemical bonds, the energy of emitted X-ray photons depends only very slightly in the atomic environment and the emitted X-ray photons are characteristic of the atom. Thus the analyses in energy of emitted X-ray photons can be used to carry out element analysis. This X-ray emission is called X-ray fluorescence (XRF). In a scanning or transmission electron microscope, the energy analysis of emitted X-ray photons is usually performed using dispersive devices, and this spectroscopy technique is called energy-dispersive X-ray spectroscopy with EDX or EDS acronyms.

3 Scanning electron microscopy

3.1 Imaging modes

Most scanning electron microscopes are equipped with both secondary and backscattered electron detectors. Fig. 3 shows the same area of surface glaze of a Chinese ceramic, imaged using these two signals. The images obtained from secondary electrons show clearly the topography of the surface while the difference of element composition between the crystals and the glassy matrix is better revealed using

FIG. 3

Surface of sauce glaze of a Chinese ceramic from Nord Song Dynasty observed using secondary electrons *(on the left)* and backscattered electrons *(on the right)*.

Credit: Philippe Sciau.

backscattered electrons. In this example, the dendritic crystals are iron oxides and so contain a heavier element (Fe) than in the glassy matrix, which is mainly constituted of silicon (Si), aluminum (Al) and oxygen (O). During the scanning, when the beam hits an iron-rich crystal, the intensity of backscattered electrons is more intense than in the case of a beam hitting the matrix. On the other hand, the secondary electrons allow highlighting that the crystals partially come out of the matrix.

By changing the energy of the incident electrons, i.e., the accelerating voltage, one can modify the investigated depth as shown in Fig. 4. In the previous example, the dendritic structures are not only located at the sample surface but also have an in-depth extension. The low accelerating voltages (2 or 5 kV) limit the depth of the analyzed zone to a few hundred nanometers and only the structures very close to the sample surface can be observed (Fig. 5). The increase of accelerating voltage (15 and 20 kV) increases the penetration depth making the deeper structures visible. However, this increase of the incident electron energy also increases the width of the zone analyzed and thus decreases the spatial resolution of the images.

An XRF signal can also be used to obtain images. Following the secondary and backscattered electron cases, the XRF signal is recorded at each pixel of the image during the scanning by the incident electron beam. Then, maps can be plotted by assigning the intensity of characteristic X-ray emission lines to each pixel. Spectral data can also be processed to quantify the elemental composition at the pixel scale and to obtain quantitative maps (Fig. 6). However, since the X-ray photons come from a bigger volume than the secondary or backscattered electrons, the spatial resolution is much lower. Accelerating voltage and current can be lowered to increase the resolution, but the accelerating voltage must be high enough to ionize atoms and the current determines the acquisition time. In the selected example, the Fe quantification was performed using the Kα line, which requires an excitation energy higher than 7 kV. In practice, it is difficult to achieve a spatial resolution better than half a micron.

3.2 Spectroscopy analysis

The spectral analysis of X-ray emission consists of counting the number of X-ray photons as a function of their energies. For this purpose, two methods were developed. The first one, called wavelength-dispersive X-ray spectroscopy (WDXS or WDS), uses an analyzer crystal to select a wavelength and to count the number of emitted photons by the sample at this selected wavelength. The spectrum is obtained by scanning the wavelength range of interest step-by-step. The spectral resolution and the sensitivity of this method are very good since only a few photons can be detected. This method is used in electron microprobes and allows very close X-ray emission lines to be separated. It is a very efficient way of achieving highly accurate quantitative analysis including trace element quantification. However, this method is time consuming and the presence of several analyzer crystals is cumbersome. Thus another method is often preferred in scanning electron microscopes.

FIG. 4

Backscattered electron images performed at different accelerating voltages from 2 kV (*top left*) to 20 kV (*bottom right*). The image modifications come from the depth variation of the analyzed volume (Fig. 5).

Credit: Philippe Sciau.

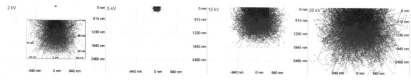

FIG. 5

Simulation of backscattered electrons *(cerise color)* and X-ray emissions *(blue color)* paths versus the accelerating voltages in iron(III) oxide. With a 2-kV accelerating voltage, the backscattered electrons come from only a few nanometers (15 nm) close to the surface while at 20 kV they come from more than 500 nm.

Credit: Christophe Deshayes.

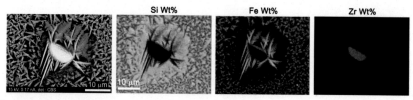

FIG. 6

Elemental maps recorded from a *brown* glaze of a Chinese ceramic (Nord Song Dynasty) showing the iron oxide crystallization around of zircon crystal (SiZrO$_4$). Detailed information concerning the Chinese *brown* glazes can be found in refs. [7, 8].

Credit: Philippe Sciau.

It consists of measuring directly the energy of emitted X-rays by means of an energy-dispersive spectrometer. This method uses multichannel detectors, which are able to discriminate energy and count the photons, thus allowing one to obtain the X-ray spectrum quickly. However, its energy resolution and sensitivity are not as good as the first method for the quantification of low-element concentration. Typically, the concentration accuracy is about 5%–10% for element concentration superior to 1%, and this method is often semi-quantitative for concentrations below this value. Schematically, two strategies can be applied.

The first involves recording an image of the area of interest especially in backscattered electron mode using the software supplied with the EDS device. Then, points or zones are selected to be scanned by the electron beam during the collection of the X-ray spectrum. The identification of emission peaks is not difficult. It is performed directly by the software; this allows fast identification of the elements present in the analyzed volume provided that the energy of incident electrons is high enough to ionize the atoms involved, which is an important point. We mentioned previously that the decrease of incident electron energy increases the spatial resolution, but this improvement is limited and depends on the atoms involved. Quantitative analysis based on peak intensities is performed using the internal software references or those added by the user. For instance, combined with Raman spectroscopy, the EDS analysis of a small number of points has emphasized the variation of the composition of

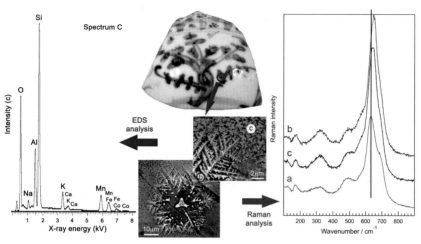

FIG. 7

SEM-EDS and Raman spectroscopy studies of a dark spot from a *blue-and-white* Chinese porcelain. The decor is characteristic of Chenghua or Hongzhi eras of Ming Dynasty *(top center)*. *Bottom center*, detail of the morphology of spinel crystals observed using backscattered electrons (15 kV, JEOL JSM 6490). *On the left*, X-ray emission spectrum from C point and on the right Raman spectra of points a, b and c. Details of the study can be found in ref. [9].

Credit: Ariane Pinto.

spinel crystals formed at the surface of blue and white porcelain during the firing [9]. Dark spots are observed at the glaze surface of some decors due to a rise of pigment during firing (Fig. 7). First, Raman spectroscopy was used to record the characteristic spectra of dendritic structures, which highlighted a significant shift of vibration modes between the crystal center (a) and outlying crystals (b and c). In a second study, the same zone was investigated by SEM-EDS after carbon deposition limited to the area of interest in order to obtain both a more accurate image of the morphology and elemental composition information. The crystal center is jacobsite type $(MnFeO_4)$, and the further we move from the center, the more the proportion of Mn increases (point c).

The second approach consists of collecting, a spectrum in each pixel of a selected area step-by-step as for the example illustrated in Fig. 8. Most of the commercial software supplied with EDS devices allows each spectrum to be processed separately and the quantitative analysis to be performed at the pixel scale. The elemental composition of specific points can be easily extracted from the data. The software also allows one to sum pixel spectra of defined area in order to determine its elemental composition. Thanks to modern detectors, the acquisition of all the spectra of an area of interest is not too time consuming, and this approach is well adapted to complex situations including several types of crystals as illustrated in Fig. 8. One hour was enough to record the 512×512 spectra of the selected area with sufficient statistics

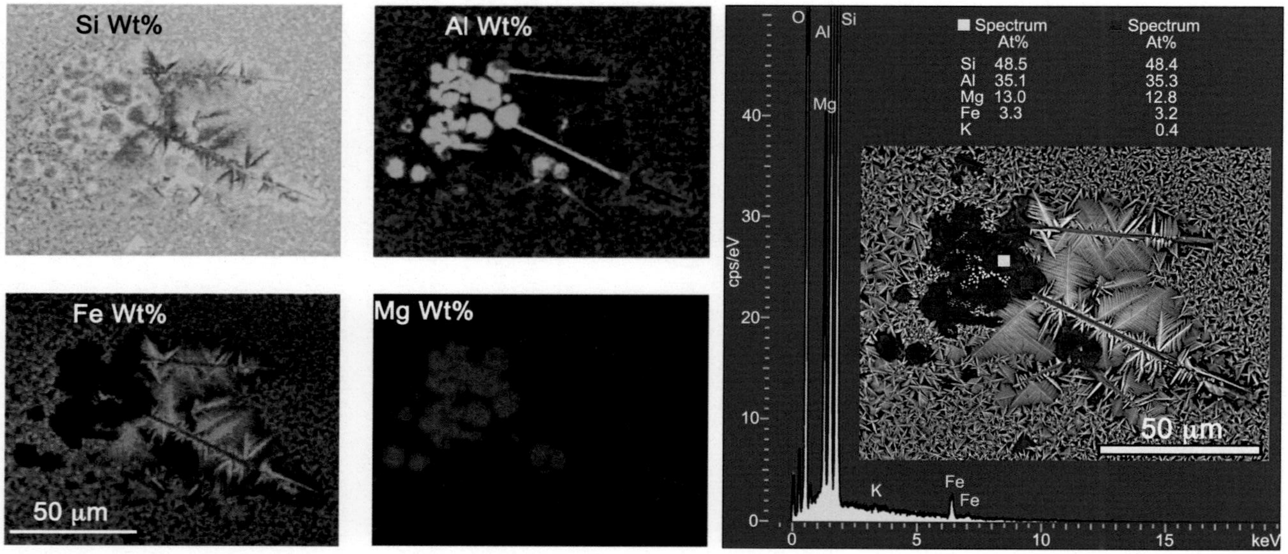

FIG. 8

SEM-EDS investigation of Chinese brown glaze from Yaozhou Song Dynasty Kiln. *On the left*, elemental maps revealing three types of crystals: Fe-rich dendrites, Al-rich needles and Mg-Al-rich pseudo-hexagonal crystals. *On the right*, the added spectra of the two areas marked on the SEM-BSE image. Quantitative results lead to a cordierite formula of $Mg_{1.4}Fe_{0.4}Al_{3.9}Si_{5.3}O_{18}$ with an accuracy of around 6% at. 15 kV, 1.4 nA, FEI Helios Nanolab 600i.

Credit: Philippe Sciau.

for an individual quantitative treatment. The elemental maps revealed the presence of three types of crystals whose composition can be specified from the extraction of pixel spectra of concerned zones. For instance, the spectra of Al- and Mg-rich zones lead to a composition corresponding to cordierite crystals ($Mg_2Al_4Si_5O_{18}$) probably containing Fe.

3.3 Diffraction

The very short wavelengths of accelerated electrons used in SEM limit the coherent Bragg diffraction to very small diffraction angle ranges. For instance, the wavelength of a 20-kV electron beam is around 0.087 Å. Thus the recording of a diffraction pattern up to 4 degree diffraction angle is equivalent to the diffraction pattern recorded up to 90 degree using Cu Kα X-ray radiation (1.5418 Å). In these conditions, the only way of recording a diffraction pattern from a bulk sample is in the reflection mode using grazing incidence. This geometry is used for the set-up of the reflection high-energy electron diffraction (RHEED) technique [10]. Thanks to the small incident angle and the short penetration length of the electrons in matter, only the atoms at the sample surface can contribute to the RHEED patterns. This technique is very efficient for characterizing the surface of crystalline samples such as a thin film deposited on a substrate [11]. However, it requires an extremely clean surface, and because of the very small angle of incidence, the spatial resolution is low. Furthermore, this technique is not well adapted to the study of complex materials found in ancient artifacts.

In addition to coherent Bragg diffraction, sharp lines, called "Kikuchi lines or Kikuchi bands", can be observed in electron diffraction patterns [12,13]. Because of the strong electron/matter interaction, a nonnegligible part of the incident electrons is scattered by the atoms close to the sample surface without significant energy loss. These scattered electrons may behave as secondary electron point-sources and strongly diverge. In a crystal, this new electron beam is diffracted by the crystalline planes and emerges from the sample surface exposed to the incident beam. This emerging beam is responsible of the formation of the Kikuchi lines. Kikuchi lines are present on RHEED patterns, but may also be observed beyond the coherent Bragg diffraction range. Kikuchi line patterns can then be obtained with high-angle incident beams. In fact, geometrical considerations show that the best contrast is achieved when the sample is strongly tilted, and when the incident beam makes an angle of about 20 degree with the sample surface. In SEM, this technique is developed under the name of electron backscatter diffraction (EBSD). The Kikuchi pattern is a characteristic of the crystallographic structures and thus can be used to identify the crystals. Each band can be indexed by the Miller indices of the corresponding diffracting planes (Fig. 9). In most cases, the indexation of three intersecting bands is enough to define the crystal orientation without ambiguity. Several commercial software programs have been developed to perform indexing based on the international crystal databases [14]. However, the technique is not so efficient to identify unknown phases. The knowledge of the phases present in the studied area is often required

FIG. 9

FIB preparation of a glazed ceramic sample (at the glaze-ceramic interface); (A) application of platinum layer to protect the surface of interest (SE image), (B) and (C) milling using a Ga⁺ ion beam to excavate a foil of about 5–10 μm across and a few microns thick (SE and BSE images), (D) securing the foil on the Cu grid (SE image), (E) thinning process up to electron transparency (BSE image), (F) final result, a thinner area has been created on the right for HRTEM analysis (SE image). Instrument: FEI Helios 600i – EDS at Centre de Microcaractérisation Raimond Castaing (Toulouse, France).

Credit: Claudie Josse.

before performing an EBSD investigation. The crystal orientation obtained relates the orientation of each sampled point to a reference crystal orientation.

For carrying out an EBSD investigation, a perfectly flat, polished and clean specimen is placed in the SEM chamber at a 70 degree tilted angle or on a suitable sample holder (Fig. 10). Then, the selected area is scanned with a focused electron beam as for the other SEM imaging techniques. For each pixel point, the Kikuchi pattern is recorded using a camera. Then, each pattern is indexed using the commercial software often provided with the detector. From the indexation results, different maps can be extracted such as a phase map, giving the spatial distribution of crystalline phases, or an orientation map, showing the spatial distribution of crystal orientation [14]. Information about strain can also be deduced from Kikuchi patterns, but it is rarely used in the field of cultural heritage material. SEM-EBSD is more and more used in addition to SEM-EDS in cultural heritage problems. For instance, SEM-EBSD was used to characterize calcium phosphate phases in a study of an inorganic consolidate for damaged carbonate stones [15]. Associated with micro-Raman spectroscopy and X-ray fluorescence analysis, the coupling SEM-EDS/EBSD also gave interesting results concerning the pigment materials used by medieval artists, allowing the identification of rare mineral mixtures [16]. A recent paper clearly outlines the SEM-EBSD potential in the structural and chemical examinations of archeological ceramics [17]. Thus it is shown that the good spatial resolution of the EBSD technique allows us to obtain elemental and phase maps permitting a good separation of clay and temper. The identification of mineral phases and intergrowths of temper particles provide essential information for clarifying clay procurement and firing techniques.

4 Transmission electron microscopy

Transmission electron microscopy (TEM) is less used than SEM in the field of cultural heritage as it is a more sophisticated facility needing a more complex sample preparation. However, it is the most powerful technique for materials investigation at

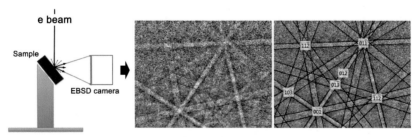

FIG. 10

Schematic diagram of EBSD set-up *(on the left)* using a suitable sample holder and EBSD pattern with its indexation *(on the right)*.

Credit: Philippe Sciau.

the nanoscale (and now subnanoscale) as it can produce atomic-resolution images and gives information about the sample chemistry and crystallography at the same time.

To inquire further into the subject, the reader is encouraged to consult the very exhaustive book written by Williams and Carter [18].

4.1 Sample preparation

If it is possible to examine a specimen in SEM without any preparation or only after simple metallization, TEM investigation always requires sample preparation. The aim of preparation is to make the sample thin enough to become transparent to the electron beam (typically less than 100 nm).

Sample preparation is crucial for TEM analysis. Indeed, the quality of the results will depend of the specimen preparation, especially sample thickness. The preparation methods available are numerous and will be chosen depending both on the object to be analyzed (material type, properties, etc.) and the type of analysis to be conducted (for instance for HRTEM or EELS techniques, which are detailed later, the thickness has to be less than 50 nm). Furthermore, practical factors such as time constraints or equipment availability also have to be taken into account.

A good preparation is already half of the work done. This is why a discussion between the people who know the specimen (curators/conservation scientists) and the TEM scientists is crucial to transfer all the knowledge about the specimen (material properties, alteration) and benefit from the microscopist's experience and skills.

Ancient materials show some special features compared to the current materials. They are often very heterogeneous at several scales, showing a macro-, micro- and nano-structuration, altered by the environment and are rare (small amount of material available, meaning you often cannot restart a preparation if you fail). It is thus indispensable to choose carefully a region of interest representative of the properties you want to examine. Indeed, TEM analysis is very local (a few microns maximum), so this choice will be determinant if you want to conclude something about the bulk object. This is why performing optical and scanning electron microscopies before TEM is essential to have relevant details concerning the morphology and the chemical composition of the object. As Williams and Carter [18] wisely put it: "know the forest before you start looking at the veins in the leaves on the trees".

A short selection of preparation methods, among the most used for ancient materials, will now be presented. To get more detailed information on the subject, the reader should look at the chapter dedicated to this topic in Williams and Carter [18] or the book of Ayache et al. [19]. Furthermore, one has to keep in mind that the field of specimen preparation is constantly evolving and that new ideas appear all the time.

The development of the focused ion beam technique (FIB) in the last two decades has revolutionized the field of specimen preparation. Nowadays numerous ancient materials are prepared using this facility [20–23]. This technique allows uniform thin foils of a wide variety of materials to be obtained with a very precise localization of the sampling area and is thus perfectly adapted to ancient materials (Fig. 9).

However, a FIB preparation can be relatively expensive (a few hundred euros per sample), and some fragile samples cannot be prepared using this technique as they would be destroyed under the Ga^+ ion beam. For rather homogeneous specimens (or those composed of particles), one easy preparation technique is to crush them into powder, put the particles into an inert solvent (like water or ethanol) and sonicate for a few minutes (to break the agglomerate) and finally put one drop on the metallic grid (different kinds of grids are available—Cu, Cu/Ni, Au—which can be covered by a carbon holey film to prevent small samples from falling). The grid is then fixed on the sample holder, which will be introduced to the microscope. Particles can also be collected directly on the grid in some cases (dust from the atmosphere, particles ejected during laser cleaning). Preparation has to be adapted to each case and this explains the resourcefulness of TEM sample preparers. For instance, Godet et al. investigated the nano-compounds responsible for the laser-induced yellowing effect [24]. The particles ejected during the laser cleaning were collected on a glass slide to be first analyzed by optical and scanning electron microscopies (Fig. 11). A copper grid was then simply rubbed gently on the slide and put into the TEM. Nanoparticles were easily observable, meaning that this preparation method is well adapted to this particular case.

If the sample is heterogeneous and we want to conserve the spatial information, different procedures are available including sandwich preparation and ultramicrotomy. For ceramic fragments covered by a slip or a glaze, a "sandwich" procedure combining mechanical and precision ion polishing is often used (described in detail in ref. [21]). Another technique principally used by microbiologists, but more and more used in the field of cultural heritage, is ultramicrotomy, which allows uniform

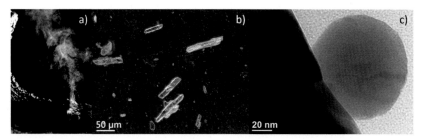

FIG. 11

For each sample, an adequate preparation has to be found: example of particles ejected during the laser cleaning of an encrusted substrate; (A) laser cleaning of a soiled stone showing a lot of particles ejected in a visible smoke, (B) ejected particles collected on a glass slide showing individual particle (OM image), (C) nanosphere observed on the surface of the particle (TEM-BF image).

Credit: Dominique Bouchardon (A), Marie Godet (B and C).

slices of multiphase materials to be obtained. The specimen is often embedded in resin using a specific mold before being cut into slices using a knife.

Finally, it should be recalled that the thinning process always has some effect on the specimen, modifying its structure and chemistry. It is thus very important to be able to recognize the artifacts introduced by each preparation method. Artifacts can be for instance: ion implantation, matter redisposition and amorphization during FIB preparation or ion polishing, distortion and breakage during mechanical preparation such as ultramicrotomy, etc. The best preparation method is the one that reveals the intrinsic nanostructure and chemistry without adding too many extrinsic defects.

4.2 The instrument

In practice, a TEM can be divided into three parts:

1. The illumination system
2. The stage of the specimen holder/objective lens
3. The imaging system

The illumination system includes the gun (electron source) and the condenser lenses, which take the electrons from the source to the sample. The electrons emerging from the sample are then dispersed by the objective lens to form a diffraction pattern in the back focal plane of the lens and recombined to create an image in the image plane. By selecting either the back focal plane or the image plane as its object, the intermediate lenses (comprising the imaging system) project either the diffraction pattern or the image on a viewing screen or currently on a camera.

The illumination system can be operated in two modes: parallel beam (TEM imaging and selected-area electron diffraction [SAED]) or convergent beam (STEM scanning imaging, convergent-beam electron diffraction [CBED]). Fig. 12 shows the different TEMs used to acquire some of the data presented below.

4.3 Electron diffraction (SAED and CBED)

We have seen that some electrons from the beam are deviated from their original trajectory after having passed through the sample. Investigating the distribution in angular deviation (corresponding to the diffraction pattern) reveals precious information on the specimen: crystallinity of the probed area (mono-crystal, polycrystal and amorphous compounds have very different diffraction patterns), crystallographic information (lattice parameters, symmetry), orientation of the specimen components relative to the beam and among themselves (texture), etc.

Diffraction patterns can be obtained in parallel or convergent beam modes:

– If the area of interest is "large" (more than a few hundred nanometers), a selection area diaphragm (SAD) aperture is inserted into the image plane of the objective lens to limit the diffraction signal to the area of interest. The diffraction pattern is then projected on the viewing screen or camera using the imaging system.

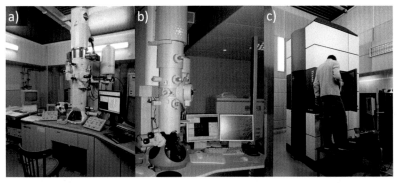

FIG. 12

Example of different TEMs used to acquire the data presented in this chapter; (A) JEOL 2100F (Institut de Minéralogie, de Physique des Matériaux et de Cosmochimie, CNRS UMR 7590, Paris, France), (B) FEI TECNAI F20 (Institut de Chimie et des Matériaux Paris-Est, CNRS UMR 7182, Thiais, France), (C) TITAN Qu-ant-em (Electron Microscopy for Materials Science, Antwerp, Belgium); note the height of Nicolas compared to that of the TEM (and Nicolas is very tall already!).

Credit: IMPMC (A), ICMPE (B), Marie Godet (C).

Sharply focused spot patterns called selected area electron diffraction (SAED) patterns are obtained.

– If the area of interest is smaller than several tens of nanometers and not beam sensitive, the electron beam can be focused to reduce the probing area. In this case, the diffracted rays leaving the specimen in a range of diffracted cones and disks are observed on the camera. If the angle of divergence is small, small disks are observed (microdiffraction mode); if the angle divergence is big, large disks having a nonuniform intensity are obtained on the so-called convergent beam electron diffraction (CBED) pattern. For more details, the reader should consult the reference text by Morniroli [25].

From a SAED pattern, much qualitative and quantitative information can be extracted such as the crystallinity, which can be assessed in just a glance.

– Amorphous: diffuse rings
– Polycrystalline (nano-sized grains): rings composed of spots showing different sizes (the broader the spot, the smaller the grain) and forms
– Monocrystalline: network of spots

The three types of crystallinity are illustrated in Fig. 13. The SAED pattern of an iron-colored lead transparent glaze is characteristic of an amorphous compound (glass). The pattern obtained from a fragment of black crust (soiling) collected on Saint-Denis Basilica (France) shows a polycrystalline nature typical of this heterogeneous system composed of a mixture of minerals, ashes, soot, etc. The last one acquired on a glazed ceramic sample (on a lead feldspar [$PbAl_2Si_2O_8$] crystal

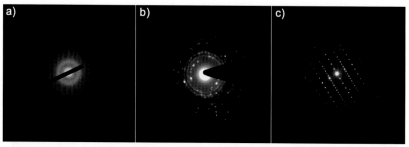

FIG. 13

SAED diffraction patterns obtained for (A) an amorphous lead glaze, (B) a polycrystalline fragment of a black encrustation collected on Saint-Denis Basilica (France), (C) a lead feldspar ($PbAl_2Si_2O_8$) monocrystal localized at the interface between a lead glaze and a white ceramic. Instrument: JEOL 2100F 200 kV for the glaze (FIB cuts) and TECNAI F20 200 kV for the black crust (powder).

Credit: Marie Godet.

localized at the interface between the glaze and the paste) reveals that the crystal is monocrystalline.

Each spot or ring position in the SAED pattern can be related to a diffracting plane in the specimen using the following relationship:

$$Rd = \lambda L,$$

where R is the distance between the direct beam and a diffraction spot or the radius of a ring, d is the corresponding interplanar spacing in the crystal, λ is the electron wavelength and L is the camera length (a constant linked to the microscope often chosen as 50 cm).

The object of this association is indexing the diffraction pattern. The ideal is to match experimental data with simulated diffraction patterns to confirm the nature of the specimen and deduce its crystallographic properties (orientation, interplanar spacings). To begin, preliminary results can also be obtained by measuring the spot/ring positions and comparing them with those of references [23]. In the case of polycrystalline compounds, a radial integration can be applied on the diffraction pattern, giving a 2D radial distribution of the diffraction planes. This method has been used, for instance, to identify hematite, spinel (close to $MgAl_2O_4$) and illite nanocrystals in pre-sigillata slips [26].

In the case of single crystals, several software such as Single Crystal (Crystal-Maker Software Ltd.) allow straightforward indexing of the diffraction patterns after entering some experimental data (distances from the center and angles).

One difficulty of electron diffraction (compared to X-ray diffraction) is that dynamical diffraction phenomenon cannot be neglected (the electron is diffracted several times in the sample). These dynamical effects contribute to the appearance of extra spots and make the spot intensity very difficult to interpret. One way of minimizing these effects is to perform precession electron diffraction (PED). In this case, the electron beam is deflected, and disoriented patterns are collected, showing reduced dynamical diffraction contributions.

Coupled to electron diffraction tomography (collection of a series of randomly oriented ED patterns at a fixed angular interval), 3D precession electron diffraction tomography has begun to be used for the investigation of cultural heritage materials [27]. This technique allows the atomic structure (space group) of single nanocrystals to be determined without prior knowledge and with improved intensity integration. Using this method, milarite-osumilite inclusions with a hexagonal symmetry (P6/mcc space group and unit cell $a=b=10.41$ Å and $c=13.84$ Å) have been identified in a blue glaze in a small pottery fragment (5–6th century BCE).

4.4 Imaging modes

TEM-imaging is linked to the spatial distribution of the beam scattered by the specimen. The electron beam can be considered as a wave and both its amplitude and phase will have changed after passing through the specimen. These modifications will be responsible for the appearance of two types of contrast in the image: amplitude contrast (bright field [BF] and dark field [DF] imaging) and phase contrast (high resolution transmission electron microscopy [HRTEM] imaging).

4.4.1 Amplitude contrast (BF and DF)

Before performing imaging, it is useful to first look at the diffraction pattern of the probed area as it gives precious information on how the specimen is scattering by allowing a straight visualization of the direct and scattered beams. To view an image, an aperture called the "objective aperture" is inserted into the back focal plane of the objective lens. This aperture is used to select either the direct beam to form a bright field image (BF imaging) or the diffracted beams to form a dark field image (DF imaging).

For both BF and DF images, the contrast is linked to:

- diffraction contrast: intensities of diffraction coming from different parts of the sample (diffracting areas appear darker in BF and brighter in DF mode)
- mass-thickness contrast: thickness and atomic number (Z) of atoms modify the contrast (thicker areas and heavy elements appear darker in BF and brighter in DF mode)

In the case of amorphous material (such as polymers), the diffraction contrast disappears, and the mass-thickness contrast will be responsible for the image contrast.

BF imaging is widely used to study the morphology of the specimen at the nanoscale. DF imaging is a powerful way to examine crystalline species as it can highlight portions of a specimen that are responsible for one or several diffraction spots (Figs. 14 and 15).

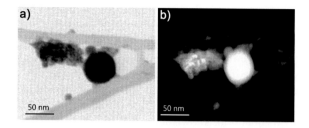

FIG. 14

BF (A) and DF (B) imaging of some particles ejected during a laser cleaning. Instrument: TECNAI F20 200 kV (ejected particles collected from a glass slide).

Credit: Eric Leroy.

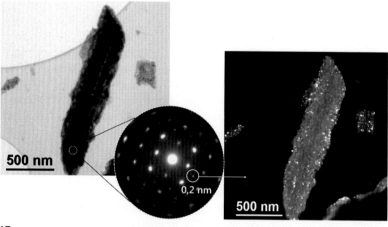

FIG. 15

Use of DF imaging to reveal nanocrystallites on the surface of a gypsum platelet ejected during laser cleaning; *left*: BF image of the gypsum platelet with a SAED diffraction pattern acquired on the dotted area; *right*: DF image corresponding to the selected diffracted beam at 0.2 nm. Instrument: TECNAI F20 200 kV (ejected particles collected from a glass slide).

Credit: Eric Leroy.

4.4.2 Phase contrast (HRTEM)

We now know that BF and DF images require that a single beam is selected using the objective aperture. A phase contrast image requires the selection of more than one beam using a large aperture. Indeed, as the transmitted and diffracted waves each have a different phase, we obtain an interference pattern called a "phase contrast" image or a high-resolution transmission electron microscopy image. The resulting phases depend on the crystallographic structure of the specimen. However, the phase-contrast image is not directly interpretable as supplementary phase modifications are created by the microscope aberrations (represented by the contrast transfer function) and sample thickness. The relationship between the image and the atomic positions is thus not straightforward. Complex image simulation is therefore required to model experimental images [18]. In the case of nano-objects too small for SAED analysis, HRTEM method can be used to obtain the diffraction pattern using a Fourier filtering of the phase contrast image (using a mask if needed to select only the region of interest) (Fig. 16).

4.5 Chemical analysis

In addition to morphological and crystallographic information obtained by imaging and diffraction respectively, chemical information can also be collected from the same area using inelastic scattering signals such as X-rays (X-ray dispersive energy [EDX] spectroscopy) or electrons that have lost some kinetic energy by interacting with the specimen (electron energy loss spectroscopy [EELS]).

Nowadays chemical analyses are usually performed in scanning mode (STEM) using a small probe instead of a parallel beam. Furthermore, using the traditional TEM mode implies condensing the beam to an appropriate size for the analysis, which can lead to misalignment of the illumination system. Besides, STEM is a spatially resolved method: it gives chemical information for each desired scanning point. As for SEM-EDX, hypermaps (often called data cube or spectrum-image in the case of TEM) can be recorded with a complete EDX or EELS spectrum for each pixel. The chemical analysis modalities in a transmission microscope are thus described in the section dedicated to STEM mode.

5 Scanning transmission electron microscopy
5.1 Principle

In this mode, TEM and scanning electron microscope (SEM) are combined to form a scanning transmission electron microscope (STEM). The electron beam is focused on a small spot (a few angstroms) and is scanned over the specimen. The signal is

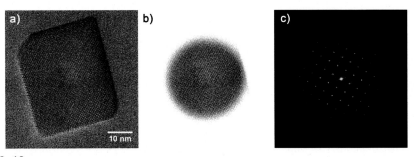

FIG. 16

(A) HRTEM image of a mullite crystal observed in a glaze and (B) application of a Fourier filtering using a mask to obtain (C) the diffraction pattern. Instrument: JEOL 2100F 200 kV on a FIB cut.

Credit: Marie Godet.

then detected using an electron detector, amplified and displayed on the computer screen. More and more TEMs can now switch over to STEM operation, and the user can now shift from TEM to STEM modes quite easily. Besides, with the development of probe aberration correctors, it is now possible to achieve atomic resolution imaging.

5.2 STEM imaging (BF, DF, HAADF)

As with the traditional TEM mode, BF and DF imaging can also be performed in STEM mode. To obtain a BF image, we use a BF detector placed in a plane conjugate to the back focal plane of the objective lens to collect only the direct beam in every scanned point. In the case of DF images, an annular dark-field (ADF) detector is used and it collects the scattered beams.

Another annular detector surrounding the ADF detector is used to collect the electrons diffracted at higher angles (>50 mrad). This high-angle annular dark field (HAADF) detector yields Z-contrast images as the intensities in this type of images are approximately proportional to Z^2. STEM-HAADF imaging is similar to SEM-BSE imaging but at the nano- and now atomic scale. As in SEM, this type of imaging is very interesting because Z-contrast images can be coupled to spectroscopic analysis using STEM-EDX and STEM-EELS.

5.3 STEM-EDX

As in SEM-EDS, STEM-EDX (the TEM community uses more frequently the acronym EDX while the SEM community prefers the acronym EDS for **E**nergy **D**ispersive **X**-ray **S**pectrometry) spectroscopy in TEM is used to detect and quantify the elemental chemical composition. As previously mentioned, this method is based on the core-shell ionization phenomenon. Under high-energy electron beam irradiation, the specimen atoms are ionized, i.e., inner shell or core electrons are ejected from the atom and create holes in the shells. To regain their stability, the excited atoms fill the holes with electrons from outer shells and this process leads to the emission of either an X-ray or an Auger electron. The X-ray energy will be equal to the energy difference between the inner and outer shells, which is characteristic of the atom encountered. Therefore collecting the energy distribution of the X-rays emitted (corresponding to an EDX spectrum) will provide the elemental composition of the specimen in the probed area. (Characteristic X-rays appear as sharp peaks in the spectrum, and the continuous background is due to bremsstrahlung X-rays generated by the interaction between the electronic beam and the nucleus.)

The probability of emitting an X-ray or an Auger electron is described by the fluorescence yield, which depends strongly on the atomic number Z. Briefly, heavy elements generate mostly X-rays (and thus EDX spectroscopy is suitable for this type of elements) whereas light elements generate mostly Auger electrons. In addition, the use of a high-energy beam (≥ 100 kV) in TEM is not favorable to ionization of light atoms. Such high energy does, however, allow core electrons of very heavy atoms such as Ag, Pb or Au to be ejected.

X-rays are emitted in an isotropic way from the specimen, but as the EDX detector collection angle is very small (0.03–0.3 sr), the collected signal intensity is often limited, and analysis can be difficult, especially as the sample is very thin. For instance, using EDX to analyze a very thin sample containing light elements is not a good idea. EELS would be more suitable in this case.

The first step when collecting an EDX spectrum is to identify the different elements present in the sample and to distinguish them from the artifacts such as escape peaks, sum peaks, and spurious X-rays emitted by some TEM components such as the pole pieces which are located next to the sample or by the metallic grid supporting the sample.

EDX point analysis, profiles and maps can be performed (Fig. 17). In particular, EDX mapping is very suitable to localize spatially the different chemical elements at the nanoscale in ancient heterogeneous materials.

If a full spectrum is recorded at each pixel of a map, we obtain a hypermap from which we can extract spectra and profiles. For instance, on Fig. 18, a hypermap of a complex core-shell titanium-based compound observed in a lead glaze has been recorded. Afterwards, spectra have been extracted from the five regions of interest identified on the hypermap and reveal that the core contains O, Ti and the crown O, Pb, Ti and Fe. In addition, two particles containing O and Fe and rectangular compounds composed of O, Al, Si and Pb have been evidenced near the core-shell compounds.

To make quantitative analysis, the Cliff-Lorimer ratio technique is almost always used by the software available with the EDX detector. As the specimen is thin, we can neglect absorption and fluorescence phenomenon using the "thin-film approximation". In this case, the ratio of the weight percent of two elements A and B is proportional to the X-ray intensities I_A and I_B following the equation:

$$\frac{C_A}{C_B} = k_{AB}\frac{I_A}{I_B}$$
$$C_A + C_B = 100\%$$

The term k_{AB} is called the "Cliff-Lorimer factor" and is a sensitivity factor linked to the accelerating voltage and the microscope (TEM and EDX spectrometer).

The equation can be extended easily to multicomponent systems and allows quantification of the weight percent of each component with an accuracy of 5%–10%.

5.4 STEM-EELS

Electron energy-loss spectroscopy (EELS) gives chemical and structural information on the specimen. An EELS spectrum is the energy distribution of the electrons that have passed through the specimen. Two regions are usually distinguished on the spectrum corresponding to different types of electronic excitations:

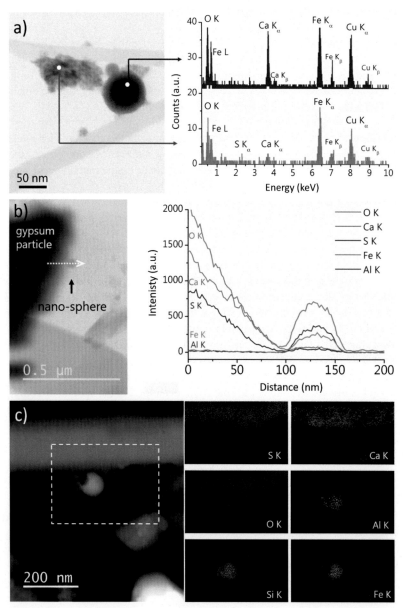

FIG. 17

Illustration of EDX (A) point, (B) profile and (C) map analysis (STEM-HAADF images). Instrument: TECNAI F20 200 kV, EDAX MeteK SDD Octave T Optima 60 mm² with no window, angle collection: 5 sr. The particles were cooled to −172°C.

Credit: Marie Godet.

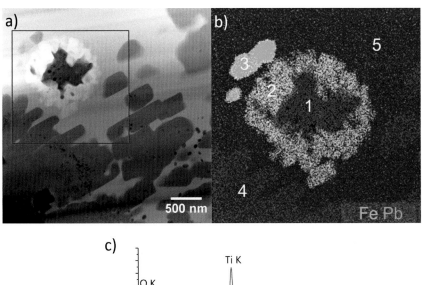

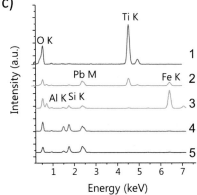

FIG. 18

EDX analysis of interface crystals in a glazed ceramic; (A) STEM-HAADF image; (B) EDX hypermaps of Ti K *(red)*, Fe K *(green)*, Pb M *(blue)* and Al K *(magenta)*, (C) EDX spectra extracted from different regions of the hypermap. Instrument: JEOL 2100F 200 kV JEOL Si(Li) detector (FIB cut).

Credit: Marie Godet.

- Low-loss region (0–50 eV): this region is dominated by the very intense zero-loss peak corresponding to the elastically scattered electrons. Other peaks observed in this area correspond to interband transitions and plasmas (collective oscillations of free electrons under the beam, often observed in high free-electron density materials such as metals).
- Core-loss region (>50 eV to a few keV): peaks observed in this area are absorption edges corresponding to electronic transitions from the inner shells to the empty states of the conduction band. Each absorption edge position is characteristic of a specific atom and of the shell involved in the transition

(K, L, M). The EELS edges also show intensity fluctuations called "fine structures" which give information on the atom speciation (coordination, valence). Investigation of these fine structures is called "energy loss near edge structure spectroscopy" (ELNES) and necessitates a subeV resolution, which is reachable using a FEG source and improvable using a monochromator.

EELS is complementary to EDX as it allows all the elements of the periodic table to be investigated including light elements (not affected by the fluorescence yield limitation) with atomic resolution and better analytical sensitivity. However, the elemental quantification is more difficult to perform than with EDX, the biggest limitation being that a very thin sample is needed (<50 nm).

During an EELS analysis, spectrum-images are most often acquired (lines or maps) from which EELS spectra and edge intensity maps can then be extracted. Fig. 19 illustrates the different possibilities of EELS analysis. In this example, iron-rich nanoparticles covering a gypsum ($CaSO_4 \cdot 2H_2O$) particle ejected during laser cleaning of a soiled stone have been investigated. A spectrum-image was acquired on the interface nanoparticle–substrate to localize the different chemical elements with a spatial resolution of 0.5 Å/pixel. From these data, edge intensity maps have been extracted from which the distribution of the elements can be seen at the subnanoscale. It was found, for instance, that the nanoparticles analyzed contain only O and Fe, and the substrate O, Ca and S (corresponding well to gypsum). EELS spectra were also extracted from the spectrum-image (with a resolution of about 1 eV) showing variations of the O K edge depending of the different areas.

In addition to spectrum-images, energy-filtered images can be collected. In this case, electrons of a specific energy are selected to form an image. Usually operated in TEM mode (EFTEM technique), it can also be operated in STEM mode to produce composition maps.

Furthermore, by using a monochromator the spectral resolution has been improved to 0.3 eV and ELNES spectroscopy has been conducted allowing magnetite (Fe_3O_4) to be identified by comparing the edge fine structures to those of iron oxide references (Fig. 20).

5.5 STEM-PACOM (precession-assisted crystal orientation mapping)

Diffraction patterns can be also collected in scanning mode. It is the base of the precession-assisted crystal orientation mapping (PACOM) technique also known commercially as the ASTAR™ technique (NanoMEGAS SPRL, Brussel). PACOM consists in collecting sequential electron diffraction patterns with a dedicated CCD camera while the specimen is scanned by an incident quasi-parallel nano beam [28,29]. Then, each individual pattern is indexed (phase and orientation determination) via comparison with previously generated reference patterns. Reference patterns are obtained from the crystallographic data of each expected phases. After this data processing, one obtains orientation and phase maps of the scanned area one.

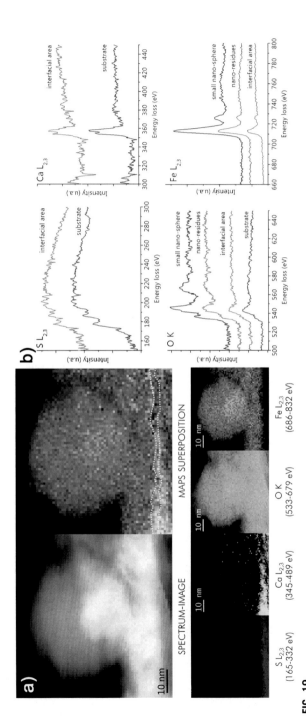

FIG. 19

EELS analysis of nanoparticles observed on an ejected gypsum particle during laser cleaning; (A) EELS data cube including a spectrum-image (0.5Å/pixel), extracted edge intensity maps and superposed (B) EELS edges (S $L_{2,3}$, Ca $L_{2,3}$, O K and Fe $L_{2,3}$) obtained on different regions of interest [24].

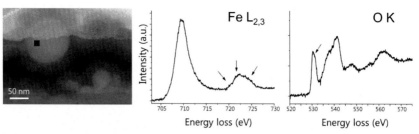

FIG. 20

ELNES Fe $L_{2,3}$ and O K edges spectra obtained on a big nano-sphere (STEM-HAADF image); *black arrows* indicate the fine structures features used for identification. Instrument: TITAN Qu-ant-em 300 kV, GIF quantum EELS detector (1 eV for EELS and 0.3 eV for ELNES with monochromator), FIB cut, cooling with liquid nitrogen.

Credit: Marie Godet.

The spatial resolution depends mainly on the effective focused beam size on the thin specimen and on the scanning step chosen. In most cases, it is around a few nanometers and can be as small as 1 nm in the best cases. The process developed by Nano MEGAS (ASTAR) allows fast acquisitions, typically between 50 and 100 images per second, which allows quick exploration of various zones of different samples. This is a major advantage in the field of cultural heritage materials where it is often necessary to analyze a large corpus in order to obtain pertinent data. Fig. 21 shows an example of an ASTAR study concerning a prehistoric flint tool [21]. Flints are sedimentary silica rocks mainly made of chalcedony (50–100 nm large quartz crystallites arranged in a fiber-like texture) in which larger isolated dendritic quartz grains (>1 μm) are trapped into the chalcedony network during the geological formation. ASTAR technique is a powerful tool to observe this specific structuration.

5.6 Beam damage

If inelastic scattering signals are very useful to get chemical information on the specimen, they also cause beam damage due to radiolysis (bond breaking), knock-on damage (atom displacement) and heating. These phenomena can change the morphology, the structure and the chemistry of the specimen. One should thus always remember that observing a specimen could change its nature.

To minimize the damage, different methods can be used:

– cool the specimen to liquid nitrogen temperature (−172°C) using a cooling specimen holder (reduce heating)
– use STEM mode (less damaging as the electron beam is moving during the scan)

Fig. 22 illustrates some beam damage observed during the STEM-EDX analysis of a gypsum particle.

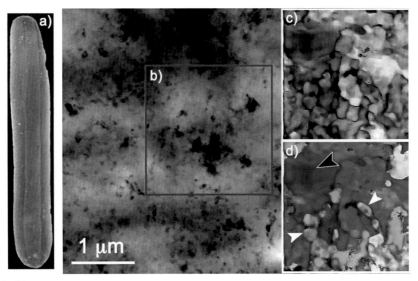

FIG. 21

ASTAR analysis of a prehistoric flint blade (A). BF image with the selected zone ($2 \times 2\,\mu m^2$) for investigation (B), (C) grain boundaries and (D) orientation maps obtained from the treatment of diffraction patterns. The *white and black arrows* indicate chalcedony fiber fragments and a dendritic quartz grain, respectively. Instrument: CM20-FEG 200 kV.

Credit: Christian Roucau.

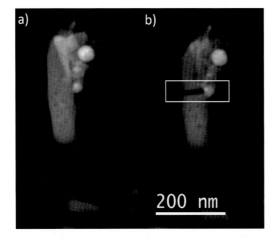

FIG. 22

Example of beam damage: gypsum rod covered by nanoparticles (A) before and (B) after an EDX line profile acquisition over the entire width of the rod (STEM-HAADF images); notice that the nanoparticle seems unaffected by the beam. Instrument: TECNAI F20 200 kV. The particles were cooled at $-172°C$.

Credit: Marie Godet.

Dose calculation can be performed to estimate the maximal dose acceptable for a system. For instance, De Seauve et al. have shown that sensitized photographic silver plates were damaged by silver chloride radiolysis in STEM mode (electron dose: 320 eA^{-2} corresponding to 0.9 Gy for a 20-s 1024×1024 image acquired on a JEOL 2100F at 200 kV) [30]. The authors applied a carbon coating on the sample and used a LN_2 cooling sample holder: no more damage was observed for doses up to a few thousand Gy. We strongly recommend checking the sensitivity of the sample under the beam while performing TEM observation to adjust the experimental parameters and minimize the irradiation effects.

6 Conclusions

SEM is a basic technique in the cultural heritage field. It is used daily in many studies of ancient artifacts. This technique is rather easy to implement and rarely requires complex sample preparations. It gives quickly relevant images and significant information about the elemental composition of sample up to the submicrometric scale. Transmission electron microscopy allows increasing the resolution to atomic level, but its implementation is more complex. Samples must be specifically prepared for the investigation, and the use of a transmission electron microscope requires a rather long training. However, as this technique is widely used in materials science, many sample preparation methods have been developed for almost all types of materials, and it is often not too difficult to find a method suitable for each case. In addition, the technique requires a very small quantity of matter, which is a significant advantage in cultural heritage analysis. The main advantage of transmission electron microscopy is that it generates an image enabling us to determine the atomic structure and to perform highly localized chemical analysis, i.e., to make individual measurements on nano-objects and undertake chemical and structural study of interfaces and extended defects. It is certainly one of the best techniques currently available for characterization at this scale. Cultural heritage materials do not differ fundamentally from other materials except that they are more heterogeneous with a more complex and imperfect structure. In addition, many of them contain nanoparticles or have a nano-scale structuration, which plays a significant role in their physical properties, or are rich in information concerning their manufacture. TEM techniques are thus well suited to investigate these types of materials especially because the developments of these in the last decades afforded both a more efficient sample preparation and faster data recording. Without TEM techniques, it would be impossible to reveal the true nature of many materials involved in ancient artifacts. These successes should encourage more and more people to use TEM techniques for characterizing diverse cultural heritage materials. The use of advanced methods is currently largely employed in physical sciences such as ASTAR technique, and structural determination from electron diffraction or electron tomography should likewise grow and open new perspectives.

Acknowledgment

The authors would like to express their sincere gratitude and deep appreciation to Peter Hawkes (CEMES) for his valuable and constructive suggestions.

References

[1] C. Colliex, La Microscopie Electronique, Presse Universitaires de France, Paris, 1998.

[2] T. Mulvey, B. Kazan, P.W. Hawkes, The Growth of Electron Microscopy, Vol. 96, Academic Press, San Diego, New York, Boston, 1996.

[3] G.D. Daniatos, Design and construction of an atmospheric or environmental SEM (part 1), Scanning 4 (1981) 9–20.

[4] G. Danilatos, Introduction to the ESEM instrument, Microsc. Res. Tech. 25 (1993) 354–361.

[5] P. Lopez-Arce, L.S. Gomez-Villalba, S. Martinez-Ramirez, M. Alvarez de Buergo, R. Fort, Influence of relative humidity on the carbonation of calcium hydroxide nanoparticles and the formation of calcium carbonate polymorphs, Powder Technol. 205 (2011) 263–269.

[6] I.C.A. Sandu, M.H. de Sà, M.C. Pereira, Ancient 'gilded' art objects from European cultural heritage: a review on different scales of characterization, Surf. Interface Anal. 43 (2011) 1134–1151.

[7] R. Wen, D. Wang, L. Wang, Y. Dang, The colouring mechanism of the Brown glaze porcelain of the Yaozhou Kiln in the Northern Song Dynasty, Ceram. Int. 45 (8) (2019) 10589–10595.

[8] P. Sciau, C. Brouca-Cabarrecq, A. Pinto, Les glaçures de céramiques chinoises colorées au fer: un matériau historique à fort potentiel en science de la matière? Technè 47 (2019) 144–149.

[9] A. Pinto, P. Sciau, T. Zhu, B. Zhao, J. Groenen, Raman study of Ming porcelain dark spots: probing Mn-rich spinels, J. Raman Spectrosc. 50 (5) (2019) 711–719.

[10] A. Ichimiya, P. Cohen, Reflection High-Energy Electron Diffraction, Cambridge University Press, Cambridge, 2004.

[11] G. Rijnders, D.H.A. Blank, J. Choi, C.B. Eom, Enhanced surface diffusion through termination conversion during epitaxial $SrRuO_3$ growth, Appl. Phys. Lett. 84 (2004) 505–507.

[12] Y. Kainuma, The theory of Kikuchi patterns, Acta Crystallogr. 8 (1955) 247–257.

[13] A.J. Wilkinson, P.B. Hirsch, Electron diffraction based techniques in scanning electron microscopy of bulk materials, Micron 28 (4) (1997) 279–308.

[14] N. Brodusch, H. Demers, R. Gauvin, Imaging with a commercial electron backscatter diffraction (EBSD) camera in a scanning electron microscope: a review, J. Imaging 4 (7) (2018) 88.

[15] E. Sassoni, S. Naidu, G.W. Scherer, The use of hydroxyapatite as a new inorganic consolidant for damaged carbonate stones, J. Cult. Herit. 12 (4) (2011) 346–355.

[16] B.H. Berrie, M. Leona, R. McLaughlin, Unusual pigments found in a painting by Giotto (c. 1266–1337) reveal diversity of materials used by medieval artists, Herit. Sci. 4 (2016) 1.

[17] D. Dietrich, T. Lampke, G. Nolze, N. Del-Solar-Velarde, D. Nickel, R. Chapoulie, L.J. Castillo Butters, The potential of EBSD and EDS for ceramics investigations case studies on sherds of pre-Columbian pottery, Archaeometry 60 (3) (2018) 489–501.

[18] D.B. Williams, C.B. Carter, Transmission Electron Microscopy, Springer US, Boston, 2009.

[19] J. Ayache, L. Beaunier, J. Boumendil, G. Ehret, D. Laub, Sample Preparation Handbook for Transmission Electron Microscopy, Springer New York, New York, 2010.

[20] P. Sciau, P. Salles, C. Roucau, A. Mehta, G. Benassayag, Applications of focused ion beam for preparation of specimens of ancient ceramic for electron microscopy and synchrotron X-ray studies, Micron 40 (5–6) (2009) 597–604.

[21] P. Sciau, Transmission electron microscopy: emerging investigations for cultural heritage materials, Adv. Imaging Electron Phys. 198 (2016) 43–67.

[22] F. Casadio, S. Xie, S.C. Rukes, B. Myers, K.A. Gray, R. Warta, I. Fiedler, Electron energy loss spectroscopy elucidates the elusive darkening of zinc potassium chromate in Georges Seurat's A Sunday on La Grande Jatte - 1884, Anal. Bioanal. Chem. 399 (2011) 2909–2920.

[23] M. Godet, G. Roisine, E. Beauvoit, D. Caurant, O. Majérus, N. Menguy, et al., Multiscale investigation of body-glaze interface in ancient ceramics, Heritage 2 (3) (2019) 2480–2494.

[24] M. Godet, V. Vergès-Belmin, N. Gauquelin, M. Saheb, J. Monnier, E. Leroy, et al., Nanoscale investigation by TEM and STEM-EELS of the laser induced yellowing, Micron 115 (2018) 25–31.

[25] J.P. Morniroli, Large-Angle Convergent-Beam Electron Diffraction (LACBED): Applications to Crystal Defects, Société Française des Microscopies, Paris, 2002.

[26] C. Mirguet, C. Dejoie, C. Roucau, P. De Parseval, S.J. Teat, P. Sciau, Nature and microstructure of Gallic imitations of Sigillata slips from the La Graufesenque workshop, Archaeometry 51 (5) (2009) 748–762.

[27] S. Nicolopoulos, P.P. Das, A.G. Pérez, N. Zacharias, S.T. Cuapa, J.A.A. Alatorre, E. Mugnaioli, M. Gemmi, E.F. Rauch, Novel TEM microscopy and electron diffraction techniques to characterize cultural heritage materials: from ancient Greek artefacts to Maya mural paintings, Scanning (2019) 1–13.

[28] S. Nicolopoulos, P.P. Das, P.J. Bereciartua, F. Karavasili, N. Zacharias, A.G. Pérez, A.S. Galanis, E.F. Rauch, R. Arenal, J. Portillo, J. Roque-Rosell, M. Kollia, I. Margiolaki, Novel characterization techniques for cultural heritage using a TEM orientation imaging in combination with 3D precession diffraction tomography: a case study of green and white ancient Roman glass tesserae, Herit. Sci. 6 (2018) 64.

[29] D. Viladot, M. Veron, M. Gemmi, F. Peiro, J. Portillo, S. Estrade, J. Mendoza, N. Llorca-Isern, S. Nicolopoulos, Orientation and phase mapping in the transmission electron microscope using precession-assisted diffraction spot recognition: state-of-the-art results, J. Microsc. 252 (1) (2013) 23–34.

[30] V. De Seauve, À l'origine des couleurs des images photochromatiques d'Edmond Becquere: étude par spectroscopies et microscopies électroniques, Thèse SACRe - Sciences, Arts, Création, Recherche — SACRe (EA 7410), ENS Paris - École normale supérieure, Paris, 2018.

UV-visible-near IR reflectance spectrophotometry in a museum environment

4

Paola Ricciardi

The Fitzwilliam Museum, University of Cambridge, Cambridge, United Kingdom

1 Introduction

Reflectance spectroscopy (or spectrophotometry, henceforth RS) in the ultraviolet, visible and near-infrared wavelength range (UV-vis-NIR, 350–2500 nm) is an analytical method that measures how materials absorb and reflect light, providing specific information on the material's color and some aspects of its components' molecular structure. The spectrum produced can be used to identify both natural and synthetic, organic and inorganic compounds including pigments, paint binders, as well as a broad range of materials used to fashion, conserve and restore archeological and historical objects and works of art. During the last 10 or so years especially, cultural heritage applications of RS have grown exponentially. The method is now a widespread routine tool for the noninvasive analysis of cultural heritage objects and the identification of artists' materials, especially in its variant with optical fibers, commonly termed Fiber Optic Reflectance Spectroscopy (FORS). Its various applications in this field were briefly summarized in a background paper published by the Analytical Methods Committee of the UK's Royal Society of Chemistry in 2016 (AMCTB No 75 [1]). This chapter updates and expands on the latter publication by providing an overview of the applications of RS to the study of cultural heritage objects, largely based on the author's own experience of using it in a museum setting.

The application of RS to the study of artworks dates back to the late 1930s, when researchers at MIT and the Fogg Museum in Harvard characterized the color of a set of specially prepared reference panels with different pigments, making reflectance measurements in the visible range (400–700 nm) [2]. Since the late 1970s, this method has been used to identify pigments on works of art. To this end, a pioneering programme of research was set up in the late 1980s at one of the Italian National Research Council's institutes in Florence (then IROE, now IFAC). A research group led by Mauro Bacci exploited the then recent development of optical fiber

103

technology to take RS out of the laboratory and "into the field," or rather into churches and museums across the city [3]. They assembled an instrument using a commercial UV-visible spectrophotometer, whose light was transmitted to an external integrating sphere by two optical fiber bundles. This setup required a close, gentle contact between the sphere and the surface of the artwork, which allowed the scattered light to be collected and detected by a photomultiplier. With such a device, the team aimed to set up "color archives" and to monitor color changes due to aging of the artworks. Among the works examined were the frescoes by Masolino and Masaccio in the Brancacci Chapel of the Chiesa del Carmine in Florence during a conservation campaign. The frescoes had been heavily damaged by a fire in 1771 and suffered further in the early 20th century due to a badly performed restoration. Bacci and his colleagues recorded more than one hundred reflectance spectra, representing nearly all the shades of color present across the frescoes. Statistical comparison of spectra collected before and after cleaning permitted the spectral properties of the "dirt" to be determined and enabled "digital restoration" of the original colors.

Since Bacci's first experiments, more than 30 years ago, much has changed in the field of heritage science. A recent publication provides a thorough review of the applications of RS in the UV-vis range to the study of artworks [4]. For this reason, and because of the numerous publications discussing the use of this technique, only a brief summary of the methodological background and the types of instrumentation most commonly used for this work is given here. For a thorough discussion of theoretical background, see Bruce Hapke's "Theory of reflectance and emittance spectroscopy" [5]. A good overview of the types of instrumentation in more widespread use is provided in a recent article by Andrew Beeby et al. [6]. Most noteworthy in recent years is the development of sturdy, portable equipment with extended sensitivity in the near-infrared range (NIR), mostly for remotesensing applications. The presence of multiple detectors within a single instrument currently allows recording of spectra in a broad range of wavelengths, usually extending from 350 nm (UV-A) to 2500 nm, i.e., to the "lower" limit of mid-infrared. It should be noted that the ~700–1000-nm wavelength range is referred to in the optics and remote sensing community as the NIR (near-infrared) and the 1000–2500-nm range is referred to as the short-wave infrared (SWIR). These terms do occasionally appear in the Heritage Science literature but are not especially common among chemists. To avoid confusion, and facilitate cross-disciplinary understanding, it may be most appropriate to refer to the entire 700–2500-nm range as near-infrared, and it is in this sense that the term is used here.

These novel instruments have now been adopted by the cultural heritage community and are increasingly preferred to the previously available spectrophotometers, usually limited to the UV and visible range or just beyond (approximately 200–1000 nm). The possibility of examining the spectral range beyond 1000 nm means that a great deal of information can be gained not only about electronic transitions (such as ligand field and band-to-band), but also about vibrational ones, whose overtones and combinations can be observed in the NIR range. The availability of this additional information considerably expands the range of artists' materials that can be reliably identified using this method. Some general comments can be made on the suitability of UV-vis-NIR RS for the study of different types of materials commonly encountered in historical and archeological objects as well as works of art, with further details given in the case studies discussed below:

1. *Historical pigments*: This method is especially useful for the identification of blue, green, red and some white pigments. Reflectance spectra of most yellow and black pigments, however, generally do not have enough distinctive features to allow identification, except in broad terms, such as iron oxides and hydroxides ("earth pigments"), and carbon-based blacks.
2. *Binding media*: These can also usually be broadly categorized, although complications can arise when their spectral features overlap with those of pigments or support materials onto which paint layers are laid.
3. *Support materials*: Paper, parchment, wood and stone can be easily differentiated when doubts exist, and FORS can be used alongside other analytical tools to study their degradation mechanisms.
4. *Other materials*: RS allows distinction between different types of textile fibers such as wool, silk, cotton and bast (e.g., linen or hemp). Many types of minerals and gemstones can also be reliably identified using this method, and recent applications of this technique to the study of historical plastics and glass objects have been published.

2 Advantages and limitations of UV-vis-NIR reflectance spectroscopy for the analysis of museum objects

Most historical and archeological objects and works of art in museum collections are, intrinsically, precious and/or irreplaceable. In addition, many of them possess characteristics which pose significant limitations to the types of investigation that can be carried out. They are, for example:

- fragile
- light-sensitive
- difficult to transport (or in remote locations)
- chemically complex and/or inhomogeneous
- not made according to a standard "recipe"
- aged

Accordingly, the analytical methods used to investigate such objects should:

- be nondestructive, i.e., leaving the object or sample intact and/or
- be noninvasive, i.e., not damaging the object in any way, which includes not requiring contact with the object, using low light and/or (laser) energy levels, etc.
- be portable
- have high spatial resolution (i.e., be able to analyze small areas and/or volumes of the sample) and/or

- be capable of working in scanning/imaging mode
- be versatile (i.e., have high analytical discrimination for a wide range of materials)
- be nontargeted, as the composition of a certain object can sometimes be a complete unknown

The main advantages of RS for the study of museum objects are its noninvasiveness and its rapidity. A good quality spectrum can be obtained in less than 10 s and viewed in real time, enabling extensive surveys to be undertaken of hundreds of areas, across an individual artwork or entire collections, in a relatively short period of time. Additionally, fiber-based instruments do not set limitations on the size and shape of the objects that can be analyzed (Fig. 1), although the use of optical fibers often limits spatial resolution to 2–3 mm. The latter can actually be an advantage in some cases, because the somewhat larger analytical "spots" analyzed by FORS are better suited to averaging the composition of complex compounds.

Due to the portability of the instrumentation, and to the fact that its application does not in most cases require contact with the object studied, FORS is especially useful for investigating objects which may be difficult to transport due to their fragility or high insurance costs (Fig. 2), and for which sampling is not commonly allowed by current conservation standards.

It is beyond this chapter to review the physical laws of reflection, refraction and scattering that provide the theoretical basis for RS, based on the optical phenomenon known as diffuse reflectance [7]. Its principles and methods of application to the study of minerals in particular have been summarized by Clark [8]. Much of Clark's extensive discussion applies equally well to other materials that are of interest to museum scientists. It should therefore be sufficient to provide here a brief list of

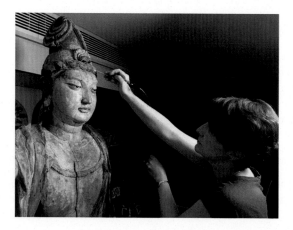

FIG. 1

Spectroscopic analysis of a large polychrome wood statue of the Buddha.

Credit: © Fitzwilliam Museum, Cambridge.

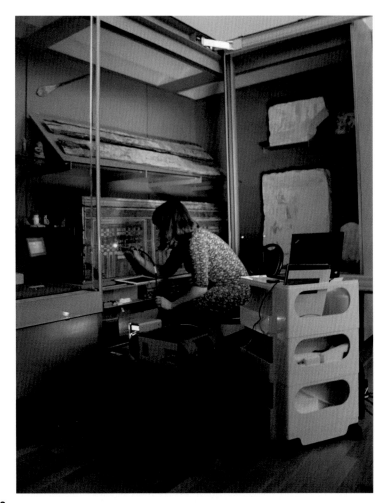

FIG. 2

Spectroscopic analysis of an Egyptian coffin, carried out without removing the object from its display case at the Fitzwilliam Museum.

Credit: © Fitzwilliam Museum, Cambridge.

the processes most likely to produce absorption bands observable in the reflectance spectra of materials. There are two general types of processes:

- Electronic transitions: including crystal field effects (giving rise, for example, to the spectral features of iron oxides); charge transfer absorptions (as seen in Prussian Blue); conduction bands (characterizing the spectra of semiconductors such as cinnabar, minium as well as cadmium pigments); and color centers (generally at wavelengths between 400 and 1000 nm);

- Vibrational features: such as hydroxyl ($-OH^-$) and carbonate (CO_3^{2-}) bands, generally at wavelengths >1000 nm.

Although visible reflectance spectroscopy is a "surface analysis" technique, best suited, for example, for the analysis of the outermost layers of an object, the NIR reflectance spectral features usually do give information on both the surface layer and the underlying materials, for example, the substrate on which paint is applied. One can distinguish between works executed on a chalk or a gypsum ground, on paper or parchment, or identify materials used for bookbindings and historical textiles.

A well-chosen measurement geometry also usually allows analysis of shiny surfaces, such as facetted gems, although no information can be obtained on metallic surfaces, which only show specular rather than diffuse reflectance. Varnish layers present on paintings do not generally interfere with the identification of pigments. However, the VIS portion of the spectrum—and therefore, the accurate measurement of color—can be influenced by the presence of a varnish or even by a binding medium. In these cases, if accurate color measurement is required, unvarnished areas can be tested and compared with reference samples made up of the same pigment bound in different media. The possibility of identifying some binding media, or at least to characterize the category to which they belong (e.g., lipidic vs. proteic media), is one of the advantages of extending the range of wavelengths probed as far as possible into the near-infrared [9–11].

As flexible and versatile as UV-vis-NIR RS is, there are of course limitations to its use. These include the difficulty in correctly identifying individual components of complex mixtures, and a relatively low spatial sampling (usually limited to 2–3 mm), which can be problematic when detailed characterization of small areas (e.g., in illuminated manuscripts) is desired. In both of these cases, the use of additional, complementary in situ methods such as Raman spectroscopy or X-ray fluorescence spectroscopy may help clarify the results (see Section 4). In some instances, due to the broadness of the absorption bands in the visible and NIR range, RS only allows general characterization of material type, rather than the precise identification of individual compounds. In these cases too, it is often necessary to utilize other, perhaps "destructive," methods to establish the exact nature of the material under study.

3 Instrumentation, setup and data processing methods

The recent rapid increase in the use of UV-vis-NIR RS for the analysis of museum objects is largely due to the continuous development of relatively inexpensive, portable, fiber-based instrumentation, which can easily be transported and used on an ever-growing variety of objects. As is often the case when scientific instruments are adopted for use in new applications—particularly in situations with significant constraints such as the study of museum objects—much of the equipment used for UV-vis-NIR RS analysis of artworks has not been purchased "off-the-shelf," but rather has been carefully designed and assembled in a laboratory for this particular purpose.

The instrument used by Bacci's group in the late 1980s, discussed earlier, was based on a commercially available spectrophotometer (a Perkin-Elmer model 552), which was then manually coupled to an external integrating sphere using two optical fiber bundles [3]. Twenty years later, the same group adapted their instrumental configuration to identify chromophores on late-14th-century stained-glass windows in situ, in the chapel of Santa Maria della Certosa at Galluzzo, in the hills just south of Florence [12]. Their novel setup commanded the use of solar light, coming in through the church's windows, as a powerful light source for producing spectra with much increased signal-to-noise ratio (Fig. 3). Due to the high transparency of church windows, measurements made using a traditional reflectance probe head

FIG. 3

Spectroscopic measurements made in situ by means of optical fibers, using sunlight as the source of illumination.

Credit: Image from M. Bacci, A. Corallini, A. Orlando, M. Picollo, B. Radicati, The ancient stained windows by Nicolò di Pietro Gerini in Florence. A novel diagnostic tool for non-invasive in situ diagnosis, J. Cult. Herit. 8 (2007) 235–241, https://doi.org/10.1016/j.culher.2007.02.001 under the terms of the Creative Commons Attribution 4.0 International License.

could gather only a very weak signal; most incident radiation was lost, as it was absorbed by or transmitted beyond the glass. A transmission measurement using sunlight as the source of illumination was an ingenious response to the challenge posed by the impossibility of placing a 100% diffuse reflective background on the outside of the church windows, located at three meters from the chapel's floor. An opaque white windowpane, which was not expected to display any characteristic absorptions in the spectral range investigated (380–780 nm), was selected to acquire a "reference" white spectrum. Bacci's setup only allowed for qualitative measurements, due to variations in sunlight intensity and spectral power distribution during the experiments. However, these were sufficient to characterize the chromophores in the stained glass.

Regardless of the specific characteristics of individual instruments, whether commercially available or custom-developed for heritage science applications, a few general comments can be made about instrumental setup and accessories.

Normally, reflectance measurements rely on calibration with a white target, which is assumed to have close to 100% diffuse reflectance throughout the wavelength range investigated. For accurate measurements, such targets are measured relative to a NIST standard, to correct for any deviation. Nowadays, a high-quality PTFE-based material such as Spectralon (developed by Labsphere) or Fluorilon (by Avian Technologies) is used, although barium sulfate ($BaSO_4$, with the same composition as the pigment barium white) also has similar characteristics and can be utilized.

The measuring geometry used for reflectance measurements can vary significantly depending on the instrument used and on the specific circumstances in which the analytical campaign takes place. Most commonly, the setup allows for both incident and reflected light to be at an angle (typically 45 degrees) from the surface normal, in order to exclude specular reflectance from the collected signal. As a result, for example, glare effects from varnished surfaces do not cause any interference. Alternatively, the light probe and the receiving fiber can be placed at 45 degrees relative to one another (typically with one of them kept parallel to the surface normal).

An additional possibility is to use a Y-shaped (or "bifurcated") fiber that conveys light from an external source onto an object's surface and collects backscattered light, transmitting it to the spectrometer. Specular reflection can be a problem in this case, but the configuration still allows for good-quality measurements in most instances. This setup also has the advantage of maximum flexibility, allowing for measurement of reflectance spectra of oddly shaped surfaces, including narrow cavities. Figs. 1, 2 and 4 illustrate how a commercial spectroradiometer, in this case a FieldSpec4 by Malvern Panalytical/ASDi, can be used in this configuration. Rugged, robust and developed to function wirelessly, this instrument can easily be deployed on site and allows for rapid collection of spectra in an extended range (350–2500 nm). The external light source can be substituted to suit the requirements of the operator; the source depicted in Fig. 4, for example, is not the manufacturer's own but rather a separately sourced one (Ocean Optics HL-2000-HP). The output from this lamp can

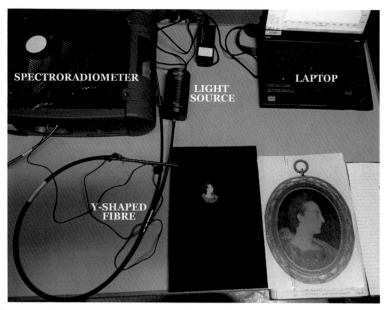

FIG. 4

Spectroradiometer for reflectance measurements in the UV-vis-NIR range (350–2500 nm), with bifurcated fiber and external light source.

Credit: © Fitzwilliam Museum, Cambridge.

be attenuated to limit the power density and therefore allows analysis of light-sensitive objects such as the early 17th-century portrait miniature depicted.

Figs. 5 and 6 illustrate two cases in which equipment has been assembled—in one case using a commercial spectroradiometer as a starting point—in an ad hoc configuration, optimized for the specific needs of the researchers. "Team Pigment," based at Durham University in the UK, specialize in the analysis of colorants in medieval manuscripts. Because of the particularly stringent requirements of such an activity, due to these objects' fragility and light sensitivity, they have developed an extremely flexible setup that allows them to carry out analyses in safe and stable conditions (Fig. 5) [6]. Three optical fibers (two for illumination and one for collection of the reflected light) are inserted in kinematic mounts held on a 3D-printed frame and kept at some distance from the object's surface, ensuring that accidental contact cannot occur. The frame itself is mounted on a vertical translation stage, which allows careful and reproducible positioning of the probe above the object and a sampled area of approximately 2 mm in diameter. The sample is illuminated by the output of a 20-W tungsten light source (also an Ocean Optics HL-2000-HP) operating at 3000 K and equipped with a shutter, optional filter and attenuator. Such system does not need the "soft but stable contact" required by some of the instruments in use by

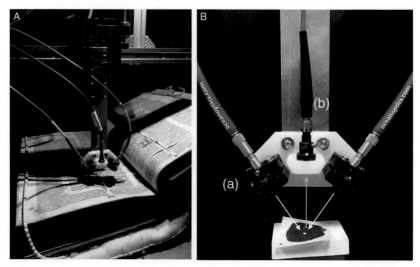

FIG. 5

"Team Pigment" FORS probe, *left*, showing head suspended from the gantry and the vertical adjustment system and, *right*, the alignment of the two spots on a test sheet of blue paint. The single bright spot in the center of the test sheet demonstrates the correct vertical alignment of the head for optimum collection efficiency.

Credit: Image from A. Beeby, L. Garner, D. Howell, C.E. Nicholson, There's more to reflectance spectroscopy than lux, J. Inst. Conserv. 41 (2) (2018) 142–153, https://doi.org/10.1080/19455224.2018.1463920 under the terms of the Creative Commons Attribution 4.0 International License.

other groups [3] and has the advantage of allowing careful adjustment of the light dose to which the analyzed area is being subjected. This makes it optimally suited to the study of fragile and light-sensitive objects.

In the Scientific Research Department at the National Gallery of Art (NGA Washington), the light source and collecting fiber of a commercial spectroradiometer (FieldSpec3 by ASD/Malvern Panalytical) have been mounted on an optical plate alongside an X-ray source and detector. The setup constitutes an integrated scanning system, which allows both reflectance and XRF measurements to be undertaken on the same area of a painting, at the same time (Fig. 6) [13]. The painting itself is positioned vertically on a high precision, computer-controlled easel, placed a few centimeters away from the instrumental setup. The XRF and FORS instruments collect spectra when triggered by a piece of software, written in-house, which also controls the lateral movement of the easel. As a result, two spatially aligned spectral data sets ("cubes") are collected for each painting, containing information on the chemical elements and compounds present in the paint layers and grounds. This setup does require significant computing power to handle and fully exploit the large amount of data collected, but provides a relatively low-cost, dual imaging modality, XRF and FORS scanner for use in a museum environment.

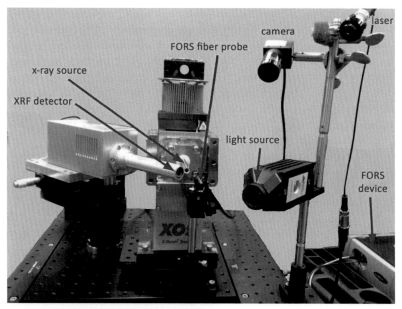

FIG. 6

(Top) a photograph of the view from the painting of the XRF and FORS point measurement collection head showing the X-ray source, X-ray detector and fiber optic reflectance sensor and light source. *(Bottom)* a photograph showing the painting on the computer-controlled easel along with the diffuse white standard and the XRF and FORS instruments.
Credit: Image from J.K. Delaney, D.M. Conover, K.A. Dooley, L. Glinsman, K. Janssens, M. Loew, Integrated X-ray fluorescence and diffuse visible-to-near-infrared reflectance scanner for standoff elemental and molecular spectroscopic imaging of paints and works on paper, Herit. Sci. 6 (2018) 31, https://doi.org/10.1186/s40494-018-0197-y under the terms of the Creative Commons Attribution 4.0 International License.

Despite the ingenious ways researchers have devised for instrument setups to be both flexible and stable, it is still often difficult to maintain a fixed geometry and be able to access all sites required for analysis on a three-dimensional museum object. This problem strongly limits the possibility of obtaining *quantitative* reflectance measurements—which would allow, for example, establishing the relative abundance of two or more different compounds in a mixture. *Qualitative* interpretation of reflectance spectra, however, can be relatively straightforward, as long as one has access to a spectral reference database. An increasing number of databases are available both online and in printed publications. However, a reference collection of one's own is often crucial, and can be developed to suit the specific, individual research requirements. A reference database may, for example, include spectra from a set of pigments bound in different media and painted on substrates such as paper, parchment or wood.

Even for a single compound, for example, a historical pigment, the unambiguity of the reflectance spectrum, and therefore the ability to unequivocally identify a material, varies according to each individual colorant. Blue pigments, for example— lapis lazuli (and its synthetic equivalent French ultramarine), azurite, indigo, cobalt blue, cerulean blue, smalt and Prussian blue—display unique features, which allow their unambiguous identification. The identification is easier and more reliable if the NIR range is included in the measurement, as many of these blue compounds have absorption bands in this region. Inorganic red pigments, on the other hand— including vermilion, red lead, cadmium red and chrome red—display very similar, S-shaped curves. One can attempt to distinguish between them by performing a simple mathematical process, i.e., by calculating the first derivative of the spectrum, which allows the precise determination of the curve's inflection point (or "transition edge"). This additional piece of information allows the unequivocal identification of red lead and chrome red, whereas vermilion and cadmium red can only be unambiguously identified by employing additional analytical methods and including the analyst's understanding of the nature and historical context of the object. Cadmium red, for example, can only be encountered in objects produced after c.1910, when the pigment was first developed. The calculation of first and second derivative curves is especially useful to determine the exact position of the (often poorly defined) absorption features of organic red colorants; this may allow a distinction between those based on alizarin, such as madder, and those containing carminic acid, such as cochineal [14,15].

When dealing with mixtures or other combinations of materials—for example, a pigment, a binding medium and the support over which they are painted—the spectral features of the various components may well overlap with each other. Mathematical methods can be usefully applied to facilitate a correct interpretation of the spectra and consequently improve the identification of specific materials. The simplest is the calculation of first derivative spectra, a technique well known to spectroscopists, which has proven to be quite useful. Further processing to separate the contributions of individual components to an overall spectrum requires some mathematical modeling. The most widely used model is based on the Kubelka-Munk theory of reflectance [16], which is used even when not all external conditions are met (e.g., in the case of collimated rather than diffuse illumination). This approach has been used to evaluate pigment quantity in paint glazes [17], to characterize pigment mixtures [18] and to "correct" fluorescence emission spectra of organic colorants [19]. A modified K-M model has been developed specifically to deal with optically rough surfaces [20].

Finally, it is worth noting that the correct interpretation of reflectance spectra obtained from complex objects may require the deployment of completely new approaches. These may range from the production of appropriate "replica" materials with subsequent compilation of ad hoc spectral databases [21–23] to the development and testing of novel methods for data processing [9,24–27].

4 Complementary methods

The variety of materials present in museum objects, the range of research questions that may instigate their analysis, as well as the inherent limitations of each analytical technique often demand the use of multiple, complementary analytical tools. As such, like every other analytical method, UV-vis-NIR RS is most efficient when used as part of a multianalytical protocol.

Once research questions have been clearly formulated, it is possible to choose what types of analytical methods one might need to employ to answer those questions. For example, to survey the pigments and painting techniques used by a busy Renaissance workshop during a determined period of time, one would need a combination of techniques that can reliably identify inorganic pigments as well as organic colorants and paint binders. Site-specific spectroscopic techniques would be helpfully supplemented by imaging methods.

To provide an initial overview of the object, most analytical protocols recommend starting a new campaign of study by applying a range of noninvasive and nondestructive techniques. These may include visual examination under magnification as well as various technical and chemical imaging methods. Specific questions that remain unanswered, or new questions that may arise from these initial analyses, can then be addressed by a combination of site-specific spectroscopic methods including FORS, as well as XRF (see Chapter 9), Raman (see Chapter 2), FTIR, XRD (see Chapter 6) and spectrofluorimetry.

FORS, for instance, has often been used, during the past 10 years or so, as one of the very first steps deployed as part of a noninvasive analytical protocol for the study of illuminated manuscripts in museums and libraries around the world [28–31]. The instrumentation's portability and the fact that it can typically be used without making contact with the surface under analysis mean that FORS is ideally suited for the investigation of fragile objects, such as manuscripts, often held in dispersed, remote locations. A case in point is the extensive research carried out by Porter and co-authors [32] who compared the purple colorants found in a series of 6th- to 10th-century codices from multiple institutions in seven different European countries.

Based on a broad survey of literature on the subject, it is clear that a combination of FORS with XRF and/or Raman spectroscopy usually proves most effective in studies that require the identification of pigments. More than three quarters of publications using FORS to identify painting materials in manuscripts exploit the complementarity of the structural information provided by this method with the elemental data gathered by XRF. Just under half of the published studies include the use of Raman spectroscopy, especially effective for the analysis of many yellow pigments, and when microscopic precision is required. Spectrofluorimety is used by a small number of research groups to complement reflectance spectroscopy, mainly for studies focusing on the identification of organic red and purple colorants, extensively used by medieval and Renaissance manuscript illuminators. FTIR and XRD, both harder to deploy in fully noninvasive mode, are only used as a complement to FORS in a small number of relevant publications. A comprehensive discussion of multi-technique protocols used for the investigation of illuminated manuscripts has been recently published [33].

5 Research questions and case studies

Scientific research in a museum environment is usually aimed at addressing a large number of diverse questions about objects in the collection. A nonexhaustive list might include:

- authorship and authentication
- composition and technology
- history and provenance
- conservation needs or concerns, including the nature of degradation processes and products as well as the identification of past treatments

Research questions regarding single objects are just as common as studies involving broader cultural or contextual queries and a commonality in museum research studies is the need for a cross-disciplinary approach. Museum scientists cannot work in isolation; their access to objects as well as their involvement in research projects is inevitably connected to their willingness to collaborate with a range of colleagues—conservators and curators in particular.

The case studies presented here illustrate the range of museum-based applications for which UV-vis-NIR RS has been successfully used in recent years. They are largely drawn from the author's own experience of working at NGA Washington and at the Fitzwilliam Museum in Cambridge, UK, and broadly cover all three areas of investigation for which scientific analyses are generally used in the heritage field (see also Chapter 1):

- Analyses to support conservation, collections care and management needs, for example, study of degradation mechanisms and deterioration products, and analytical surveys to help with documentation;
- Analyses to aid curatorial research on the date, authorship and provenance of objects; to uncover or clarify hidden, disfigured or obscured details on artifacts; or to add information on the development of cultural, social or economic trends;
- Development of novel analytical methods and protocols, as well as data processing techniques, to support these research activities.

5.1 Cross-disciplinary research on medieval and Renaissance illuminated manuscripts

Since 2012 the Fitzwilliam Museum in Cambridge has been undertaking in-depth cross-disciplinary research on medieval and Renaissance illuminated manuscripts, within the context of the MINIARE research project (Manuscript Illumination: Non-Invasive Analysis, Research and Expertise, www.miniare.org). The MINIARE project protocols combine codicological, art historical, and noninvasive analytical investigation and rely on the expertise of a heterogeneous group of experts, including curators, conservators and heritage scientists. The project has resulted in a significant number of publications, including the proceedings of an international conference, in the form of two printed volumes [34] and a collection of open-access journal articles [35]; and in a major exhibition, staged in 2016 to celebrate the Fitzwilliam Museum's bicentenary, which was accompanied by an illustrated catalogue [31] and a digital resource [36].

The importance of UV-vis-NIR RS within the MINIARE project is well illustrated by the cross-disciplinary investigation of a richly illuminated 13th-century manuscript, the so-called Breslau Psalter (Fitzwilliam Museum, MS 36-1950) [30]. The Psalter's 294 pages contain 28 full-page miniatures, 10 large historiated initials, 168 small miniatures and a variety of figural and ornamental motifs in the margins. The volume's extensive decorative programme poses challenges in terms of the attribution of its numerous illuminations to specific artists, as well as the need to process an enormous amount of analytical data to help clarify its method of production and to inform its long-term preservation. The first phase of our investigation focused on the volume's full-page miniatures and historiated initials, a total of 38 images. The images were studied using methods from palaeographical, textual, historical, art historical and codicological research, and the results interpreted vis-à-vis the images and spectra obtained by photomicroscopy, near-infrared imaging, UV-vis-NIR RS, and X-ray fluorescence spectroscopy. Because of the quantity of information needed to characterize such a large number of complex and colorful images, we chose to use UV-vis-NIR RS to carry out an extensive survey of painted areas on these illuminations. Over 600 FORS spectra were collected within a few days, allowing the identification of most of the pigments used in the 38 illuminations. Sample spectra from fol. 39v of the Psalter (Fig. 7) are shown in Fig. 7B.

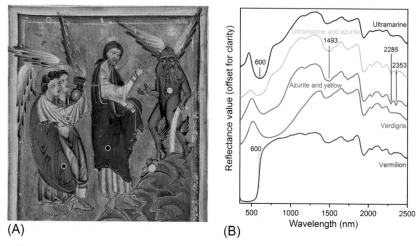

FIG. 7

(A) *First Temptation of Christ*, detail from fol. 39v of the so-called Breslau Psalter (Fitzwilliam Museum, MS 36-1950. Silesia, Breslau, c.1255–1267). (B) Sample FORS spectra from *blue*, *green* and *red* areas marked in (A), showing characteristic absorption features *(solid lines)* and transition edges *(dotted line)*.

Credit: © Fitzwilliam Museum, Cambridge.

FORS analysis identified ultramarine in most blue areas, thanks to its characteristic symmetrical absorption centered at 600 nm and subsequent rise in reflectance with an inflection point at 700–720 nm. Selected dark and "dull" blue areas showed evidence for the presence of both ultramarine and azurite, which the artist may have mixed, or possibly layered one over the other. Azurite, characterized by absorption bands at 1493, 2285 and 2353 nm, was also found in a few green areas: FORS analysis revealed greens produced by a mixture of azurite and a yellow pigment, which could not be identified based on its reflectance features alone. Green areas were more commonly painted with verdigris, characterized by a broad absorption centered at 700 nm followed by a slow rise in reflectance, peaking around 1320 nm. Vermilion was identified in all red areas, thanks to its sharp transition edge at 595–600 nm.

FORS analyses were also instrumental in understanding similarities and differences between flesh tones painted throughout the manuscript. Flesh tones are usually highly complex mixtures, and not all individual components could be precisely identified. Using reflectance spectra, however, they could be separated into groups whose overall features revealed different patterns in the use of materials.

Three different yellow pigments were used in the few yellow-colored areas found within the manuscript. FORS analysis alone was unable to identify these and was therefore supplemented by XRF and Raman spectroscopy, revealing the use of orpiment, a yellow earth or ochre and an organic yellow on different pages.

Overall, technical analyses of illuminations in the Psalter suggest the presence of six distinct pigment "palettes," allowing a preliminary division of the miniatures into groups. When combined with the results of codicological and art historical research, they allow attribution of the 38 major illuminations in the Psalter to no less than nine different artists, probably working with a number of assistants, demonstrating the benefits of an integrated, cross-disciplinary approach to improve our understanding of such an exceptionally complex manuscript.

There are other notable examples of how the use of RS has significantly advanced our understanding of manuscript illumination practice during the Middle Ages and the Renaissance. These include the first-ever analytical identification of the systematic use of egg yolk as a paint binder, by a group of 15th century Florentine painter-illuminators [37]; and the discovery of the extensive use of Egyptian Blue—whose manufacturing technology is thought to have been entirely unknown to medieval practitioners—in a lavish, 10th century copy of Hrabanus Maurus' *Laudibus Sanctae Crucis* [38].

5.2 Getting it right: Identification of gemstones in historical jewelry

One of the less-well-known aspects of the Fitzwilliam Museum's holdings is a collection of jewelry and metalwork dating back to between 1850 and 1940. These objects were created during an intensely productive period of design, and include pieces made by some of the most notable jewelers and metalwork craftsmen of the time, as well as objects by famous architects and artists whose work in silver and gold is less well known. In 2018, a special display and publication highlighted 70 pieces of jewelry and metalwork belonging, or on loan to, the Fitzwilliam Museum. The pieces were designed by some 30 named designers/workshops and exhibited a broad range of styles [39].

Technical analyses were undertaken in 2017 on a selection of objects to confirm—or, if necessary, correct—the identification of the jewels' gemstones. UV-vis-NIR RS proved to be an especially straightforward, rapid way of providing reliable information to curatorial colleagues who were researching the museum's jewels. The most challenging aspect of this particular campaign of analyses proved to be the difficulty in finding appropriate reference spectra for some of the gemstones. It was only by expanding bibliographical searches to include publications outside the heritage science field, in particular gemmological journals and reference books, that a reliable reference database could be assembled for direct comparisons with the spectra collected in the Fitzwilliam's analytical laboratory. RS is one of the most reliable analytical methods adopted by gemmologists in recent years, supplementing "classic" diagnostic tools, such as microscopy, specific gravity, and refractive index among others [40].

Spectra collected from two of the jewels (Fig. 8) are shown in Fig. 9, to illustrate the usefulness of this type of analysis and the direct, practical contribution of scientific examination to understanding and reconstructing object "biographies."

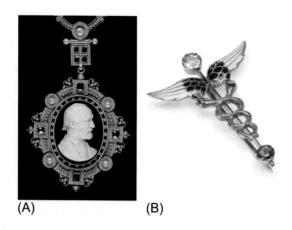

FIG. 8

(A) Pendant with a cameo bust of Giuseppe Mazzini, Italian, c.1872–1882 (Fitzwilliam Museum, M.7 & A-1983). (B) Caduceus Brooch, made by Ferdinand Giuliano, London, c.1903–1910 (Fitzwilliam Museum, M.5 & A-1983).

Credit: © Fitzwilliam Museum, Cambridge.

The lavish pendant and necklace with a cameo bust of Giuseppe Mazzini shown in Fig. 8A had been previously catalogued as containing "green and red" stones, which were readily identified as emeralds and rubies, respectively (top two spectra in Fig. 9). The spectra of emeralds (a green-colored variety of beryl, $Al_2Be_3[Si_6O_{18}]$) are strongly characterized by the presence of a significant number of narrow absorption bands in the NIR range, due to the presence of water (H_2O) molecules within the stone's structure. Their specific positions relate to thermal conditions at the time the crystals were formed [41]. The visible portion of the spectrum gives instead some indication on the chromophore ion(s), in this case verified by XRF as being both chromium and vanadium. Chromium, occasionally accompanied by iron, is also the chromophore responsible for the bright color of rubies, a red variety of corundum (Al_2O_3). The most characteristic feature of the ruby spectrum is not linked to the absorption phenomena, but rather to the intense emission of light at approximately 693 nm in response to the visible-light excitation [42].

Spectroscopic analysis allowed correction of information about an early 20th-century enameled gold caduceus brooch (Fig. 8B), originally catalogued as including a "white sapphire" (Al_2O_3) and a citrine (a variety of quartz, SiO_2, colored yellow by Fe ions). The gemstones were instead found to be a colorless topaz (a fluorine-containing aluminum silicate, $Al_2(F,OH)_2SiO_4$) and an amber-colored grossular (i.e., a calcium garnet, $Ca_3Al_2[SiO_4]_3$). Colorless stones, such as the topaz included in this jewel, cannot typically be identified with UV-vis reflectance spectroscopy [4], because of the lack of spectral features in this range. The extended spectral range that we were able to probe, however, revealed a wealth of sharp absorption bands due to

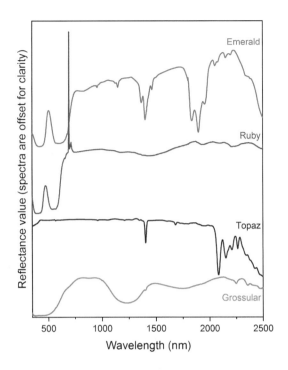

FIG. 9

FORS spectra of gemstones: emerald and ruby from pendant with cameo bust (Fig. 8A); topaz and grossular from caduceus brooch (Fig. 8B). Spectra collected with a FieldSpec4 spectroradiometer.

Credit: © Fitzwilliam Museum, Cambridge.

—OH and possibly to Al-OH bonds within the crystal structure, allowing reliable identification of this stone [43]. Pure grossular, on the other hand, would not be expected to exhibit any electronic absorption bands in the entire wavelength range probed, due to the lack of spectrally active transition series elements. Natural stones, however, commonly contain a range of cation substitutions, giving rise to numerous absorption features. These can be broad—especially if caused by Fe^{2+} and Fe^{3+} ions—and their positions can change significantly, depending on differences in structure between the sites occupied by the substituting ions [44]. In the case of the Fitzwilliam's brooch, the amber-colored stone was identified as a grossular based on the presence in its reflectance spectrum of broad absorptions around 880 and 1230 nm. These are due to a significant iron content, relatively high for this type of stone and confirmed by XRF analysis. As a clear demonstration of the immediate impact of the technical analysis, both stones embellishing this stunning brooch could be correctly identified in the catalogue published to celebrate the Fitzwilliam's 2018 jewelry display [39, p. 145].

5.3 Recovering lost pigments and revealing construction techniques of medieval polychrome wood sculpture

The Fitzwilliam Museum holds a small but exceptional collection of approximately 50 medieval wood sculptures, largely polychrome or formerly polychrome, made across Western Europe between about 1300 and 1550. For the most part extremely fragile, most of the sculptures have never been exhibited and are largely unknown both to the public and to scholars. In 2017, an in-depth technical study of the sculptures was initiated, aiming to reconstruct their material histories, including identification of original functions and polychromatic schemes, subsequent modifications and current conservation needs.

One of the sculptures tested is especially enigmatic. Probably produced in Germany or Austria, and currently dated to the 15th century based on curatorial expertise, even its subject is unclear. It is most commonly thought to represent a male Saint, although the raised position of its right hand suggests that it might depict the Risen Christ (Fig. 10A). Imposing in size, measuring over 1 m in height, the sculpture is carved in the round from a single piece of Swiss pine, with parts of the base and other areas having been replaced and adapted later on. There are scant traces of polychromy, mostly present toward the rear of the sculpture, including remains of blue, red and flesh-colored paint, occasionally surviving only in minuscule fragments under drapery folds and within the object's deepest recesses.

Due to the fragmentary nature of the extant polychromy, and to its presence in hard-to-reach positions, FORS was an obvious choice for the initial noninvasive characterization of the paint layers, followed by micro-sampling for more in-depth investigation. Despite the small size and complex stratigraphy of the remaining paint layers, reflectance spectroscopy provided extremely helpful insights. The following discussion is based on an unpublished technical report by Helen Howard and Paola Ricciardi.

In particular, observation under magnification allowed the identification of blue paint under a fold in the drapery between the figure's legs (Fig. 10B). Several FORS spectra were collected at a number of narrowly spaced locations within this area and suggested the presence of azurite, indigo, smalt, zinc white and Prussian blue (Fig. 11). Subsequent examination and analysis of microscopic paint samples confirmed the presence of three layers of paint within the blue robe: an original azurite layer followed by two phases of repainting.

The presence of small amounts of indigo in the original layer can only be hypothesized based on an absorption maximum at 650–660 nm, in spectra which also show evidence for azurite, identified through its characteristic features at 1491, 2286 and 2351 nm. The latter two absorption bands partially overlap with CH_2 stretches at approximately 2305 and 2350 nm. These are typical of lipidic binding media, such as the drying oil that was probably used here. FORS analysis also identified indigo clearly in other areas of the figure's drapery, including at the right side of Christ's chest. A cross-section sample taken from the remains of blue paint in a recess of this

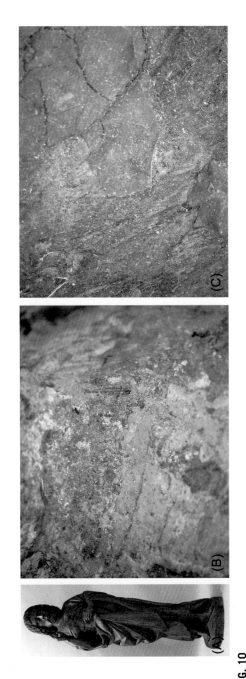

FIG. 10

(A) *The Risen Christ*, polychrome wood, Austrian or German?, possibly 15th century (Fitzwilliam Museum, M.24–1929). (B) Photomicrograph of area between the sculpture's legs, under the front drapery fold, showing fragmentary *blue paint* layers (original magnification 20 ×).
(C) Photomicrograph of area at the base of the sculpture's proper right foot, showing remains of parchment (original magnification 25 ×).

Credit: © Fitzwilliam Museum, Cambridge.

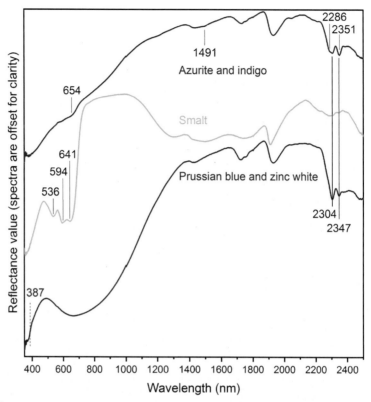

FIG. 11

Sample FORS spectra from areas of *blue paint* on the sculpture (see Fig. 10B), showing characteristic absorption features *(solid lines)* and transition edges *(dotted line)*. Spectra collected with a FieldSpec4 spectroradiometer.

Credit: © Fitzwilliam Museum, Cambridge.

area shows indigo combined with lead white, applied directly over a white ground, suggesting that indigo was indeed part of the robe's original palette.

The first layer of overpaint contains smalt, whose glassy particles were clearly visible under UV illumination and which can be easily identified by FORS, based on characteristic features at ∼536, 594 and 641 nm, in addition to a broad absorption between ∼1250 and 1850 nm. The second layer of overpaint contains zinc white and Prussian blue bound in oil, the latter giving rise to the sharp absorption features mentioned above. The reflectance spectrum of zinc white is characterized by a transition edge around 380 nm, whereas Prussian blue displays a broad, deep absorption with a maximum at 660 nm and extending into the NIR range.

FORS analysis was also instrumental in clarifying the nature of an unusual material, observed in a cavity on Christ's chest and behind the figure's proper right foot (Fig. 10C). Initial observation under magnification suggested the material to be paper or textile. One hypothesis was that these might be remains of a scroll, possibly held in the figure's proper left hand and flowing down toward his feet. The spectroscopic analyses, however, revealed the typical features of parchment (Fig. 12), which can easily be distinguished spectrally from paper, wood and a range of textiles including silk, wool and linen. Strips of parchment and linen were in widespread use throughout Renaissance Europe as a way to cover joins, imperfections or faults in wooden panels used as supports for easel paintings [45]. It is probably for this same reason that these fragments of parchment were employed during the construction of this sculpture.

6 **Where next?**

This chapter has hopefully given readers a taste of what an exciting array of research questions around museum objects can be successfully approached using UV-vis-NIR RS. Far from being exhaustive, the case studies presented include just a few of the materials for whose characterization RS can play an important role. Paintings—on wall, panel and canvas—have always been, and will continue to be, the focus of the majority of FORS research in the heritage field. In addition to illuminated manuscripts, gemstones set in items of jewelry, and medieval polychrome wood sculpture, researchers have also profitably started to use FORS to investigate historical textiles [46,47] as well as colorants in ceramics and glass [48,49]. Future work will undoubtedly reveal the full potential of this method for the characterization of wooden objects such as African combs [50], particularly in terms of wood speciation [51]; patinas found on outdoor bronze monuments [52,53]; and the collagen content of archeological and historical bone [54]. An increased use of statistical and mathematical modeling will indubitably allow augmentation and refinement of the information one can gather from FORS spectra, such as the tensile strength of historical silk [55] and the lignin content of historical paper [56], to name just two.

A popular topic of discussion and the cause of innumerable problems for our planet, plastic materials are also commonly characterized using NIR RS. Widely used in the recycling industry, commercial NIR spectroscopy systems typically cover only a portion of the near-infrared range (from c.1000 to 1800 nm or from 1600 to c.2400 nm) and can successfully identify up to 30 different types of plastics, if appropriate standards are measured and suitable mathematical modeling implemented for interpretation of the spectra.

One should not be surprised to learn that museum conservators and heritage scientists, too, have an interest in methods that can identify synthetic plastics rapidly and noninvasively. These materials are frequently found in museum collections, in a variety of artifacts, but are often unstable and can deteriorate rapidly and catastrophically. Identifying problems and acting promptly to safeguard their integrity are crucial to the preservation of recent artifacts of all kinds. Different types of plastics, however, deteriorate in different ways. The correct and reliable identification of the basic polymer of a museum object is therefore essential for choosing the correct specialist storage environment and conservation treatment. Of the wide range of polymers that can be found in museum collections, five types appear to be particularly

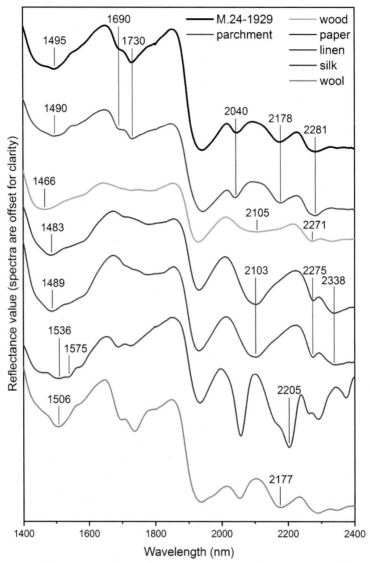

FIG. 12

Sample FORS spectra of *(from the top)* parchment, wood, paper, linen, silk and wool, with indication of selected spectral features. A spectrum captured in the area behind the sculpture's foot (Fig. 10C) is also shown for comparison. Spectra collected with FieldSpec series spectroradiometers at the Fitzwilliam Museum and NGA Washington.

Credit: © Fitzwilliam Museum, Cambridge.

problematic, namely PVC (polyvinyl chloride), polyurethane, cellulose acetate, cellulose nitrate and rubber. The first three are commonly encountered in everyday objects, and can usually be identified by commercial NIR spectrometers. Cellulose nitrate is identifiable with a simple chemical spot test. Rubber, however, displays many modifications and is the most problematic. It is also not recyclable, which may be a reason why manufacturers of scientific instruments have not so far focused on its correct identification.

In the past few years, research has been undertaken on the noninvasive, reliable characterization of historical plastics by means of NIR spectroscopy coupled with statistical analysis. Promising results have been published, showing the possibility of obtaining accurate predictions for almost 90% of polymers included in an ad hoc database created for the development of a statistical classification model; this approach shows potential for useful improvement [57].

For continued positive developments and advancement toward improved characterization of all types of materials found in cultural heritage objects, it remains crucial to explore synergies with other research fields, particularly gemmology, remote sensing, and quality control for human food as well as animal feed, crops and timber. Profitable industries inevitably drive instrumental development; the cultural heritage community can and should continue to adopt those improvements and adapt them to suit its specific requirements, while bringing to the fore new and interesting challenges of its very own.

Acknowledgments

I owe my thanks to a number of people who contributed, directly or indirectly, to the production of this chapter. I was first introduced to UV-vis-NIR RS, and to its use for the study of museum objects, by John K. Delaney (Senior Imaging Scientist at NGA Washington), from whom I learned most of what I know about this topic. With Susanna Pancaldo, Anna Mazzinghi and Trevor Emmett, John read drafts of this document and provided insightful feedback. Trevor, Stella Panayotova and Helen Howard contributed directly to the technical analyses presented here. Helen Ritchie, Vicky Avery, Jo Dillon, Elsbeth Geldhof, Jennifer Marchant and Sophie Rowe provided curatorial and objects conservation expertise. Marcello Picollo and Maurizio Aceto have answered innumerable questions about FORS spectra during the years. Julie Dawson and Suzanne Reynolds incessantly provide support and encouragement, for which I could not be more thankful.

Dedication

I dedicate this chapter to the memory of Jenny Marchant, amazing colleague and friend, a fearless conservator who was never afraid to pick up the skills she needed to undertake scientific work herself. Jenny had a keen interest in reflectance spectrophotometry and could often be found analyzing pigments on Egyptian coffins. Her infectious smile and boundless energy will always be missed.

References

[1] Analytical Methods Committee AMCTB No 75, UV-visible-NIR reflectance spectrophotometry in cultural heritage: background paper, Anal. Methods 8 (2016) 5894–5896, https://doi.org/10.1039/c6ay90112c.

[2] N.F. Barnes, Color characteristics of artists' pigments, J. Opt. Soc. Am. 29 (5) (1939) 208–214, https://doi.org/10.1364/JOSA.29.000208.

[3] M. Bacci, V. Cappellini, R. Carlà, Diffuse reflectance spectroscopy: an application to the analysis of artworks, J. Photochem. Photobiol. B 1 (1987) 132–133, https://doi.org/10.1016/1011-1344(87)80016-7.

[4] M. Picollo, M. Aceto, T. Vitorino, UV-vis spectroscopy, Phys. Sci. Rev. 4 (4) (2018) 20180008, https://doi.org/10.1515/psr-2018-0008.

[5] B. Hapke, Theory of Reflectance and Emittance Spectroscopy, second ed., Cambridge University Press, Cambridge, 2012, https://doi.org/10.1017/CBO9781139025683.

[6] A. Beeby, L. Garner, D. Howell, C.E. Nicholson, There's more to reflectance spectroscopy than lux, J. Inst. Conserv. 41 (2) (2018) 142–153, https://doi.org/10.1080/19455224.2018.1463920.

[7] J.P. Blitz, Diffuse reflectance spectroscopy, in: F.M. Mirabella (Ed.), Modern Techniques in Applied Molecular Spectroscopy, John Wiley & Sons Inc., Hoboken, NJ, 1998, pp. 185–219.

[8] R.N. Clark, Spectroscopy of rocks and minerals, and principles of spectroscopy, in: A.N. Rencz (Ed.), Manual of Remote Sensing, vol. 3, Remote Sensing for the Earth Sciences, John Wiley and Sons Inc, New York, 1999, pp. 3–53. Also available online at https://archive.usgs.gov/archive/sites/speclab.cr.usgs.gov/PAPERS.refl-mrs/refl4.html.

[9] A. Jurado-López, M.D. Luque de Castro, Use of near infrared spectroscopy in a study of binding media used in paintings, Anal. Bioanal. Chem. 380 (2004) 706–711, https://doi.org/10.1007/s00216-004-2789-5.

[10] M. Vagnini, C. Miliani, L. Cartechini, P. Rocchi, B.G. Brunetti, A. Sgamellotti, FT-NIR spectroscopy for non-invasive identification of natural polymers and resins in easel paintings, Anal. Bioanal. Chem. 395 (2009) 2107–2118, https://doi.org/10.1007/s00216-009-3145-6.

[11] P. Ricciardi, J.K. Delaney, M. Facini, J.G. Zeibel, M. Picollo, S. Lomax, M. Loew, Near infrared reflectance imaging spectroscopy to map paint binders in situ on illuminated manuscripts, Angew. Chem. Int. Ed. 51 (23) (2012) 5607–5610, https://doi.org/10.1002/anie.201200840.

[12] M. Bacci, A. Corallini, A. Orlando, M. Picollo, B. Radicati, The ancient stained windows by Nicolò di Pietro Gerini in Florence. A novel diagnostic tool for non-invasive in situ diagnosis, J. Cult. Herit. 8 (2007) 235–241, https://doi.org/10.1016/j.culher.2007.02.001.

[13] J.K. Delaney, D.M. Conover, K.A. Dooley, L. Glinsman, K. Janssens, M. Loew, Integrated X-ray fluorescence and diffuse visible-to-near-infrared reflectance scanner for standoff elemental and molecular spectroscopic imaging of paints and works on paper, Herit. Sci. 6 (2018) 31, https://doi.org/10.1186/s40494-018-0197-y.

[14] C. Bisulca, M. Picollo, M. Bacci, D. Kunzelman, UV-vis-NIR reflectance spectroscopy of red lakes in paintings, in: Proceedings of the 9th International Conference on NDT in Art, Jerusalem, Israel, 25–30 May, 2008. https://www.ndt.net/search/docs.php3?id=6120.

[15] B. Fonseca, C. Schmidt Patterson, M. Ganio, D. MacLennan, K. Trentelman, Seeing red: towards an improved protocol for the identification of madder- and cochineal-based

pigments by Fiber optics reflectance spectroscopy (FORS), Herit. Sci. 7 (2019) 92, https://doi.org/10.1186/s40494-019-0335-1.

[16] P. Kubelka, F. Munk, An article on optics of paint layers, Fuer Tekn. Physik 12 (1931) 593–609 (translated by S.H. Westin, 2004).

[17] G. Dupuis, M. Menu, Quantitative evaluation of pigment particles in organic layers by fibre-optics diffuse-reflectance spectroscopy, Appl. Phys. A Mater. Sci. Process. 80 (2005) 667–673.

[18] G. Dupuis, M. Menu, Quantitative characterisation of pigment mixtures used in art by fibre-optics diffuse-reflectance spectroscopy, Appl. Phys. A Mater. Sci. Process. 83 (2006) 469–474.

[19] C. Clementi, B. Doherty, P.L. Gentili, C. Miliani, A. Romani, B.G. Brunetti, A. Sgamellotti, Vibrational and electronic properties of painting lakes, Appl. Phys. A Mater. Sci. Process. 92 (2008) 25–33, https://doi.org/10.1007/s00339-008-4474-6.

[20] A.B. Murphy, Modified Kubelka–Munk model for calculation of the reflectance of coatings with optically-rough surfaces, J. Phys. D. Appl. Phys. 39 (2006) 3571–3581, https://doi.org/10.1088/0022-3727/39/16/008.

[21] M. Aceto, A. Agostino, G. Fenoglio, A. Idone, M. Gulmini, M. Picollo, P. Ricciardi, J.K. Delaney, Characterisation of colourants on illuminated manuscripts by portable fibre optic UV-visible-NIR reflectance spectrophotometry, Anal. Methods 6 (2014) 1488–1500, https://doi.org/10.1039/c3ay41904e.

[22] A. Cosentino, FORS spectral database of historical pigments in different binders, e-conserv. J. 2 (2014) 53–65.

[23] M.J. Melo, P. Nabais, R. Araújo, T. Vitorino, The conservation of medieval manuscript illuminations: a chemical perspective, Phys. Sci. Rev. 4 (8) (2018) 20180017, https://doi.org/10.1515/psr-2018-0017.

[24] D. Lichtblau, M. Strlič, T. Trafela, J. Kolar, M. Anders, Determination of mechanical properties of historical paper based on NIR spectroscopy and chemometrics – a new instrument, Appl. Phys. A Mater. Sci. Process. 92 (2008) 191–195, https://doi.org/10.1007/s00339-008-4479-1.

[25] A.R. Pallipurath, J.M. Skelton, P. Ricciardi, S. Bucklow, S.R. Elliott, Multivariate analysis of combined Raman and fibre-optic reflectance spectra for the identification of binder materials in simulated medieval paints, J. Raman Spectrosc. 44 (2013) 866–874, https://doi.org/10.1002/jrs.4291.

[26] A. Pallipurath, R. Villő Vőfély, J.M. Skelton, P. Ricciardi, S. Bucklow, S.R. Elliott, Estimating the concentrations of pigments and binders in lead-based paints using FT-Raman spectroscopy and principal component analysis, J. Raman Spectrosc. 45 (2014) 1272–1278, https://doi.org/10.1002/jrs.4525.

[27] A.R. Pallipurath, J.M. Skelton, P. Ricciardi, S.R. Elliott, Estimation of semiconductor-like pigment concentrations in paint mixtures and their differentiation from paint layers using first-derivative reflectance spectra, Talanta 154 (2016) 63–72, https://doi.org/10.1016/j.talanta.2016.03.052.

[28] M. Aceto, A. Agostino, M. Gulmini, E. Pelliz, V. Bianco, A protocol for non-invasive analysis of miniature paintings, in: A. Miranda, M. João Melo (Eds.), Medieval Colours: Between Beauty and Meaning, Instituto de História de Arte, Lisbon, 2011, pp. 231–241.

[29] M. Aceto, A. Agostino, G. Fenoglio, M. Gulmini, V. Bianco, E. Pellizzi, Non invasive analysis of miniature paintings: proposal for an analytical protocol, Spectrochim. Acta A Mol. Biomol. Spectrosc. 91 (2012) 352–359, https://doi.org/10.1016/j.saa.2012.02.021.

[30] P. Ricciardi, S. Panayotova, A holistic, noninvasive approach to the technical study of manuscripts: the case of the Breslau Psalter, in: A. Nevin, T. Doherty (Eds.), The Non-invasive Analysis of Painted Surfaces: Scientific Impact and Conservation Practice, Smithsonian Institution Scholarly Press, Washington, DC, 2016, pp. 25–36, https://doi.org/10.5479/si.19492367.5.

[31] S. Panayotova (Ed.), Colour. The Art and Science of Illuminated Manuscripts, Harvey Miller/Brepols Publishers, London and Turnhout, 2016.

[32] C. Porter, M. Aceto, E. Calà, A. Agostino, G. Fenoglio, A. Idone, M. Gulmini, Looking for lichen, fooled by folium and tricked by Tyrian: a brief tour and new research on purple in manuscripts, in: S. Panayotova, P. Ricciardi (Eds.), Manuscripts in the Making. Art and Science, Vol. 2, Harvey Miller/Brepols Publishers, London and Turnhout, 2018, pp. 64–77.

[33] P. Ricciardi, C. Schmidt Patterson, Science of the book: analytical methods for the study of illuminated manuscripts, in: S. Panayotova (Ed.), A Handbook of the Art and Science of Illuminated Manuscripts, Harvey Miller/Brepols Publishers, London and Turnhout, 2020, pp. 51–87.

[34] S. Panayotova, P. Ricciardi (Eds.), Manuscripts in the Making. Art and Science, 2 vol, Harvey Miller/Brepols Publishers, London and Turnhout, 2017-2018.

[35] P. Ricciardi (Ed.), Manuscripts in the making, art and science, Conference held in Cambridge, UK, December, 2016. https://www.springeropen.com/collections/Fitzwilliam2017. (Accessed 7 May 2020).

[36] Illuminated: Manuscripts in the Making, www.fitzmuseum.cam.ac.uk/illuminated (Accessed 7 May 2020).

[37] P. Ricciardi, M. Facini, J.K. Delaney, Painting and illumination in early Renaissance Florence: the techniques of Lorenzo Monaco and his workshop, in: D. Saunders, M. Spring, A. Meek (Eds.), The Renaissance Workshop, Archetype Publications, London, 2013, pp. 1–9, ISBN: 9781904982937.

[38] S. Panayotova, P. Ricciardi, Painting the Trinity Hrabanus: materials, techniques and methods of production, Trans. Camb. Bibliogr. Soc. 16 (2) (2017) 227–261.

[39] H. Ritchie, Designers & Jewellery 1850–1940: Jewellery and Metalwork from the Fitzwilliam Museum, Philip Wilson Publishers and the Fitzwilliam Museum, 2018.

[40] S. Karampelas, L. Kiefert, Gemstones and minerals, in: H. Edwards, P. Vandenabeele (Eds.), Analytical Archaeometry: Selected Topics, The Royal Society of Chemistry, London, 2012, pp. 291–317, https://doi.org/10.1039/9781849732741-00291.

[41] D.L. Wood, K. Nassau, Infrared spectra of foreign molecules in beryl, J. Chem. Phys. 47 (1967) 2220, https://doi.org/10.1063/1.1703295.

[42] Y. Liu, T. Lu, T. Mu, H. Chen, J. Ke, Color measurement of a ruby, Color. Res. Appl. 38 (5) (2013) 328–333, https://doi.org/10.1002/col.21743.

[43] R.N. Clark, T.V.V. King, M. Klejwa, G.A. Swayze, High spectral resolution reflectance spectroscopy of minerals, J. Geophys. Res. 95 (B8) (1990) 12653–12680, https://doi.org/10.1029/JB095iB08p12653.

[44] M.R.M. Izawa, E.A. Cloutis, T. Rhind, S.A. Mertzman, J. Poitras, D.M. Applin, P. Mann, Spectral reflectance (0.35–2.5 µm) properties of garnets: implications for remote sensing detection and characterization, Icarus 300 (2018) 392–410, https://doi.org/10.1016/j.icarus.2017.09.005.

[45] K. Dardes, A. Rothe (Eds.), The Structural Conservation of Panel Paintings, Part Two: History of Panel-Making Techniques, Getty Conservation Institute, Los Angeles, CA, 1998. http://hdl.handle.net/10020/gci_pubs/panelpaintings.

[46] E. Richardson, G. Martin, P. Wyeth, X. Zhang, State of the art: non-invasive interrogation of textiles in museum collections, Microchim. Acta 162 (2008) 303–312, https://doi.org/10.1007/s00604-007-0885-x.

[47] J.K. Delaney, P. Ricciardi, L. Glinsman, M. Palmer, J. Burke, Use of near infrared reflectance imaging spectroscopy to map wool and silk fibres in historic tapestries, Anal. Methods 8 (2016) 7886–7890, https://doi.org/10.1039/c6ay02066f.

[48] I. Reiche, S. Röhrs, J. Salomon, B. Kanngießer, Y. Höhn, W. Malzer, F. Voigt, Development of a nondestructive method for Underglaze painted tiles – demonstrated by the analysis of Persian objects from the nineteenth century, Anal. Bioanal. Chem. 393 (2009) 1025–1041, https://doi.org/10.1007/s00216-008-2497-7.

[49] C. Fornacelli, A. Ceglia, S. Bracci, M. Vilarigues, The role of different network modifying cations on the speciation of the Co^{2+} complex in silicates and implication in the investigation of historical glasses, Spectrochim. Acta A Mol. Biomol. Spectrosc. 188 (2018) 507–515, https://doi.org/10.1016/j.saa.2017.07.031.

[50] P. Edqvist, African Hair Combs: Exploring Material Technology Using Fibre Optics Reflectance Spectroscopy, Project report. Available online at https://www.academia.edu/12959032/African_Hair_Combs_Exploring_Material_Technology_Using_Fibre_Optics_Reflectance_Spectroscopy2013 (Accessed 7 May 2020).

[51] T.F. Emmett, P. Edqvist, The use of VIS-NIR fibre optic reflection spectroscopy (FORS) in the speciation of wood: potentials and problems in the analysis of African hair combs, in: Poster presented at the 1st International Conference on Science and Engineering in Arts, Heritage and Archaeology (SEAHA), University College London, 14–15 July, 2015. Available online at http://www.seaha-cdt.ac.uk/wordpress/wp-content/uploads/2015/07/Emmett.pdf. (Accessed 7 May 2020).

[52] E. Catelli, G. Sciutto, S. Prati, Y. Jia, R. Mazzeo, Characterization of outdoor bronze monument patinas: the potentialities of near-infrared spectroscopic analysis, Environ. Sci. Pollut. Res. 25 (2018) 24379–24393, https://doi.org/10.1007/s11356-018-2483-3.

[53] E. Catelli, L. Lyngsnes Randeberg, H. Strandberg, B. Kåre Alsberg, A. Maris, L. Vikki, Can hyperspectral imaging be used to map corrosion products on outdoor bronze sculptures? J. Spectr. Imaging 7 (2018) a10, https://doi.org/10.1255/jsi.2018.a10.

[54] M. Sponheimer, C.M. Ryder, H. Fewlass, E.K. Smith, W.J. Pestle, S. Talamo, Saving old bones: a non-destructive method for bone collagen prescreening, Sci. Rep. 9 (2019) 13928, https://doi.org/10.1038/s41598-019-50443-2.

[55] P. Garside, P. Wyeth, X. Zhang, Use of near IR spectroscopy and chemometrics to assess the tensile strength of historic silk, e-Preservation Sci. 8 (2011) 68–73.

[56] M. Strlič, M. Cassar, J. Kolar, D. Lichtblau, M. Anders, T. Trafela, L. Cséfalvayová, G. de Bruin, B. Knight, G. Martin, J. Palm, N. Selmani, M.C. Christensen, NIR/Chemometrics approach to characterisation of historical paper and surveying of paper-based collections, in: J. Bridgland (Ed.), Preprints of the 15th ICOM-CC Triennial Conference, 22–26 September 2008, New Delhi, India, Vol. I, Allied Publishers Pvt Ltd., New Delhi, 2008, pp. 293–300.

[57] V. Šuštar, J. Kolar, L. Lusa, T. Learner, M. Schilling, R. Rivenc, H. Khanjian, D. Koleša, Identification of historical polymers using near-infrared spectroscopy, Polym. Degrad. Stab. 107 (2014) 341–347, https://doi.org/10.1016/j.polymdegradstab.2013.12.035.

Neutron and X-ray tomography in cultural heritage studies

5

Eberhard Lehmann and David Mannes
Paul Scherrer Institute, Villigen, Switzerland

1 Introduction: The aim of cultural heritage studies with tomography methods

Objects of our cultural heritage are manifold in composition, size and age. All of them should be analyzed only nondestructively in order to preserve them for coming generations. An analysis can answer many questions regarding age, condition, purpose, manufacturing methods, provenance and authenticity.

With the help of tomography methods, it becomes possible to analyze the whole volume of an object (3D) or to derive at least specific slices (2D) of it lying transverse to the transmission direction of the applied radiation. This can be done in an acquisition process involving many discrete projections of the object by transmitted radiation and subsequent reconstruction by means of mathematical procedures.

Tomography with X-rays was developed in the last century when powerful computers became available to handle large amounts of data in a reasonable time for data reconstruction and visualization.

The mathematical and methodological details had already been devised much earlier by Radon [1–3] and were first applied by G. Hounsfield (UK) and A. Cormack (USA) during the early 1970s [4]. The major application of X-ray tomography has been the observation of the human body in medicine [4]. Shortly after, the method was introduced as nondestructive testing method in research and industry for components of different sizes. Nowadays, using synchrotron light source facilities [5] and micro-focus X-ray sources, research activities in many fields become possible on the microscopic scale.

Tomography with neutrons arrived much later because there is the need for digital image acquisition to feed into the reconstruction algorithm in order to derive the information about the third dimension of the object under investigation. These digital neutron detectors were only developed in the 1990s [6,7] (previously, only film-based single frame images could be obtained in a mode analogous to X-radiography). Now, both neutron and X-ray tomographies are similar standard imaging techniques, and are available at several neutron research centers (see Section 8 and Table 1).

133

Spectroscopy, Diffraction and Tomography in Art and Heritage Science. https://doi.org/10.1016/B978-0-12-818860-6.00009-X

Table 1 Facilities for neutron tomography, available in principle for cultural heritage studies.

Country	Site	Institution	Facility	Neutron source	Spectrum	Power (MW)	Status
Australia	Sydney	ANSTO	DINGO	OPAL reactor	Thermal	20	Operational
Germany	Munich-Garching	TU Munich	ANTARES	FRM-2 reactor	Cold	25	Operational
Germany	Munich-Garching	TU Munich	NECTAR	FRM-2 reactor	Fast	25	Operational
Germany	Berlin	HZB	CONRAD	BER-2 reactor	Cold	10	Shut down end of 2019
Hungary	Budapest	KFKI	NORMA	WWS-M reactor	Cold	10	Operational
Hungary	Budapest	KFKI	NRAD	WWS-M reactor	Thermal	10	Operational
Japan	Kyoto	Kyoto University	Imaging beamline	MTR reactor	Thermal	5	Standby
Japan	Tokai	JAEA	Imaging beamline	JRR-3M reactor	Thermal	20	Standby
Japan	Tokai	JAEA	RADEN	JPARC spallation	Cold	0.5	Operational
Korea	Daejon	KAERI	Imaging beamline	HANARO reactor	Thermal	30	Standby
Russia	Dubna	JINR	Imaging beamline	IBR-2M pulsed reactor	Thermal	2	Operational
Switzerland	Villigen	PSI	NEUTRA	SINQ spallation	Thermal	1	Operational
Switzerland	Villigen	PSI	ICON	SINQ spallation	Cold	1	Operational
UK	Oxfordshire	Rutherford Lab	IMAT	ISIS spallation	Cold	0.3	Operational
USA	Gaithersburg	NIST	BT-2	NBSR reactor	Thermal	20	Operational
USA	Gaithersburg	NIST	NG-6	NBSR reactor	Cold	20	Operational
USA	Oak Ridge	ORNL	CG-1D	HFIR reactor	Cold	85	Operational
South Africa	Pelindaba	NECSA	SANRAD	SAFARI reactor	Thermal	20	Standby
France/EU	Grenoble	ILL	D50	ILL reactor	Cold	58.3	Operational

While the acquisition and reconstruction principles of X-ray and neutron tomography are very similar, the results are distinct, principally because of the differing interaction mechanisms of these two kinds of radiation with matter. This behavior will be described and the consequences explained in Sections 3 and 4.

Because these two kinds of radiation are able to penetrate matter and the investigated materials are more or less transparent to them, the result of a tomography study is a volume data set which comprises information on the inner content with the spatial resolution of the detection system used. By means of suitable specialized software, it is possible to analyze the data in more detail by slicing, segmenting, measuring and counting of features and to visualize the results by hiding or illuminating zones and analyzed features.

In this way, both tomography techniques have a high impact in investigations of cultural heritage objects because they are completely nondestructive and noninvasive when used appropriately. It depends, of course, on the material and size of the object, and on which method is best suited for the specific investigation. The relevant criteria are described in Sections 3 and 4.

Fig. 1 shows an example of the study of an old Tibetan Buddha sculpture by neutron tomography. It was filled, sealed and sanctified about 500 years ago. The aim of the study was to analyze the inner content of the object without opening—and to compare the filling with assembly practices, which are common today. Opening is not only forbidden for religious reasons as it would desecrate the assembly, but also to avoid alteration, damage and disturbing the arrangement of the devotional objects within.

FIG. 1

Results of a tomography study with neutrons: a Buddha sculpture "Kunga–Peljor" (about 20cm high), private CL Tibet collection Therwil, Switzerland: *left*: virtual slice through the object; *right*: virtually extracted devotional objects from the sculpture inside—the "life tree" and four ceramic objects (so-called tsha-tshas).

The example shows that all positions inside the object are accessible and a detailed analysis becomes possible based on the characteristic material properties with respect to the applied kind of radiation. In this particular case, X-ray tomography would fail most likely due to the limited transmission through the metallic cover and the low contrast for the hidden material (e.g., wood, ceramics, textile and other organics).

Studies of cultural heritage objects have been performed at the author's institute (Paul Scherrer Institute, Switzerland) for local Swiss museum's collections, but also with objects provided from institutions abroad. While the major competence is in the field of neutron tomography, in-house facilities for X-ray imaging were used for comparison. Because the number of investigations is quite large, the area of cultural heritage studies can never be covered completely by this article. However, we provide examples of similar studies within our reference list.

It has to be mentioned that neutron and X-ray tomography are able to visualize and measure dimensionally the size and the shape of most components of an object. However, there are also many other techniques which help museum experts to analyze their collections, described in other chapters of the book. Therefore this chapter will help and guide potential applicants to decide whether a tomographic study should be performed, and if so, which kind and under what conditions. The seven case studies below can be taken as examples and templates.

2 Tomography as a general method

Here is a definition from Wikipedia [8]: "Tomography is imaging by sections or sectioning, through the use of any kind of penetrating wave". The method is used in radiology and many other research fields. X-ray tomography is the commonest due to its manifold uses in hospitals. The tool is still under development and subject to improvement.

The principle acquisition process in tomography is given in Fig. 2: the emitted radiation from the source penetrates the object to be investigated and the attenuated radiation is registered by a radiation sensitive detector, commonly as a discrete 2D pixel matrix.

This transmission process is described to first order by the Beer-Lambert law, where the intensity in front of the sample $I0$ is related to the detected attenuated intensity behind the sample I by:

$$I = I_0 e^{-\Sigma d} \tag{1}$$

where the sample thickness d in beam direction and the so-called attenuation coefficient Σ are relevant. Σ is, on the one hand, an inherent material property (caused by the sensitivity of the particular material to the kind of radiation) and otherwise linearly dependent on the material density. Unfortunately, all information about the attenuation in beam direction z is integrated over the whole material thickness d while in the directions across (dimensions x and y in Fig. 2) a pixel-wise value

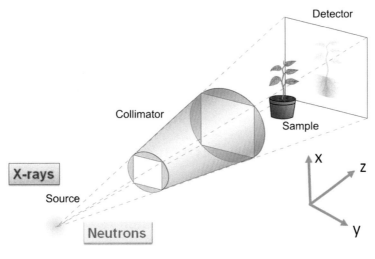

FIG. 2

Setup for the transmission imaging of one projection of the sample.

can be derived. Obtaining the values of Σ along a path through the sample in the z direction is the challenge in tomography. At the end of a tomography study, we will have the data for $\Sigma(x, y, z)$ in the whole sample volume. Because this is only possible in a discrete digital format in order to apply an algorithm for the reconstruction of the values in the z direction, we will have the matrix of attenuation coefficients Σijk.

The standard approach in X-ray and neutron tomographies, known as computed tomography (CT), is structured into three steps, as demonstrated in Fig. 3.

The basic data acquisition is done by acquiring many projections of the sample from different viewing angles over 180 degree (parallel beam) or 360 degree (cone beam). If possible, the source and the detector are rotated simultaneously around the sample with the sample position fixed (typical approach in medical radiology when the patient should not be moved). If the source is very large (as it is in the case of neutron sources or synchrotron beam lines), the sample has to be rotated either around its vertical or horizontal axis across the beam. Depending on the required or intended precision in spatial resolution or achievable image quality, several hundred projections are taken.

The second step, where the necessary mathematical procedures are applied, is the reconstruction of data in the horizontal direction z as the inversion of Eq. (1). The resulting slices $\Sigma(y, z)$ can be stacked in the x direction and the whole $\Sigma(x, y, z)$ volume data set is made available for further treatment. The most common method for volume reconstruction is the "Filtered Back Projection (FBP)" method, which is based on the "Inverse Radon Transformation". Because all mathematical details are well described in the literature [3], we will not repeat them here, but focus on

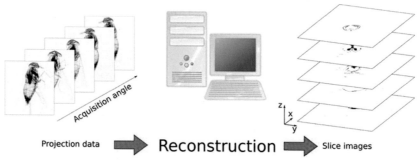

FIG. 3

Schematic workflow of a tomography study.
*Taken from J.H. Hubbell, S.M. Seltzer, X-Ray Mass Attenuation Coefficients, (2004), https://www.nist.gov/pml/
x-ray-mass-attenuation-coefficients (Created September 17, 2009, Updated December 11, 2019).*

the methodology, which is more relevant for the study of cultural heritage objects. Even though FBP is already demanding in computation power and calculation time for large data sets, there are still more demanding methods like Iterative Algorithms [9], which are applied if the image data quality needs to be enhanced further. However, these require a priori information about the sample.

It is quite important which beam geometry from the source is delivered to the sample. A parallel beam illuminates all parts of the investigated object equally without geometrical distortion. Only projections over 180 degree rotation are needed due to repetitions in the mirrored case. On the other hand, a conical beam allows some magnification if the sample and the detector are at suitable distance. Here, the full 360 degree range of sample rotation has to be investigated. In this case, the final limit in spatial resolution is given by the spot size of the source. The reconstruction algorithm for the cone beam case is much more demanding than for the parallel beam case, but implemented and available in modern tools, e.g., [10].

Most of the studies presented in this chapter were done in (nearly) parallel beam geometry in a manner where the source was far away from the sample and the detector was arranged very close to and behind the sample. The "collimator" in Fig. 2 is needed to cover the beam in the limits of sample and detector (also for radiation protection reasons). There is an important number which describes the collimator properties: the collimation ratio L/D, where L is the distance between source and detector and D the diameter of the source. Values higher than 500 already can be considered as "quasi-parallel" where FBP of the parallel beam geometry option is applicable.

The third step in tomography data processing is the "visualization" one. Even if it is possible to handle the Σijk data manually, there are modern software tools available (both freeware and commercial) which are able to perform many analysis steps in a comfortable manner. It has to be taken into account that the volume data sets are often large, depending on the number of voxels that are defined by the detector size. Currently employed neutron imaging detector systems (camera-based) operate with 2048×2048 pixels in the (x, y) plane and have a 16-bit dynamic range. The tomography data volume ends up then in a $2048 \times 2048 \times 2048$ voxel matrix, corresponding to 17.2 GB, which has to be stored and handled simultaneously in the main memory of the computer. This is not a significant problem for modern computers—but it was nearly impossible two decades ago.

However, with increasing detector area and higher spatial resolution, the data volume will increase again in cubic order.

Software tools like Avizo (Thermo Fisher Scientific) and the Volume Graphics product range (Volume Graphics GmbH) [11,12] have many analysis options available. By comparing similar values of Σ_{ijk} structures and their surfaces can be defined and accordingly visualized.

Because the voxel position and the voxel size are well defined, the sizes of zones with similar values can be derived and measured in three dimensions. With the definition of subzones they can be identified and handled (separated, moved, hidden) specifically. Similar features can be counted, extracted and measured in their distribution. The example of wall-thickness determination is given in Fig. 4 for a famous object from the Roman period, discussed in more detail in Section 6.5. All these options (which are under further development in interaction with the users) have been applied successfully in the studies described in this chapter.

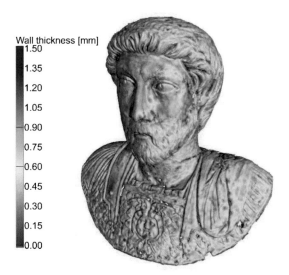

FIG. 4

Result of a tomography study of the golden bust of Marcus Aurelius, museum Avenches, Switzerland, about 25 cm high, where the thickness of the material layer is derived with 0.1 mm precision (in color code) for all positions at the object at the same time.

Before starting a tomography acquisition, which takes a few hours, depending on beam intensity and required resolution, simple projection data should be taken from the two main directions. The tomography run should only be started if the transmission or the required contrast is good enough (5% to 95% transmission). Otherwise, the reconstruction might fail or artifacts will be induced, which hinders a serious interpretation afterwards.

3 Neutron interaction with matter

Neutrons as free particles can only be made available in nuclear reactions as nuclear fission or spallation. In the first case, a nuclear reactor where a stationary process via chain reactions of fission with U-235 or Pu-239 is running has to be set up. Reactors for neutron production differ in principle from power plants: their purpose is to deliver as many as possible free neutrons while the generated power should be as low as possible.

Spallation neutron sources are driven by a powerful accelerator where charged particles (mostly protons) are sent to a target with high mass density. Depending on the particle energy and the target material, about 10–15 neutrons are set free per spallation event. Whereas the nuclear reactor is running (in most cases) continuously, the spallation source can be operated in a pulsed mode with frequencies of the order of about 25 Hz.

Although there are initiatives for smaller sources, also accelerator driven, the most prominent neutron imaging facilities are located at either fission or spallation sources. A summary of the existing user facilities is given in Section 8, Table 1.

All primarily generated free neutrons have a high initial energy of 1 MeV or higher. They are not very useful for neutron research because the interaction probability differs only little from material to the next. Much more interesting are the neutrons in the energy range of meV, corresponding to wavelengths (according to de Broglie's relation) of few angstroms—i.e., the order of the interatomic spacing in condensed matter.

To slow down the fast neutrons over nine orders of magnitude in energy, many subsequent nuclear collisions within a "moderator" (light or heavy water) are necessary. While light water stops the neutrons more efficiently, heavy water does not absorb neutrons as much and a better distribution of thermal (or cold) neutrons around the initial source can be obtained.

Modern research neutron sources deliver neutrons to many different beam ports and serve many neutron instruments, mainly for neutron scattering and imaging. Fig. 5 shows the example of the spallation source SINQ, PSI, Switzerland. In the case of SINQ, a second moderator with liquid deuterium at 25 K slows down the neutrons further to energies of only about 5 meV. The advantage of these cold neutrons is a higher image contrast for most of the materials under investigation compared to thermal neutrons.

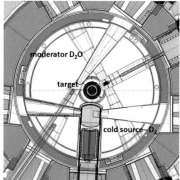

FIG. 5

The Spallation Neutron Source SINQ (*left*: overview; *right*: layout of the target region and the D2O moderator around, including the beam port exits).

The two most important nuclear reactions with relevance for neutron imaging are neutron capture and neutron scattering in the samples. Both attenuate the incident beam and are considered equally on first order. The scattered neutrons might return to the resulting image and can disturb the data as a background. However, there are methods for scattering corrections available. Generally, the "total attenuation cross-sections Σtot" as sum of all contributing interactions are taken for data analysis in neutron imaging. Because the interaction takes place exclusively with the atomic nuclei, the electrons in the shell are "ignored".

If we compare Σtot for all the elements and their isotopes, we can find huge differences, as demonstrated in Fig. 6. (Here, the ratio of Σtot and the material density is given—the so-called "mass-attenuation coefficient".) There is no "golden rule" behind the data because the nuclear interaction is very random from one isotope to the next. Only with knowledge concerning a particular material behavior, can the data in the neutron tomography process be interpreted in the right manner.

As also indicated in Fig. 6, there is a high interaction probability for light elements like H, Li, B, C, etc., but much less attenuation in the heavy elements such as Pb, Bi, Au, Cu or Pd. This explains why the object in Fig. 1 (with its bonze cladding) can be penetrated by neutrons easily but with the wooden pillar visible at the same time (due to the hydrogen content).

4 X-ray interaction with matter

X-rays were discovered in 1896 by C. Röntgen, well before neutrons which were discovered by J. Chadwick in 1932. Because the generation of X-rays at useful intensity and quality is much easier than for neutrons, X-ray imaging is much more readily available than imaging with neutrons. The most prominent X-ray application is still

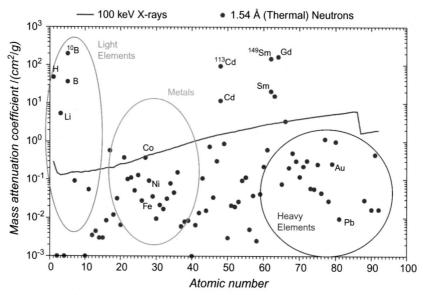

FIG. 6

Mass-attenuation coefficients for the observed material according to their atomic number in the interaction with thermal neutrons *(dots)* and 100 keV X-rays for comparison *(line)*. The data are for natural isotopic abundances unless the isotope is labeled specifically. Note that different isotopes of the same element have very different neutron mass attenuation because of the different number of neutrons in the nucleus. The step in the X-ray attenuation for heavy elements occurs when 100 keV X-rays can ionize the K-level of the element.

Data from https://www.ncnr.nist.gov/instruments/bt1/neutron.html (Accessed October 2020); J.H. Hubbell, S.M. Seltzer, X-Ray Mass Attenuation Coefficients, (2004), https://www.nist.gov/pml/x-ray-mass-attenuation-coefficients (Created September 17, 2009, Updated December 11, 2019).

medical radiography to image inner parts of the human body to investigate disease, bone fractures and other health issues.

As is the case with neutrons, there are also two different processes in the interaction of X-rays with matter: photon scattering and photon absorption, which both attenuate an incoming beam of X-rays. These two processes are very dependent on the photon energy as shown in Fig. 7 for aluminum. However, there is also a different behavior with respect to the mass number of the element under investigation, as indicated in Fig. 6. Because the interaction takes place only with the electrons of the materials, no isotopic effects are visible. The higher the number of electrons, the higher are the interaction probability and attenuation contrast.

As a consequence of this interaction behavior, metals are much less transparent for X-rays while thick layers of organic materials (wood, human body, wet materials) are much more penetrable than with neutrons.

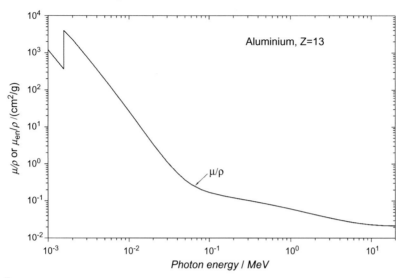

FIG. 7

Mass-attenuation coefficients for aluminum as a function of the photon energy; in the most relevant energy region (0.01 to 1 MeV), a very high slope is visible (consider the log scale); data taken from [13].

Having both kinds of radiation available, it has now to be decided which one is better suited to investigate archaeological finds or museum objects. In the best case, both beams can be applied and the two images obtained can be compared or combined in order to get even more information about the material distribution inside the object. In the case of the SINQ imaging facilities, X-ray tube systems have been integrated into the NEUTRA and ICON stations and data fusion (the combination of data sets from different techniques) is enabled [14].

5 Tomography facilities at PSI

From the 15 instruments at SINQ, accessible within the user program, two are dedicated to neutron imaging: NEUTRA [15] for thermal neutrons and ICON [16] for cold neutrons.

Sharing the beam time with other users, the BOA beam line [17] with polarized neutrons and the POLDI [18] facility (high-intensity beam spot) can be used in addition for imaging purposes.

There is a high flexibility with respect to the field-of-view (FOV) and the spatial resolution, given mainly by means of the detector performance. The neutron beam size (up to 40 cm in diameter) has never been a limiting factor. However, the sample penetration due to the amount of material in beam direction and its attenuation behavior for the incident radiation controls the setting up.

Most of the detection systems for neutrons and X-rays are camera based, where the light from a suitable scintillator screen, excited by the incoming radiation, is observed pixel-wise. By means of different lens systems, the FOV can be selected from 35 cm down to 5 mm while the pixel size is between 0.2 mm and 2 μm. The selection of the best possible setup then depends on the sample size and material composition.

For tomography studies, it is necessary to obtain the number of projections required from the different viewing angles over either 180 degree or 360 degree, depending on the reconstruction method. A well-aligned rotation table performs the rotation about the vertical axis in synchronization with the detector system. In addition, flat field and dark current images are required, showing, respectively, the beam flux/scintillator efficiency/camera sensitivity variation across the FOV and the intrinsic thermal noise signal in the camera.

Although both imaging facilities are dedicated for use in imaging with neutrons, there are X-ray tubes available, which can be used alternatively under similar conditions. This allows both data fusion and a comparative approach as will be demonstrated in Section 6.2 for a sword retrieved from Lake Zug.

6 Examples of tomography studies for cultural heritage objects

Objects of cultural heritage importance may be from: a museum collection, recent excavations, a museum workshop or even a private collection. The purpose varies from one study to another. In some cases, the internal content needs to be investigated in order to understand the production method and the conservation status. In others, processes like corrosion need to be analyzed for the purpose of treatment planning. In this chapter, we can only show a few examples of previous studies, which should be taken as templates—both to understand the principle and also the limitations of the method.

There are some requirements to be considered before a full tomography run can be started. First, the object should be sufficiently transparent from all directions. This can be checked by simple radiography images, taken from two perpendicular directions. Second, the object should not become activated (be made radioactive) through neutron exposure for a long period. Because a 20-s exposure is very unlikely to induce too much activity, but provides a reasonable signal for gamma spectroscopy, a check after one projection can be used to decide about the content and activation risk. Co, Ag or Hg in large amounts should not be exposed for hours due to the long half-life of the activation products.

6.1 Renaissance Bronzes from the Rijksmuseum Amsterdam, The Netherlands

During a very long maintenance period in the Rijksmuseum in Amsterdam (2003–2013), the opportunity was taken to perform a systematic study of the collection of bronze sculptures from the Renaissance period [19]. The museum owns and exhibits many important examples, which were manufactured about 500 years ago, but became available for analysis while not on public display.

As part of his PhD project [20], van Lang, the leading curator of the museum, shipped 14 objects from Amsterdam to the neutron imaging facilities at PSI with the aim of deriving volumetric data by means of neutron tomography investigations.

It was verified by simple transmission measurements in radiography mode that the neutron penetration should be high enough to investigate even bigger samples than the field-of-view (40 cm) of the initial neutron beam. In such cases of taller objects, they were examined at two or three different heights and combined later into one volume data set.

Trials of alternative X-ray studies failed because the metal layers of copperalloys (containing, besides Cu, varying amounts of Zn, Sn and Pb) were up to centimeters in thickness, far beyond the X-ray penetration depth. Therefore neutron imaging with thermal neutrons was found to be the only method to derive the required volumetric information without distortions such as beam hardening and starvation artifacts.

The following information can be derived from the tomography data:

- wall thickness of the bronze at any arbitrary position of the object
- distribution of inner structures (casting remains, distance holders, stabilizers)
- repair and refurbishment work (with other than the base material)
- structure of the inner surfaces (from the original casting process, not further finished)
- corrosion traces and surface finishing work, lacquer or resin at the surface

All these material science questions were supported and combined by standard investigations such as neutron diffraction (see Chapter 9) to learn about the crystalline properties of the material.

The data were published in different formats: as an overview in [20] and in the form of animated views for YouTube, also mentioned in detail in [21]. The authors are willing to provide the raw data for further detailed analysis on demand subject to the conditions of use.

An example of a sample from the Rijksmuseum collection is given in Fig. 8. The sculpture is quite hollow and the empty areas can be visualized, showing the shape of the casting core used in manufacturing and then removed mechanically. The object and its noninvasive segmentation and analysis can be seen in an animated movie: https://youto.be/rPM_kS8Zodg.

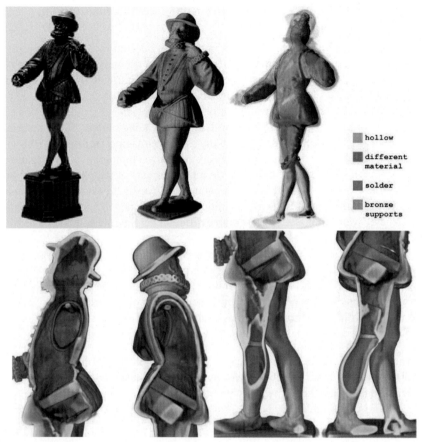

hollow

different
material

solder

bronze
supports

FIG. 8

Neutron tomography study of the "striding nobleman," a Renaissance bronze sculpture
from Rijksmuseum, Amsterdam, estimated production time 1580–1600, Inv.-No. BK-16083:
top: photo, outer tomography view, visualization of the inner empty space; *bottom*: two
artificial slices of the *upper* and *lower part* of the object, respectively. The animation of the
volume can be found on https://youto.be/rPM_kS8Zodg.

6.2 The sword from the Lake Zug, the "Oberwiler Degen"

In 2010, a heavily corroded sword was found during diving activities in Lake Zug,
Switzerland. In comparison to similar such weapons, the sword was attributed to the
period of the second half of the 15th century. Before the final preservation of the
object was started, the museum in Zug agreed to perform nondestructive investigations with X-rays and neutrons in tomography mode. These measurements were performed at the NEUTRA facility at the Swiss spallation neutron source at PSI.

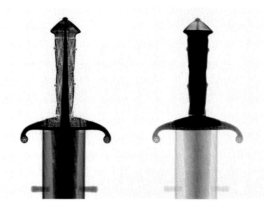

FIG. 9

Comparison of the radiography data (*left*: 250 kV X-rays; *right*: thermal neutrons) of the upper part of the sword from lake Zug (the visible Al tube below was used to fix the object during the tomography run).

Source: D. Mannes, F. Schmid, J. Frey, K. Schmidt-Ott, E. Lehmann, Combined neutron and X-ray imaging for non-invasive investigations of cultural heritage objects, Phys. Procedia 69 (2015) 653–660, https://doi.org/10.1016/j.phpro.2015.07.092.

Already the radiography data (Fig. 9) demonstrated the equivalence of X-rays and neutrons regarding the image data quality, but also the differing contrast of the components of the sword assembly: iron, wood and other metals for decoration and fixing of the structure.

The tomography data documented the impressive complementarity of the two applied methods, as illustrated in Fig. 10. While the neutron data describe the status of the preserved wooden handle in very great detail and give a clear picture of the situation of the blade regarding corrosion, the X-ray data enable one to see the intricate metallic inlay decoration in the wood. Through the use of data fusion, both tomography data sets were merged in order to get a complete picture of the object. These data were then used to build a design template for the construction of a perfect copy of the sword [22], now exhibited in the museum in Zug together with the original. The movie of the virtual disassembly and later assembly can be watched here [23].

6.3 Corrosion studies on iron samples

Corrosion is a serious topic when dealing with archaeological findings or other cultural heritage objects. Neutron imaging can be an important tool for the investigations of such corrosion. Fig. 11 illustrates the example of a heavily corroded nail from the Roman period, and shows how well regions with varying degrees of corrosion can differentiated with neutron imaging. This is due to the fact that the corrosion does not only change the density but also the elemental composition, mainly through

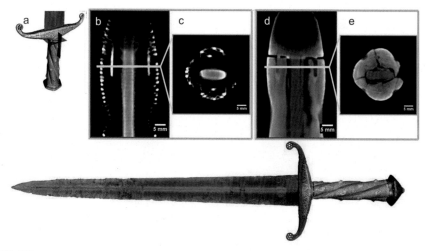

FIG. 10

Results of a combined neutron and X-ray tomography investigation: sections through the reconstructed X-ray (B and C) and neutron (D and E) CT-data sets yield only partial information; only by combining neutron and X-ray CT data the full information is available as seen in the 3D visualization of the object (A and F).

Data from E. Lehmann et al., Bronze Sculptures and Lead Objects tell Stories about their Creators, in: N. Kardjilov, G. Festa (Eds.), Neutron Methods for Archaeology and Cultural Heritage, Springer, 2017, pp. 19–41. ISBN: 9783319331614, modified layout.

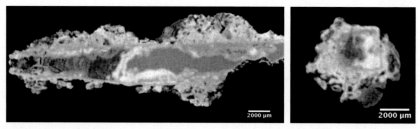

FIG. 11

Sections through the neutron CT-data set of a heavily corroded nail from the Roman period found in Switzerland.

Source: D. Mannes, E. Lehmann, A. Masalles, K. Schmidt-Ott, A.V. Przychowski, K. Schäppi, F. Schmid, S. Peetermans, K. Hunger, The study of cultural heritage relevant objects by means of neutron imaging techniques, Insight 56 (2014) 137–144.

increases in the hydrogen and chlorine content. These elements can cause an increase of the local attenuation when compared to the sound metal matrix and can hence be discerned.

The higher sensitivity of neutrons for corrosion products can be used to study the effectiveness of conservation treatments, especially when combined with X-ray

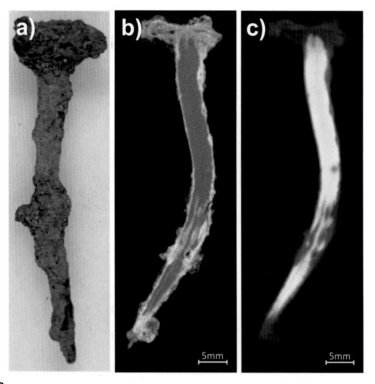

FIG. 12

Photograph of a heavily corroded nail from the Roman period in Switzerland (A), sections through the reconstructed neutron (B) and X-ray (C) CT-data set.

Source: M. Jacot-Guillarmod, K. Schmidt-Ott, D. Mannes, A. Kaestner, E. Lehmann, C. Gervais, Multi-modal tomography to assess dechlorination treatments of iron-based archaeological artifacts, Herit. Sci. 7 (2019) 29, https://doi.org/10.1186/s40494-019-0266-x modified.

investigations, which are, in such a case, only sensitive to density changes. This complementary behavior can be seen in Fig. 12, which shows sections through neutron (Fig. 12B) and X-ray CT data (Fig. 12C) at an identical position. While the sound metal gives a relatively modest signal in the neutron section, the Cl-containing corroded regions are visible as more attenuating, bright regions. The X-ray data show the inverse behavior, where the corroded regions are less attenuating, due to lower density, while the sound metal is highly attenuating and appears therefore as a bright region in the image. By combining neutron and X-ray data using a bivariate histogram (Fig. 13), it is not only possible to improve the segmentation of different regions within the tomography data set but also to assign different materials and material compositions to these regions. This allows one to map different materials and corrosion phases within the 3D volume data, and to show the depth of penetration and effectiveness of conservation treatments. A detailed description of the project can be found in [24].

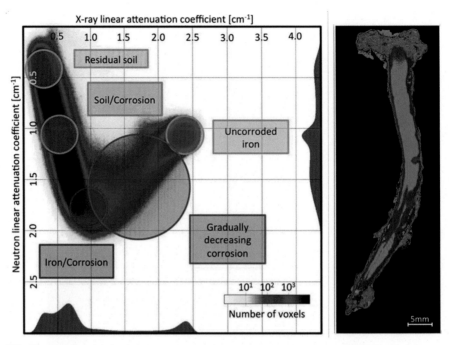

FIG. 13

The bivariate histogram, combining X-ray and neutron histograms *(left)*, allows to segment and map different materials in the volume data set *(right)*.

Source: D. Mannes, E. Lehmann, A. Masalles, K. Schmidt-Ott, A.V. Przychowski, K. Schäppi, F. Schmid, S. Peetermans, K. Hunger, The study of cultural heritage relevant objects by means of neutron imaging techniques, Insight 56 (2014) 137–144. modified.

6.4 Documentation of the corrosion condition of the lead sculpture "El Violinista" by Pablo Gargallo

The sculpture "El Violinista" by the Catalan artist Pablo Gargallo consists of a multitude of lead sheets which are fixed onto a carved wooden core sculpture. In recent years, the sculpture has started to show signs of corrosion (bloated spots on the surface, white lead carbonate powder below the pedestal). In order to address this problem and to determine the best way to save this artwork, the extent of the corrosion the sculpture had to be assessed by an investigation by means of neutron tomography. Lead, which is used as shielding material for X-rays, is relatively transparent to neutrons, which makes neutron-based methods ideal for such an investigation.

Besides the corrosion, the shape of the inner wooden core as well as the points, where the lead sheets were fixed onto the core and to each other were in the scope of the study. Fig. 14 shows some of the results; the wooden core sculpture can be seen in Fig. 14B, while Fig. 14C shows a mapping of the corroded areas (in red) as well as the fixation points such as nails and soldering (in blue). All of this information is needed

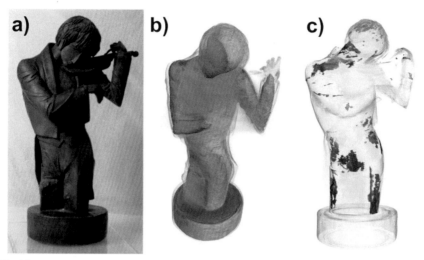

FIG. 14

The lead/wood sculpture "El Violinista" by P. Gargallo; (A) photograph; (B) visualization of the wooden core based on the neutron CT-data; (C) mapping of corroded areas *(red)* and fixation points *(blue)* based on the neutron-CT.

Source: D. Mannes, E. Lehmann, A. Masalles, K. Schmidt-Ott, A.V. Przychowski, K. Schäppi, F. Schmid, S. Peetermans, K. Hunger, The study of cultural heritage relevant objects by means of neutron imaging techniques, Insight 56 (2014) 137–144.

to develop a concept for the preservation of the sculpture. The mapping of the corrosion is needed to ascertain which parts need restoration; the fixing points have to be known, so that the sheets can be dismantled with the least effort and risk of deformation or destruction. As gas emissions from the wooden core were held responsible for the initiation of the corrosion process, the shape of the inner core had to be assessed. With the help of a surface model, it is possible to recreate an exact copy from chemically inert material, on which the preserved sheets could be repositioned after the conservation treatment.

A detailed description of the project can be found in [25,26].

6.5 The golden bust of Marcus Aurelius

During construction work in the region where the former Roman settlement named Aventicum was located, a well-preserved golden head was found in 1939. This hollow bust was later attributed to the emperor Markus Aurelius (living between CE 121 and 180)—Fig. 15. The size is about 80% of that of a normal human head and the wall thickness is a few millimeters or less. Chemical analysis has shown that the gold content is higher than 90%. With the gold weight of 1589.06 g, it represents the largest currently known portrait of a Roman emperor.

FIG. 15

Photographic views of the bust of the Roman emperor Marcus Aurelius, found in 1939 during excavations in Avenches, Switzerland.

With the help of noninvasive techniques, it was intended to research the following aspects: How was the bust made: from only one piece or composed from several pieces? How is the material density and the wall thickness distributed over the entire object? Are there indications for repair work or corrosion in the bulk of the object?

A first trial using X-rays failed because the transmission of millimeters of Au is very limited (attenuation coefficient μ for 100-keV photons is of the order of $36\,\text{cm}^{-1}$). In comparison, the transmission by thermal neutrons is much better ($\Sigma = 6.2\,\text{cm}^{-1}$). Therefore neutron tomography was used to perform a full 3D volumetric scan of the object. Five hundred and twelve projections were taken over 180 degree of sample rotation around its vertical axis (parallel beam approximation) and the voxel matrix was determined after volume reconstruction using the FBP method. Slices of the object at arbitrary positions are possible and the material thickness can be derived along the slice path. This method is, of course, very time consuming if the local gold layer thickness is to be determined over the whole object. Therefore it is very helpful to have a layer thickness tool such as that provided by the software from Volume Graphics [11]. This was applied to the volume data of Marcus Aurelius bust.

Next to the data in Fig. 4, we present here another view from the wall thickness tool, together with one of the projections of the neutron tomography run. Although the simple radiography provides an impression of the material thickness, as indicated in the grey levels, the real numbers can only be derived from the color-coded wall thicknesses. There are large thickness variations (from 0.1 mm to 1.5 mm) with two

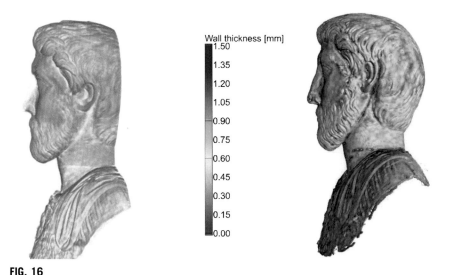

Wall thickness [mm]
1.50
1.35
1.20
1.05
0.90
0.75
0.60
0.45
0.30
0.15
0.00

FIG. 16

One profile view neutron imaging projection of the object *(left)* and the tomography view with the local wall thickness, given in a color code, of the Marcus Aurelius bust from Avenches, Switzerland.

very thin regions: a horizontal one at the end of the neck and another between the neck and beard. Although direct visual inspection did not indicate material differences and fitting traces, the neutron tomography data suggest that the bust might have been fabricated in several pieces and then joined together. To build such a high-quality bust in one piece seems to be a very complicated procedure (Fig. 16).

Since the neutron imaging data of the Marc Aurelius bust remain available, future discussions and consideration by experts are possible based on the neutron tomography study.

6.6 Study of the content of Buddhist bronze sculptures—The Buddha Shakyamuni Bhumisparsha Mudra

For about thousand years, Buddhist sculptures have been produced as hollow cast religious specimens that are filled in a special procedure with various samples of spiritual importance at the time of their creation. Such ancient objects are collected today in museums and private collections worldwide (in addition to their sites in abbeys and temples). Such objects were originally constructed in Tibet, India and China among other countries.

With the opportunity to look deeper into such objects, mainly Buddha representations and Stupas, several interesting aspects can be investigated. However, the main boundary condition is NOT TO OPEN such objects due to the loss of its spiritual value and the destruction of the assembly of the content of the sculpture,

FIG. 17

The Buddha Shakyamuni, Bhumisparsha Mudra from central Tibet and made towards the end of the 15th century *(photo left)*, was studied with X-rays *(middle)* and neutrons *(right)*.

preserved over many centuries. The filling procedure of holy objects is still common in Buddhist religion today and it is very interesting to compare both the ancient approaches with "modern" consecration using sacred materials and the filling materials used.

We were able to demonstrate that X-ray imaging mainly fails to study the inner content of the metallic samples due to the high attenuation of the cover and the low contrast for organic inner material assemblies [27]. On the other hand, neutron imaging was found to fit perfectly to the investigation of the Buddha objects and their inner content. The reason for this advantage is that the neutron attenuation properties of the materials concerned are better matched to the task (see Fig. 6).

We were able to study about 70 ancient Buddhist sculptures from Swiss museums and private collections with success. Here, we can only present two examples (Figs. 17 and 18 and Fig. 1). A further discussion and some more cases are shown in [28,29].

6.7 Comparative study of pearls using X-ray and neutron tomography

Pearls have formed an important constituent of jewelry for centuries due to their durability and brilliance. Therefore they can be attributed to the cultural heritage in many countries and in history. The value of a pearl is mostly determined by its size and quality, and whether it is naturally grown or cultured. It is very difficult to distinguish between them because the outer layer has about the same appearance. A destructive investigation is not very useful due to the damage to the sample. Therefore investigations with X-rays and thermal neutrons have been performed to study the inner structure of pearls with different origin and history.

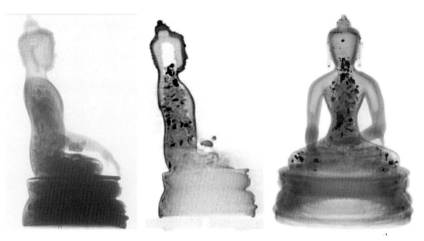

FIG. 18

The Buddha Shakyamuni, Bhumisparsha Mudra, central Tibet, end of 15th century, in neutron radiography and tomography mode; *middle*: vertical slice; *right*: semitransparent viewing option of the volume data.

Given by the limited size of the pearls (millimeters to centimeters only), high-resolution imaging techniques have to be applied to provide reasonable details in the observations. While X-ray tomography is enabled by the magnification option in a conical beam of a micro-spot source, for high-resolution neutron imaging only a few facilities are available. Here, we provide data taken at the micro-tomography setup at the ICON beam line in PSI [30].

Fig. 19 compares tomography slices of the same sample, taken with 100 kV X-rays and cold neutrons, respectively [31].

For a simple inspection of samples it is often enough to perform radiography, both with neutrons and X-rays. However, deeper understanding of the material composition is best obtained with tomography data.

7 Future trends and developments

We have demonstrated that neutron tomography is a method of equal importance for the study of cultural heritage objects, compared to the more common X-ray methods. In many cases, it complements studies with X-rays, but sometimes it is the only successful approach for noninvasive material analysis.

There are still developments on the way, which can be applied to cultural heritage studies too in the future. The spatial resolution has now reached the order of several micrometers [32]. Of course, this can be applied to correspondingly small objects of some millimeters only. More sophisticated techniques like "neutron grating interferometry" deliver information on the spatial distribution of structures below 1 μm from

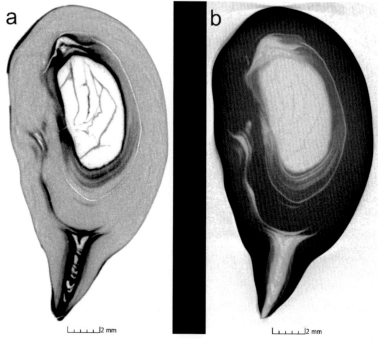

FIG. 19

Tomography slice data of a cultured pearl, *left*: neutrons; *right*: X-rays. Because neutrons have a high sensitivity for hydrogen, the organic material parts are visible at the positions where the sample seems to be empty in the X-ray data. This example documents the high complementarity of both methods for the analysis of the inner content and composition of pearls.

the so-called dark-field contrast, without, however, actually resolving the structures [33].

The direct combination of neutron and X-ray data (see, e.g., Fig. 13) in a "data fusion" process is still under development, mainly in the definition and implementation of suitable analysis tools [14].

Time-dependent studies, e.g., sample treatments (impregnation, desalting, drying, etc.), can be performed up to a frame rate of about 100 Hz, if needed. The slower the process, the more the frame rate can be reduced while the image quality per frame can be correspondingly increased.

Different neutron energies can be used for investigations. As a guideline, cold neutrons provide the highest contrast while thermal neutrons give a higher transmission through the sample. For very dense and bulky objects, the use of fast neutrons can be considered. Unfortunately, there are fewer facilities available than for thermal neutrons. Fast neutrons cannot provide as good a resolution and image quality as thermal neutrons due to the detector's properties.

A strong interaction between the museum experts plus responsible persons from the collections and the operators of the tomography facilities is needed to decide which particular method should be applied for the best possible investigation.

8 Tomography facilities for cultural heritage studies and how to access them

Table 1 provides information about existing facilities for neutron imaging worldwide. Most of them are organized according to a "user lab approach", where beam time is provided after submission and accepting of a scientific or technical proposal. There are some more facilities with less performance as communicated in the ISNR/IAEA database [34].

References

[1] G.T. Herman, Fundamentals of Computerized Tomography: Image Reconstruction from Projection, second edition, Springer, 2009, ISBN: 9781852336172.

[2] J. Radon, Über die Bestimmung von Funktionen durch ihre Integralwerte längs gewisser Mannigfaltigkeiten, Berichte über die Verhandlungen der Königlich-Sächsischen Akademie der Wissenschaften zu Leipzig, Mathematisch-Physische Klasse 69 (1917) 262–277. [Reports on the proceedings of the Royal Saxonian Academy of Sciences at Leipzig, mathematical and physical section], Translation: J. Radon, P.C. Parks (translator), On the determination of functions from their integral values along certain manifolds, IEEE Trans. Med. Imaging 5 (4) (1986) 170–176, https://doi.org/10.1109/TMI.1986.4307775.

[3] A.C. Kak, M. Slaney, Principles of Computerized Tomographic Imaging, Society of Industrial and Applied Mathematics, 2001.

[4] A.M. Cormack, G.N. Hounsfield, G.N. Hounsfield, Biographical, 2020. https://www.nobelprize.org/prizes/medicine/1979/hounsfield/biographical/. (Accessed October 2020).

[5] A.C. Thompson, J.L. Lacer, L.C. Finman, E.B. Hughes, J.N. Otis, S. Wilson, H.D. Zeman, Computed tomography using synchrotron radiation, Nucl. Instrum. Methods Phys. Res. 222 (1984) 319–323.

[6] E. Lehmann, H. Pleinert, L. Wiezel, Status of the installation of a new neutron radiography facility at the spallation neutron source SINQ, in: World Conference on Neutron Radiography 5, Berlin, DGZfG, 1997, pp. 440–447, ISBN: 3931381080.

[7] B. Schillinger, W. Ludwig, C. Rausch, U. Wagner, R. Gehbard, B. Haas, 3D neutron tomography in material testing and archaeology, in: World Conference on Neutron Radiography 5, Berlin, DGZfG, 1997, pp. 688–693, ISBN: 3931381080.

[8] Tomography. https://en.wikipedia.org/wiki/Tomography. (Accessed October 2020).

[9] P.P. Bruyant, Analytic and iterative reconstruction algorithms in SPECT, J. Nucl. Med. 43 (2002) 1343–1358.

[10] A.P. Kaestner, MuhRec – a new tomography reconstructor, Nucl. Instrum. Methods Phys. Res. 651 (2011) 150–160, https://doi.org/10.1016/j.nima.2011.01.129.

[11] Volume Graphics, 2020. https://www.volumegraphics.com/.

[12] Avizo (software), 2020 (This page was last edited on 13 February 2020, at 17:46 (UTC)) https://en.wikipedia.org/wiki/Avizo_(software).

[13] J.H. Hubbell, S.M. Seltzer, X-Ray Mass Attenuation Coefficients, 2004. https://www.nist.gov/pml/x-ray-mass-attenuation-coefficients. (Created September 17, 2009, Updated December 11, 2019).

[14] A.P. Kaestner, J. Hovind, P. Boillat, C. Muehlebach, C. Carminati, M. Zarebanadkouki, E.H. Lehmann, Bimodal imaging at ICON using neutrons and X-rays, Phys. Procedia 88 (2017) 314–321, https://doi.org/10.1016/j.phpro.2017.06.043.

[15] E.H. Lehmann, P. Vontobel, L. Wiezel, Properties of the radiography facility NEUTRA and SINQ and its potential for use as a European reference facility, Nondestruct. Test Evaluat. 16 (2001) 191–202, https://doi.org/10.1080/10589750108953075.

[16] A.P. Kaestner, S. Hartmann, G. Kühne, G. Frei, C. Grünzweig, L. Josic, F. Schmid, E.H. Lehmann, The ICON beamline – a facility for cold neutron imaging at SINQ, Nucl. Instrum. Methods Phys. Res. A 659 (2011) 387–393.

[17] BOA, Beamline for Neutron Optics and other Approaches, 2020. https://www.psi.ch/de/sinq/boa. (Accessed October 2020).

[18] POLDI, A Materials Science Time-Of-Flight Neutron Diffractometer, 2020. https://www.psi.ch/de/sinq/poldi. (Accessed October 2020).

[19] F. Scholten, M. Verber, From Volcan's Forge, Bronzes from the Rijksmuseum, Amsterdam 1450–1800, Daniel Katz Ltd., London, 2005, ISBN: 0954505824.

[20] R.J.C.H.M. van Langh, Technical Studies of Renaissance Bronzes: The use of neutron imaging and time-of-flight neutron diffraction in the studies of the manufacture and determination of historical copper objects and alloys, PhD Thesis, TU, Delft, 2012, ISBN: 9789071450495.

[21] E. Lehmann, et al., Bronze Sculptures and Lead Objects tell Stories about their Creators, in: N. Kardjilov, G. Festa (Eds.), Neutron Methods for Archaeology and Cultural Heritage, Springer, 2017, pp. 19–41, ISBN: 9783319331614.

[22] D. Mannes, F. Schmid, J. Frey, K. Schmidt-Ott, E. Lehmann, Combined neutron and X-ray imaging for non-invasive investigations of cultural heritage objects, Phys. Procedia 69 (2015) 653–660, https://doi.org/10.1016/j.phpro.2015.07.092.

[23] Paul Scherrer Institut, The sword from Oberwil (ZG, Switzerland): A combined X-ray- and neutron-tomography study, 2015. https://youtu.be/N5wb_n7BJyY.

[24] M. Jacot-Guillarmod, K. Schmidt-Ott, D. Mannes, A. Kaestner, E. Lehmann, C. Gervais, Multi-modal tomography to assess dechlorination treatments of iron-based archaeological artifacts, Herit. Sci. 7 (2019) 29, https://doi.org/10.1186/s40494-019-0266-x.

[25] D. Mannes, E. Lehmann, A. Masalles, K. Schmidt-Ott, A.V. Przychowski, K. Schäppi, F. Schmid, S. Peetermans, K. Hunger, The study of cultural heritage relevant objects by means of neutron imaging techniques, Insight 56 (2014) 137–144.

[26] A. Masalles, E. Lehmann, D. Mannes, Non-destructive investigation of "the violinist" a lead sculpture by Pablo Gargallo, using the neutron imaging facility NEUTRA in the Paul Scherrer Institute, Phys. Procedia 69 (2015) 636–645.

[27] E. Lehmann, S. Hartmann, M. Speidel, Investigation of the content of ancient Tibetan metallic Buddha statues by means of neutron imaging methods, Archaeometry 52 (3) (2010) 416–428, https://doi.org/10.1111/j.1475-4754.2009.00488.x.

[28] M. Henss, E. Lehmann, The scanned Buddha: neutron radiography and tomography of Tibetan Buddhist metal images, Orientations 47 (2016) 77.

[29] M. Henss, E. Lehmann, The scanned Buddha: hidden treasure revealed, in: 6th International Conference on Tibetan Archaeology & Arts (ICTAA VI), Hangzhou, October, 2015.

[30] D. Mannes, C. Hanser, M. Krzemnicki, R.P. Harti, I. Jerjen, E. Lehmann, Gemmological investigations on pearls and emeralds using neutron imaging, Phys. Procedia 88 (2017) 134–139.

[31] C.S. Hanser, M. Krzemnicki, C. Grünzweig, R.P. Harti, B. Betz, D. Mannes, Neutron radiography and tomography: a new approach to visualize the internal structures of pearls, J. Gemmol. 36 (2018) 54–63.

[32] P. Trtik, J. Hovind, C. Grünzweig, A. Bollhalder, et al., Improving the spatial resolution of neutron imaging at Paul Scherrer Institut – the neutron microscope project, Phys. Procedia 69 (2015) 169–176.

[33] R. Harti, M. Strobl, B. Betz, K. Jefimovs, et al., Sub-pixel correlation length neutron imaging: spatially resolved scattering information of microstructures on a macroscopic scale, Sci. Rep. 7 (2017) 44588, https://doi.org/10.1038/srep44588.

[34] T. Bücherl, Facilities – Neutron Imaging Facilities around the World, 2020. http://www.isnr.de/index.php/facilities/facilities-worlwide.

X-ray diffraction

Mark Dowsett[a,c], Rita Wiesinger[b], and Mieke Adriaens[c]

[a]*Department of Physics, The University of Warwick, Coventry, Warwickshire, United Kingdom*
[b]*Institute of Science and Technology in Art, Academy of Fine Arts Vienna, Vienna, Austria*
[c]*Department of Chemistry, Ghent University, Ghent, Belgium*

1 Introduction

The smallest unit of a substance which retains its chemical properties is a molecule. Whether the molecule consists of one atom, as it may do in the case of a pure element, or two or more atoms as is the case of a compound, solids created from these building blocks exist in one of two forms at normal temperature and pressure. In the first, the molecules are arranged in three dimensional space with some average separation, but individual spacings and orientations which vary in a random manner. We call such materials amorphous and they have little or no long range order (order on a scale larger than the molecule itself). Silica (SiO_2) can exist in this form (glass). In the second, molecular spacing and relative orientation remain essentially constant over volumes containing many molecules. These materials are known as crystalline and exhibit long range order over many times the molecular dimensions. Silica can also exist in this form—as quartz, for example. Crystalline materials can be further subdivided into polycrystalline and single crystal. In the former, this high level of organization may persist over a volume with linear dimensions of many nanometers or micrometers (a crystallite) bounded by a nm-scale discontinuity beyond which the three dimensional pattern re-emerges in a different crystallite with the same structure but perhaps rotated or tilted or both with respect to the first. Metals are often polycrystalline and deliberately induced changes in the sizes, shapes and relative orientations of the crystallites have a profound effect on the metal properties (ductility, hardness, malleability and so forth). For this reason, measurements of the metal's crystalline structure can reveal much about the intentions and methods used by the manufacturer of an artifact. Single crystal materials retain their long range order throughout their entire volume. Most gemstones are single crystals. (An exception is opal which is a near-amorphous form of polycrystalline quartz with crystallite sizes of the order of $\lesssim 10$ nm [1].) Artificial single crystals can be made with the only practical limit on their size being the dimensions of the apparatus in which they

Spectroscopy, Diffraction and Tomography in Art and Heritage Science. https://doi.org/10.1016/B978-0-12-818860-6.00011-8

are produced. Thus the basis of the silicon semiconductor industry is silicon single crystals ≥ 300 mm diameter and up to 2000 mm long at the time of writing.

Polycrystalline and single crystal materials usually contain minute imperfections in the structure. These can, for example, take the form of defects (missing atoms, atoms out of position or extra atoms) or departures from the theoretically ideal interatomic spacing (strain). The characterization of crystal imperfections can be used to give information on manufacturing methods, origins, forgeries and much more. Naturally occurring crystalline materials almost always contain imperfections whereas artificially produced ones may be close to perfect.

In heritage science, diffraction techniques and specifically X-ray diffraction (XRD) are used primarily, but not exclusively to study crystalline solids. However, more generally XRD has also been used to study amorphous solids, liquids and gases since its invention [2]. Where the crystal structure is unknown, XRD can be used to determine it, revealing the relative separations and orientations of the atoms in their repeating matrix with an accuracy potentially <1 pm. XRD is particularly synergistic with techniques sensitive to elemental composition (as opposed to structure) such as particle induced X-ray emission (PIXE) (see Chapters 10 and 11) energy dispersive spectroscopy (EDS) (see Chapter 3) or X-ray fluorescence (XRF) (see Chapter 9). A knowledge of which elements are present is especially helpful in unravelling XRD data from mixtures of compounds. Where a measured XRD pattern is that of a known material or, as is most common in heritage science, a mixture of materials, the pattern helps to identify the presence of particular compounds (e.g., corrosion products) and gives information on the size ranges of the crystallites, any characteristic distortion of the crystal structure, layer structure, alloy composition and so on [3, 4].

2 A brief description of X-rays and their interaction with matter

X-rays were discovered by Wilhelm Röntgen in 1895 [5] and so-called by him because their nature was (briefly) unknown. By 1912 or so it was suspected that they were a high energy version of visible light—i.e., a form of electromagnetic radiation. Indeed, early in that year, von Laue, Knipping and Friedrich squeezed an informal experiment into the busy schedule of Knipping's X-ray equipment and, with superb economy, verified the nature of X-rays, demonstrated XRD, and discovered X-ray crystallography [6, 7]. By the late 1930s a very complete foundation for describing the interaction of X-rays and matter had been built, particularly in the context of XRD. This took Maxwell's classical description of electromagnetic radiation [8] as a wave and modeled its interaction with electrons bound into quantum states in atoms to predict scattered intensities and other quantities. The description is still in use [9]; firstly because of its excellent results, secondly because of the historical continuity with earlier wave theories of light from which phenomenology and formalism can be borrowed wholesale, and perhaps finally because of the appealing way in which a wave theory presents visible analogies (e.g., with water waves).

Yet, as early as 1905, Einstein had proposed that electromagnetic radiation was a stream of packets of energy—*photons*—and was able to explain, for example, the photoelectric effect which contradicted the predictions of wave theory [10]. Indeed X-rays (like visible light) are detected as photons; they can be counted and the size (in Joules or electron volts) of each energy packet can be measured. In the visible, these energy packets are all the same size (in energy) for a given color. For X-rays the energy determines the size; in an 8 keV beam of X-rays, each photon carries 8 keV of energy. By analogy with light, a beam where all the X-ray photons have the same energy is called *monochromatic* and is characterized by a single wavelength (but see Chapter 1, Section 4.3).

Between 1928 and 1950 the theoretical basis for describing all electromagnetic phenomena in terms of photon interaction with charged particles [quantum electro-dynamics (QED)] was developed [11–14]. It became clear that the apparent wave-like nature of the radiation is a consequence properties and behavior of photons. This means that we can continue to use a wave description where it is convenient but that QED can be used to enhance the theoretical description of XRD where classical the-ory fails, for example in accurately predicting the scattering from heavy atoms [15]. In what follows, we will describe X-rays as photons or waves depending on context (see also Chapter 1, Section 2.3, especially comments on QED).

Whereas a visible photon (see Chapter 1) has an energy of a few eV, X-ray pho-tons have energies starting at a few 100 eV and extending to 100,000s of eV. The energies typically used in XRD are in the range 5–25 keV. In this range, the wave-length associated with the X-ray (~0.25–0.05 nm) is similar to the separation between atoms in a solid and, in particular, to the separation between the planes of atoms in a crystalline material. Just as reflection of visible white light from the surface and oil-water interface of an oil film or the wing scales of a butterfly [16] creates the effect of iridescence because of the presence of layers or other features separated by a distance of the order of the visible wavelength, so the planes of atoms in a crystal reflect particular X-ray wavelengths in specific directions allowing the crystalline structure to be probed. When a beam of X-rays impinges on a crystalline material sharp peaks in X-ray intensity—*diffraction peaks* or *reflections*—can be measured as a function of angle with respect to the incident beam (the *diffraction pattern*). This angle (the *scattering angle*) is easy to predict for a given plane sepa-ration if the X-ray energy is known. Conversely, the plane separation can be com-puted from the angle.

Photons are scattered by charged particles and it is the electrons in matter which are primarily responsible for scattering X-ray photons rather than the atomic nuclei. Analysis of X-ray fluxes scattered from matter shows how the electrons are distrib-uted on average within the material. Since most electrons are localized in atoms, the scattered X-rays also contain information on the way the atoms are arranged in space. There are two kinds of scattering event which can occur: coherent and incoherent scattering. In coherent scattering, the energies of the scattered and incident photons are the same (equivalently, the wavelengths of the incident and scattered beam are the same). If that is the case, then the scattered radiation flux can produce a diffraction pattern. For incoherent scattering (e.g., Compton scattering), the energy of the scattered photon is changed from that of the incident one and can take a range of values. There is no possibility of forming a diffraction pattern, although spectroscopy on the emitted X-rays provides several analytical methods synergistic with XRD.

XRD is therefore study of the patterns in the scattered X-ray intensity which result following coherent scattering. To see where these patterns come from, especially in the case of a crystalline material, we need to look at the 3-D geometry of a crystal and examine how this is described.

3 Crystal structure

3.1 Lattice point and point lattice

A perfect crystal structure can be linked to a 3-D array of points which make up a point lattice. As should become clear below, these lattice points are hypothetical positions in space and not to be confused with the chemical composition of the crystal. The points inside the body of the point lattice are all identical. Each one is surrounded by the same number of nearest neighbor lattice points, with the same separations, in the same directions. The surface of a point lattice is an obvious exception to this—there are no lattice points (belonging to that crystal) outside the boundary. A lattice is defined by the way any lattice point would have to be moved to coincide with its nearest neighbors. The translational directions correspond to three crystallographic axes, usually labelled a, b, and c separated by angles α, β, and γ as shown in Fig. 1. These six quantities are known as the lattice parameters.

Fig. 2 shows the effect of repeated translation of the lattice point in Fig. 1 to fill the space with its point lattice. As first demonstrated by Auguste Bravais (1811–63) there are 14 different possible spatial arrangements of lattice points [17]. These are known as the Bravais lattices. They differ according to the levels of symmetry they possess which is determined by whether any of the distances a, b, or c is equal and whether any of α, β, or γ is equal or equal to 90° or both. Bravais lattices are depicted

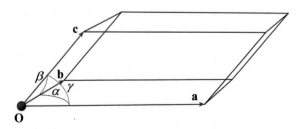

FIG. 1

One cell in a point lattice. The point at O when moved along the vectors **a**, **b** or **c** arrives at a new point in the lattice. This particular lattice is triclinic: $a \neq b \neq c$, $\alpha \neq \beta \neq \gamma \neq 90°$, and has the least possible symmetry.

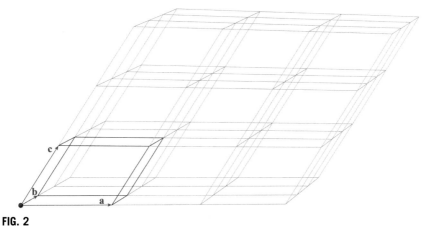

FIG. 2

A triclinic point lattice formed by the repeated translation of the point (or the cell) shown in Fig. 1. A lattice point exists at each intersection. Each cell owns one lattice point.

in most text books on crystallography and on the internet with varying degrees of clarity (and are often inaccurately shown and described in the latter). Clear descriptions of their properties are to be found in Refs. [18–20].

Fig. 3A shows a slightly more complicated Bravais lattice, but one having more symmetry—the face centered cubic lattice. The blue spheres denote the lattice points (not atoms) while the gray rods are just there to guide the eye.

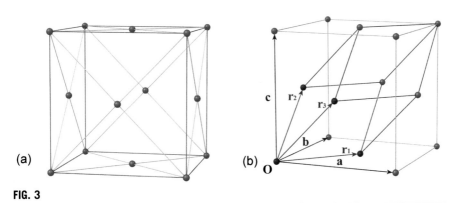

(a)   (b)

FIG. 3

An example of a Bravais lattice—the face centered cubic lattice. (A) The arrangement in space of the lattice points (blue spheres). The grey rods are there to guide the eye and have no other significance. (B) The lattice point of the simplest unit cell for the face centered cubic lattice is the red sphere at O and its edges are also in red, forming a rhombohedral shape. The (usually) more convenient cell consists of the four red spheres and the orthogonal axes a, b, and c.

The pattern of eight lattice points at the corners of a cube in Fig. 3A with a further six at the centers of the faces will (for an inorganic material) be several hundred picometers on a side, typically, and the pattern will repeat tens to billions of times in each direction in three dimensions to form the crystal.

3.2 The unit cell

In order to fully describe a crystal lattice it is useful to define the structure with the smallest number of lattice points which, when repeated in a regular pattern in space, will replicate the lattice. For example, the unit cell of the triclinic system is shown in Fig. 1 and the simplest unit cell for of the face-centered-cubic lattice is shown in red in Fig. 3B where the red bars show the edges of the rhombohedral cell. Both of these so-called primitive cells contain one lattice point. Translation of the corner lattice point along any of the vectors r_1, r_2, or r_3 will move it to another lattice point. The magnitudes of the three vectors are equal, and the angles between them are all 60°. However, it is often more convenient to refer to the whole unit of a Bravais lattice as the unit cell with the translation axes for the face-centered cubic (FCC) cell being **a**, **b**, and **c** as shown in Fig. 2B by the blue arrows. The lengths of **a**, **b**, and **c** are equal, and the angles between them are all 90°. This cell contains four lattice points—one at the corner of the cube plus the three nearest face centered points [9, 17]. With that definition, six translations are needed to fill space with the lattice: **a**, **b**, and **c** themselves plus the "face centering" translations (movement which puts the points at the face centers) (**a** + **b**)/2, (**b** + **c**)/2 and (**c** + **a**)/2.

3.3 The basis and the crystal lattice

Associated with every lattice point is one structure made of one or more atoms—the basis. The basis has the same stoichiometry as the molecule of the substance, i.e., it contains one or more complete molecules. The basis is repeated at each lattice point. In a pure elemental material the basis may be one atom (but is not necessarily so) or it may be as complex as a molecule or cluster of molecules containing tens or even thousands of atoms. The relationship between the crystal structure and the point lattice is:

$$\text{point lattice} + \text{basis} = \text{crystal structure}$$

For example, copper metal has a FCC structure (Fig. 3) obtained through the bonding of many identical copper atoms in a regular 3-D array with one copper atom per lattice point. Halite (NaCl) also has an FCC structure, but the basis is now one Na^+ ion and one Cl^- ion as shown in Fig. 4 which is shown to scale using ionic radii from Ref. [21].

This basis is shown in Fig. 4A, arranged on the four lattice points in Fig. 4B and after the six translations in Fig. 4C. (The pale colored atoms belong to the next repeat of the structure.) We have arbitrarily chosen to place the Na^+ ions on the lattice points. We could have chosen the Cl^- or a point half way between the two. The result

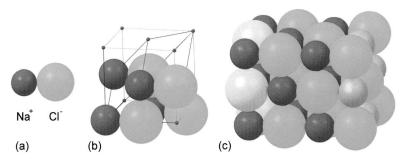

FIG. 4

The FCC lattice of halite (NaCl), common salt. (A) The basis: the ions are scaled to their correct ionic radii (Na$^+$ 116 pm, Cl$^-$ 167 pm) with respect to the lattice parameter ($a = 562.7$ pm). (B) The four lattice points are populated by eight ions—four each of Na$^+$ and Cl$^-$. (C) The full Bravais FCC structure made by six translations of the four lattice points. The pale colored atoms at the ends belong to the next repeat of the structure. The structure can also be described as two interpenetrating FCC lattices—one for the Na$^+$ and one for the Cl$^-$ ions.

would be identical. The result is two interpenetrating FCC lattices, one populated by Na$^+$ and the other by Cl$^-$ and displaced with respect to each other by $a/2$ along each axis. In the body of a crystal, each Na$^+$ ion is surrounded by six Cl$^-$ ions at the corners of a regular octahedron and, equally, each Cl$^-$ ion is surrounded by six similarly placed Na$^+$ ions.

The pure carbon compound buckminsterfullerene—C$_{60}$ [22]—also forms crystals with an FCC structure at room temperature [23], but in this case the basis is the C$_{60}$ molecule, and the crystal is built from the more or less spherical molecules as the basis, each containing 60 atoms and spatially organized in the FCC array with their surfaces effectively in contact.

To show how the basis and its orientation with respect to the cell can significantly change the crystal structure, even when the Bravais lattice remains the same, we show the example of sphalerite (ZnS) in Fig. 5, again drawn to scale. This configuration is adopted by numerous substances and is known as the zinc blende structure. (Note that although Zn is twice the mass of S, its 2+ ionic radius is around half that of S^{2-}.)

In the basis (Fig. 5A) the axis through the two ions is aligned along the diagonal of the cube. In the unit cell (Fig. 5B) note how this orientation has resulted in the lower left sulfur ion becoming surrounded by four zinc ions at the corners of a regular tetrahedron. After the six repeat translations we get one unit of the full FCC structure which again consists of two interpenetrating FCC lattices (one for Zn^{2+}, one for S^{2-}) but this time having the displacement along the cube diagonal. In the bulk of the crystal each S^{2-} ion is surrounded by four Zn^{2+} ions at the corners of a regular tetrahedron. Conversely, each Zn^{2+} is similarly surrounded by four S^{2-} ions.

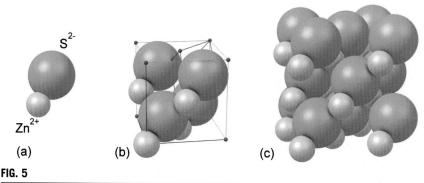

FIG. 5

The FCC structure of sphalerite (ZnS) also known as the zinc blende structure. (A) The basis of $Zn^{2+}S^{2-}$ to scale regarding separation and ionic radius. (B) The 4 lattice points of the unit cell populated with the basis. (C) After the six repeats at **a**, **b**, **c**, (**a** + **b**)/2, (**b** + **c**)/2 and (**c** + **a**)/2. Again there are two interpenetrating FCC lattices but the relative displacement is along the diagonal of the cube, not its sides.

The point lattice has its symmetry and the basis also has symmetrical properties. A full description of crystal structure takes both of these into account using the concept of space groups [9, 18, 20]. Taking account of all the possible symmetries and their combination in the 14 Bravais lattices results in a total of 230 space groups.

4 X-ray (and other) diffraction

4.1 Coherence and interference

X-ray photons used for diffraction measurements are scattered principally by the electrons in the material. In the scattering event (Fig. 6) a photon may lose or gain energy, i.e., increase or decrease its wavelength. These effects are Compton and inverse Compton scattering, respectively (see Chapter 1, Section 1.4.2). Alternatively, it may be scattered with no change in energy (Thomson scattering [9]). The first two possibilities result in *incoherent scattering*. The last is *coherent scattering*. Coherent scattering gives rise to the interference effects which create the peaks in a diffraction pattern. Incoherent scattering cannot result in interference or diffraction peaks but contributes to the background signal.

Interference is discussed in Chapter 1 in the context of Young's two slit experiment. Interference due to the scattering of X-rays by atoms is different in two respects only: there are (typically) order 10^{16} to 10^{20} sources of scattered radiation rather than just two and the scattering can take place in any direction and is therefore 3-dimensional. Interference phenomena may be described either by associating a wave with the beam of monochromatic X-rays, or by taking the weighted sum over the multiplicity of possible paths and phases associated with a *single* photon in its

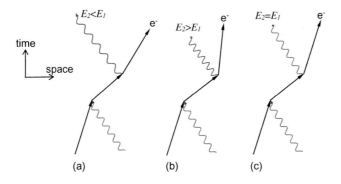

FIG. 6

Photon-electron scattering events (Feynman diagrams). The electron concerned may be free or bound in an atom. A photon (wavy line) with energy E_1 scatters off an electron (solid line). (A) The photon is scattered with a lower energy and the electron gains energy (Compton scattering), incoherent. (B) The photon is scattered with a higher energy and the electron loses energy (inverse Compton scattering), incoherent. (C) Neither the photon nor the electron energies change in the scattering event (Thomson scattering), coherent.

journey from the source to the detector. What one must *not* do is to conflate the two descriptions (as one often sees in thesis introductions). Interference may be described as taking place between many coherent scattered waves, or as the result of a summation over all the possible paths (path integration) a single photon might take. It has nothing to do with interaction between photons (scattered photons do not "interfere with each other"). The path integration method is beyond our scope except in the descriptive sense [24] although it is the more physically correct approach. So here, we discuss coherence and interference a little more using a wave description.

As we mentioned, coherent scattering implies that the scattered radiation has the same energy as the incident beam, and therefore the same wavelength. Coherence is also a property of the incident beam; the narrower the bandwidth of the beam the more coherent (loosely: describable as a single wave) it becomes. Fig. 7A and B shows how coherent waves can combine to produce constructive or destructive interference, whereas incoherent waves combine to form noise (Fig. 7C). Statistically, there is a high probability that a large number of incoherent waves will partially cancel so the noise tends to be small compared to a signal produce by constructive interference.

4.2 Crystallographic planes and Bragg's law

Looking back at Fig. 2, it is apparent that there are sets of parallel planes passing through the lattice points. For example, there is one set of four in the plane of the page, another set more or less "horizontal" and yet another slanted at an angle to the latter. However, these are just the most obvious. There are many other sets of

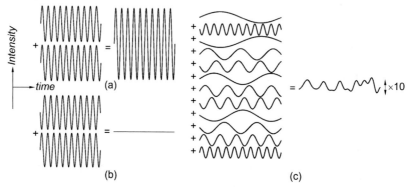

FIG. 7

Schematic representation of interference between overlapping waves: (A) The addition of two (or more) in-phase waves (crests align with crests) gives constructive interference. (B) If the waves are 180° out of phase (crests align with troughs) then the waves sum to zero. (C) The addition of many randomly incoherent waves results in noise.

planes running at different angles to those identified and usually passing through a lower density of lattice points. These planes are profoundly significant to the determination of crystal structure when using XRD or any other diffraction technique; indeed the discussion of both Bragg's law and the Laue equations which follows applies equally to XRD, electron diffraction and neutron diffraction with the distinction that typical electron wavelengths are very small compared to the other radiation so the diffraction angles are also very small in comparison.

Although it is really the electrons associated with the atoms arranged around the lattice points which do the scattering (for XRD and electron diffraction), and although these are also found above and below the hypothetical planes, the overall effect can be described simply by considering the planes themselves. In the autumn of 1912 whilst walking along the Backs in Cambridge [25], W.L. Bragg realized that any plane can be treated as having the same effect on X-rays as a partially silvered mirror does on visible light [26] (Fig. 8); some of the intensity is reflected, some is transmitted and some is absorbed. (Absorption gives rise to other analytical techniques—see Chapter 9.) Nevertheless, an individual plane is a very weak mirror; for example, the combination of reflection and absorption at each plane in a copper crystal means that 8 keV X-rays diminish to around 1/3 of their intensity at the surface after travelling ~20 μm through the material, or traversing >55,000 atomic planes making the reflection coefficient at each plane <0.002%. This is important for at least three reasons: (i) The information content of an XRD measurement of a surface therefore comes from an average of the top several microns (decreasing with rising atomic number and atomic density and increasing with X-ray energy and angle of incidence). (ii) Measurements taken in transmission through a powder

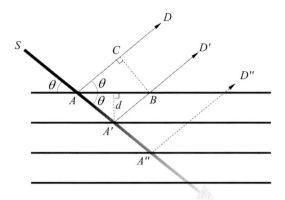

FIG. 8

A variation on the usual two-plane schematic describing scattering from parallel planes a distance d apart in a crystal. The input beam is attenuated by reflection and adsorption at each plane. Constructive interference will occur when the length difference between reflected paths such as SAD or SA'D' is an integer number of wavelengths. The diagram may be interpreted in terms of interfering waves or as being representative of a few of the infinite possible paths which might be taken by a single photon.

or single crystal require a material thickness large enough to present the beam with sufficient material to make a measurable diffraction pattern, but not so great that the beam is seriously attenuated. This thickness may be as much as 0.3–0.5 mm [8] but with modern instruments it can be just a few microns. (iii) The reflected intensities from each successive plane in a set of hundreds of parallel planes will effectively be equal, so that a relatively simple argument based on interference due to scattering from two adjacent planes can be used to describe the first order properties of the diffracted beam.

Constructive interference will occur between reflected paths when the difference in path length is $n\lambda$, where n is an integer and λ is the X-ray wavelength, i.e., a whole number of wavelengths. Considering Fig. 8, we see that the path difference δP between SAD and SA'D' (for example) is given by

$$\delta P = AA' + A'B - AC$$

Since the angle of incidence θ is equal to the angle of reflection, $AA' = A'B$ so

$$\delta P = \frac{2d}{\sin\theta} - \frac{2d\cos^2\theta}{\sin\theta}$$

$$= \frac{2d}{\sin\theta}(1 - \cos^2\theta)$$

$$= 2d\sin\theta$$

where θ is the angle of incidence. Adding the condition for constructive interference, we arrive at Bragg's law

$$n\lambda = 2d\sin\theta \tag{1}$$

For values of θ where Bragg's law is true, all the parallel planes with spacing d will contribute to constructive interference. For small changes in θ either side, the interference will rapidly become destructive because so many planes contribute and so the peaks will be sharp.

4.3 The Laue equations and Miller indices

An alternative, but equivalent, formulation of the law of diffraction was given by Max von Laue in 1912 [27]. This approach results in the three Laue equations which also illustrate the method of Miller indices [28] used to label different crystallographic planes. Because corresponding atoms in repeats of the basis are separated by the same distance as the lattice points, we can initially pretend that the lattice points themselves are the scatterers (this will not give the correct scattered intensity). For example, Fig. 9 shows two adjacent lattice points at P and P' with a separation x along a row of lattice points in some arbitrary direction $\mathbf{x}$ through the lattice.

The path difference δP between rays SPD and $S'P'D'$ is

$$\delta P = AP' - BP$$
$$= x(\cos\psi_x - \cos\phi_x)$$

For the paths to interfere constructively, the path difference must be an integer number of wavelengths $n\lambda$, say. Then

$$n\lambda = x(\cos\psi_x - \cos\phi_x) \tag{2}$$

Since we use the crystallographic axes $\mathbf{a}$, $\mathbf{b}$, and $\mathbf{c}$ to describe crystal geometry (Figs. 1–3), we can write three examples of Eq. (2), one for each axis:

$$h\lambda = a(\cos\psi_a - \cos\phi_a)$$
$$k\lambda = b(\cos\psi_b - \cos\phi_b) \tag{3}$$
$$l\lambda = c(\cos\psi_c - \cos\phi_c)$$

where h, k, and l are integers, a, b, and c are the spacings of lattice points along the axes, and the angles are measured from the corresponding axis. For constructive interference, all three equations must be simultaneously true. These are one form of the Laue equations.

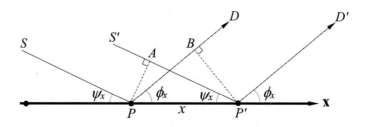

FIG. 9

Scattering from a row of lattice points equispaced by a distance x.

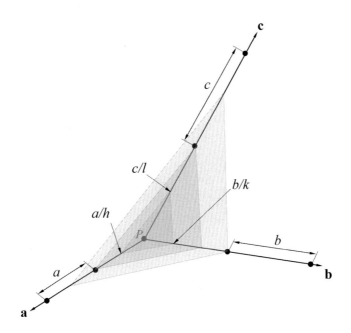

FIG. 10

Suppose the Laue equations are satisfied for $h = 2$, $k = 3$, and $l = 2$ in Eq. (3). Then in the corresponding incident and scattering direction, scattering from P at the origin would be one wavelength out of phase with scattering from points at a/h, b/k, and c/l. These three points define a plane which is a member of a set of equispaced parallel planes extending throughout the crystal. Scattering from any pair of adjacent planes would also be one wavelength out of phase and scattering from alternate planes, two wavelengths, etc.

Of particular interest are the integers h, k, and l. If the Laue equations are simultaneously satisfied, then there exist three points on the **a**, **b**, and **c** axes located at a/h, b/k, and c/l where the path difference would be one wavelength (Fig. 10).

These three points define a plane. Everywhere on that plane reflects with a phase difference of one wavelength with respect to a parallel plane through P. A family of planes can be constructed, all parallel to the first plane and with their intersections on the **a**, **b**, and **c** axes a/h, b/k, and c/l from the previous plane. Scattering from alternate planes would have a path difference of two wavelengths, from every third plane, three wavelengths, and so on. These planes are defined by their hkl values, or Miller indices, 232 in the case shown in Fig. 10. The distance between the planes (i.e., along a line normal to the planes) we write d_{hkl}. When the Laue equations are simultaneously true, the incident and scattered paths will be at an angle θ to the (232) family of planes and Fig. 8 can be taken as being a section through them. Then Bragg's law becomes plane-family specific:

$$\lambda = 2d_{hkl}\sin\theta \tag{4}$$

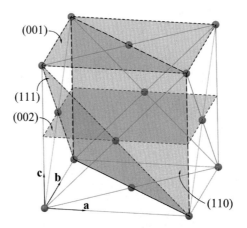

FIG. 11

Some possible plane families with their Miller indices shown with reference to the FCC lattice.

Returning to Fig. 9 for a moment, suppose $\psi_x = \phi_x$ then the path difference $\delta P = 0$. Then Eq. (2) gives $n\lambda = 0$. The wavelength cannot be zero so n must be. Suppose this occurs for the **a** axis, then $h = 0$, and a/h tends to infinity. Skipping to Fig. 10 and sliding the a/h intersection along the **a** axis towards infinity, we see that this condition corresponds to a plane family parallel to the **a** axis. This is the $(0kl)$ set of planes. Such a condition can occur for one axis or any two out of the three axes in Eq. (3) so, for example, we might have constructive interference for a family of planes parallel to the plane containing the **a** and **b** axes for $l = 1$. This is the (001) plane family. Some of the planes with their Miller indices are illustrated in Fig. 11 with reference to the FCC lattice of Fig. 3.

4.4 Some nomenclature and notation

Families of parallel planes extending through a crystal are denoted with curved parentheses, thus (hkl). For example, in Fig. 11, the (100) planes are those parallel to, and a distance a from, the bounding plane on the right of the cube $(d_{hkl} = a)$ (also Fig. 12A). Corresponding reflections in a diffraction pattern are labelled hkl without parentheses. Planes, which have negative intercepts on a crystallographic axis are labelled with a bar over the corresponding index; e.g., $(hk\bar{l})$ has a negative intercept $-c/l$ on the **c** axis.

Fig. 12B shows the $(\bar{1}\bar{1}1)$ plane, for example. Directions of crystal planes, defined as the direction normal to the plane, are written $[hkl]$, i.e., with square brackets. For example, in Fig. 12C, the $[111]$ direction is orthogonal to the (111) planes along the diagonal of the cube.

The more symmetry a crystal structure has, the more planes have an identical arrangement of atoms on them so that they will reflect the same intensity. For

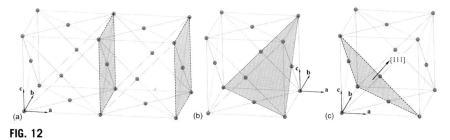

FIG. 12

(A) The (100) planes. (B) The $(\bar{1}\bar{1}1)$ plane. (C) The [111] direction.

example, the 6 cubic faces of the FCC lattice as shown in Fig. 13, (001), (100), (010), $(\bar{1}00), (0\bar{1}0), (00\bar{1})$ have an identical arrangement of atoms. If the crystal were rotated 90°or 180° about any of the axes, its symmetry would not change. Planes of identical symmetry like this are said to belong to the same *form* and are written {*hkl*}. So, Fig. 13 shows the {100} planes. In a similar way, the directions [001], [010], [100], $[\bar{1}00], [0\bar{1}0], [00\bar{1}]$ have equivalent symmetry and <100> refers to all these directions. For more on this, see for example [19, 20].

Finally, we mention that crystal planes within a hexagonal Bravais lattice are often described using four Miller indices rather than three [19].

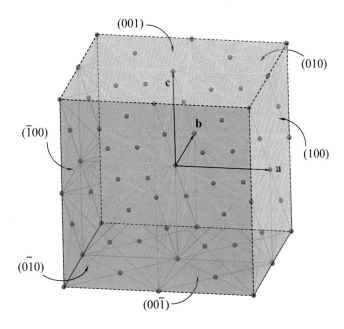

FIG. 13

One of the sets of planes having an identical arrangement of atoms and form factor for the FCC lattice.

4.5 Symmetry, missing reflections and intensity

Consider the (001) and (002) planes in Fig. 11. For the values of θ and λ for which the 001 plane spacing d_{001} gives a path difference of a whole number of wavelengths, the distance between the 001 and 002 planes will give rise to a path difference of $\lambda/2$. Since the atomic population of the 002 and 001 planes is identical, just laterally displaced, the reflected intensity from adjacent 001 and 002 planes will be identical in intensity but 180° out of phase. So the interference will be destructive and the FCC lattice has no {100} reflections. A similar argument applies to the {110} form and others for the FCC lattice. Indeed, for all the Bravais lattices, the symmetry of the lattice determines the maximum number of constructive reflections which is allowed. For cubic lattices an underlying rule for allowed reflections can be written down simply as shown in Table 1. For other systems see, for example, [18] Chapter 5 et seq.

Table 1 is far from giving the whole story. The actual intensity of a reflection, and, indeed whether it is even detectable depends on many factors a few of which we now describe qualitatively.

Firstly, the reflected intensity will depend on the number of lattice points (and therefore repeats of the basis) on a given plane. Planes with small hkl values tend to have a higher areal density of atoms than those with large values, so those planes will tend to give stronger reflections.

Secondly, every atom in the periodic table has its own scattering factor for X-rays depending on the X-ray wavelength, the number of electrons (atomic number, Z) and the atomic structure. The scattering power of an atom increases with Z so if the basis consists of a mixture of similar numbers of heavy and light atoms, the heavy atoms will do most of the scattering, provided the X-ray energy does not lie close to an absorption edge (see Chapter 9) for the atom. The intensity scattered into an hkl reflection will depend on the sum of the intensities from each atom in the basis. Because the atoms are arranged in a structure (the basis) with its own symmetry around a lattice point, there will be phase differences in the coherent scattering depending on how far above or below the theoretical plane through the lattice point the atom happens to be. There will therefore be partially destructive interference in a reflection which will control its intensity. Moreover, circumstances can arise where an allowed reflection is not seen either because it is too faint and lost in the background, or because it is cancelled entirely. Each hkl reflection has its own *structure factor*, which shows how the intensity from each atom contributes to the final intensity. This is why a diffraction pattern not only allows the Bravais lattice of a

Table 1 Allowed reflections for cubic lattices.

Lattice	Allowed combinations of h, k, l
Simple/primitive cubic (cubic P)	Any
Body centered cubic (BCC, cubic I)	$h + k + l =$ even
Face centered cubic (FCC, cubic F)	h, k, l either all even or all odd

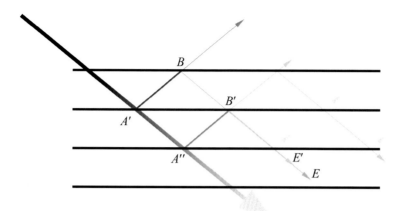

FIG. 14

Multiple reflections at *hkl* planes leading to extinction (attenuation of the incident beam).

substance to be identified, but also the relative positions of all the different atoms in the basis amongst far too much other structural information to be covered in a short chapter.

A final point (but by no means *the* final point) concerns the reflection process itself. In Fig. 8, a ray like $A'D'$ will be weakly reflected from the top plane, $A''D''$ from the second plane in, and so on. It is easy to see that repeated reflections of reflections will result in a huge range of possible paths for an entrant photon as indicated in Fig. 14.

Now, something we have not mentioned so far is that each reflection results in a phase shift of 90° with respect to the incident beam. Since all the emergent beams contributing to a diffraction pattern as shown in Fig. 8 will suffer the same shift, their phase *differences* and the resulting interference are unaffected. However, ray paths such as $A'BE$ and $A''B'E'$ (Fig. 14) which suffer with two reflections will be phase shifted by 180° with respect to the incident beam and will interfere destructively with it. The incident intensity is therefore weakened. This phenomenon, which occurs in a perfect crystal, or a perfect part of a crystal, is called primary extinction. The effect, and its influence on the diffraction pattern becomes much more complicated if the crystal is imperfect [19].

5 Instrumentation and measurement

5.1 Overview

An X-ray diffractometer consists of a source of X-rays, X-ray optics to condition the beam according to the type of instrument, a goniometer or sample manipulator for mounting and orientating the sample with respect to the beam and a detector. There

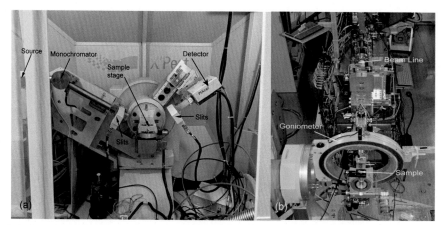

FIG. 15

(A) A laboratory diffractometer, the Panalytical X'pert Pro MPD, photo courtesy of Dr. David Walker, The University of Warwick. (B) Infrastructure, the XMaS beam line (BM28), the UK's CRG at the ESRF.

are two broad classes of instrument: laboratory instruments including portable devices (e.g., Fig. 15A) and infrastructural machines (e.g., Fig. 15B). Laboratory instruments are usually compact with a footprint of around desktop size or smaller if they are portable. When running they are typically enclosed in a lead shielded cabinet with lead glass doors.

Infrastructural instruments are very large, typically tens of meters long, with lead shielded rooms ("hutches") housing different elements of the instrument (e.g., Fig. 15B). Infrastructural facilities are generally owned by individual countries or multinational collaborations such as the European Synchrotron Radiation Facility (ESRF) since the cost of construction typically exceeds $1 Bn. These are based on synchrotrons or X-ray free electron laser sources (XFELs). A synchrotron usually has around 30 beam lines distributed tangentially around a particle storage ring with different optics on each beam line conditioning the X-rays (or other electromagnetic radiation) for a particular range of applications—see, for example [29, 30]. An XFEL typically has around five beam lines, each dedicated to a specific area of chemistry or physics which can take advantage of very short (femtosecond), high intensity $(10^{18}\,\mathrm{W\,cm^{-2}})$ coherent X-ray pulses [31].

Laboratory instruments deal with the vast bulk of XRD based heritage science and visits to synchrotrons are only required when the capabilities of a laboratory instrument are inadequate. Reasons for an application for synchrotron beam time may include:

• Noninvasive analysis—laboratory instruments have small sample stages and can certainly not accommodate an artifact such as a painting or an intact ceramic.

Synchrotron beam lines can allow whole artefacts to be analyzed in selected areas. However, portable equipment may often be mounted very close to a large artifact in such a way that no sample stage is required.

- High beam intensity—a synchrotron beam line can operate at around 1000–10,000 times the intensity of a laboratory instrument, delivering some 10^{12} photons s^{-1} of monochromatic beam or more. Given fast enough detectors, this means that many samples can be analyzed in seconds or minutes rather than hours or days. This is not without risk. Whereas the flux or power density (see Chapter 1) available on a laboratory instrument is very unlikely to cause detectable sample alteration, users are advised to check the stability of their samples at synchrotron intensities. [The XFEL operates as an extreme case of the "nondestructive" envelope described in Chapter 1. XFEL pulses are around 1 fs long and each pulse can contain as many photons as the synchrotron can deliver per second. Structurally characteristic diffracted X-rays are emitted less than 1 ps before the sample (which can be a nanovolume or less) is blown apart.]

- Small area analysis and imaging—many heritage-related samples have curved or rough surfaces and are inherently unsuited to analysis in a laboratory diffractometer because the surface topography introduces a continuous height error which leads to peak shifts and broadening in the pattern. With synchrotron beam intensities a pattern can be obtained from a surface area of mm^2 to μm^2 depending on the beam line and flat areas can be selected for analysis. With a beam at the lower end of the spot size range, if the sample is scanned across the beam in a mechanical raster (see Chapter 1) XRD patterns can be collected from each pixel for the study of heterogeneous surfaces.

- Measurement of a process in real time—the high beam intensity coupled with a fast parallel detector allows changes in surface chemistry of a sample to be observed as a function of process and time; the beam is so intense that adequate fluxes of diffracted X-rays can reach the detector after passing through polymer windows and liquid films over the sample. Therefore, processing, corrosion and coating studies can be carried out in environmental cells filled with liquids or gasses, and/or maintained at elevated temperatures. Electrochemical properties, gas composition and surface spectroscopy are amongst many other parameters which can be recorded in parallel. Conventional "before and after" measurements on a laboratory diffractometer are still advisable though. These guide the user to the most informative part of the pattern to monitor in real time and give warning where the synchrotron beam is affecting the results (e.g., through radiolysis of an electrolyte).

Currently there have been few, if any, applications of XFELs in heritage science. However, reasons to access the highly specialized capabilities of an XFEL might be: structural studies of ancient biomolecules, examination of single virus particles and crystallography of samples on the scale of a few hundred nanometers or less.

5.2 Single crystal, powder, and surface powder diffraction

Single crystal, powder and surface powder diffraction are the three most frequently encountered ways of using XRD. They are usually carried out using dedicated laboratory instruments, but many synchrotron beam lines have beam conditioning, sample manipulation and detector alternatives which allow them to operate in all three modes.

There are two basic XRD methods for the structural analysis of single crystals:

Firstly, the original method developed by Laue and co-workers [6], which used broadband X-radiation from a tungsten target. The instruments are known as Laue cameras or Laue diffractometers and are still used, principally for producing 2D (spot) patterns from single crystals (Fig. 16).

Different wavelengths across the band will be constructively scattered by specific d spacings so that the complete crystal structure can be probed using a 2D detector. Laue used photographic film and some instruments still do [32], but more modern instruments use X-ray cameras which deliver digital images of the spot patterns produced [19]. However, most modern single crystal analyzers use monochromatic X-ray beams which give higher angular resolution and, importantly, can be focused to microbeam proportions so that a structural analysis of single crystals <100 μm linear dimensions can be carried out. The output from such an instrument and

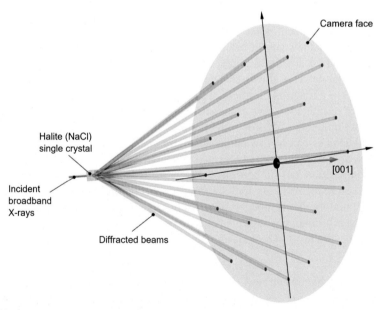

FIG. 16

Schema of Laue diffraction from single crystal halite with the broadband beam incident in the [001] direction. Laue used film as the detector and this is still sometimes used. A digital camera designed for X-ray detection would be more common today.

associated software starts with a 3-D map of the electron density in the crystal from which very complete structural details can be deduced.

Single crystals are often unavailable or inappropriate for a particular study. For example, corrosion on metals often grows as a layer (or multilayer) of small polycrystals, usually with a wide range of orientations, and pigments in artworks are often found in the form of ground crystalline material in a binder. The techniques of powder diffraction can be used here for both structural measurements and substance identification [3, 18, 33, 34].

The central principal of powder diffraction is that scattering should take place from a large number of randomly orientated crystalline fragments or small crystals. For each crystallite, sets of planes which are reflecting constructively would produce diffracted rays which would intersect the inner surface of a sphere concentric with the analyzed area or volume in a set of spots. The randomness of orientation, however, ensures that the axes of the crystallites are rotated with respect to each other. The component of rotation around the beam direction produces a rotation in the spot pattern so that the spots from each crystallite coalesce into rings. The other two orthogonal components of rotation bring the full range of planes in the structure into the diffraction condition. The net result is that X-radiation is scattered from the sample in a set of conical shells concentric with the beam (Fig. 17A and B). These are

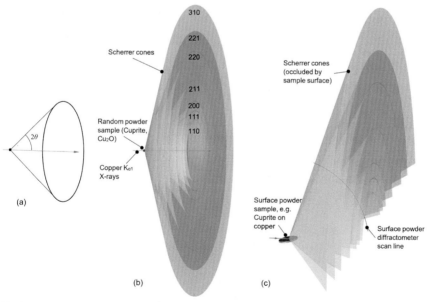

FIG. 17

(A) Geometry of a diffraction (Scherrer) cone from a powder sample. The cone is concentric with the beam with an apex half-angle (the scattering angle) of 2θ. (B) Schematic of the diffraction cones from a sample of cuprite powder for copper $K_{\alpha 1}$ wavelength X-rays. The reflections are labeled and drawn with the appropriate scattering angles. (C) The same diffraction cones for a thick cuprite layer on copper assuming random orientation.

known as Scherrer cones. The result is similar to what would be seen from a single crystal rotating about three orthogonal axes.

The sample may be in the form of a loose powder packed into a thin walled tube, or even held between two layers of adhesive tape or polymer film, either side of a thin washer. Neither the tube, nor the tape should be made of anything which contributes to the pattern (many varieties of transparent sticky tape, and polymer films are crystalline). The beam is fired through the sample and into a dump and the diffraction is detected as a series of concentric rings where the cones intersect the (flat) active surface of a 2D detector. If the detector is orthogonal to the beam the rings will be complete circles or circular arcs. Otherwise, they will be elliptical.

Surface powder diffraction is almost identical in principle except that the powder is distributed over a surface. The "powder" might be a corrosion layer on its native substrate, for example. Only that part of the diffraction pattern reflected away from the surface will be visible for the sample will block the rest. Fig. 17C shows how the Scherrer cones are occluded by the sample surface. A laboratory instrument such as that shown in Fig. 15A could be used to measure the pattern in which case the sample is usually rotated during measurement while the detector moves along the scan line cutting the cones (Fig. 17C). The sample rotates to change the incident angle θ and the detector is rotated to sample a scattering angle 2θ. [In the case of a 1D or 2D detector (see below) the normal to the center of the detector surface would be kept at this angle.] The scanning can be done, for example, by keeping the sample stationary and rotating the source and the detector at the same rate, or by keeping the source stationary, and rotating the detector at twice the rate of the sample surface (Fig. 18).

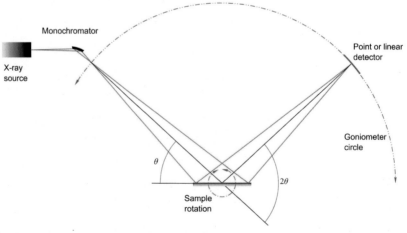

FIG. 18

Simplified schematic of a surface powder diffractometer showing the relative positions of the source, sample and detector in Bragg-Brentano geometry. Either the sample stays fixed while the source scans clockwise (say) and the detector scans anticlockwise at the same rate, or the source is fixed and the sample rotates anticlockwise (say) while the detector also rotates anticlockwise but at twice the rate.

If a 2D detector is used, and this subtends a large enough range of scattering angle, it is possible to acquire data without angular scanning but if the crystalline layer on the surface has grown with a preferred orientation (i.e., the polycrystals are not randomly orientated) some planes may not be brought into the diffraction condition and reflections may be missing.

5.3 X-ray sources

In a laboratory instrument a typical X-ray source for XRD is made by accelerating electrons from a hot filament in a vacuum up to 40 keV or so (37% of the speed of light) in an X-ray tube and smashing them into a heavy metal target (Fig. 19). This kind of source was invented by Coolidge in 1913 [35] but has undergone a vast amount of development since. As the electrons interact with the target several processes take place of which the most important in this context are:

- Heavy deceleration of the electrons as they plough through the electron fog in the metal. This results in the emission of X-rays across a wide range of energies (broadband emission). Such X-rays are known as Bremsstrahlung.
- Release of tightly bound (core level) electrons from the metal atoms through collisional processes leaving atoms in an unstable state with one or more missing electrons (core level holes). The filling of the holes from more weakly bound states or the conduction band results in the emission of X-rays of characteristic energies (narrow band emission). [These are the same X-rays as are used in EDS in the electron microscope and XRF spectroscopy to identify the atoms comprising a sample (Chapters 3 and 9). Here, however, they will be used as the X-ray source.]
- A large amount of power dissipation in the form of heat, which would melt or even vaporize the metal target if it were not very efficiently cooled.

Most modern laboratory instruments use characteristic X-rays produced when K-level holes are filled by electrons from the L_{II} and L_{III} sub shells (K_α radiation). The rest of the emission is removed by filtering or by monochromation using a silicon or germanium crystal to diffract a narrow band of wavelengths through an aperture.

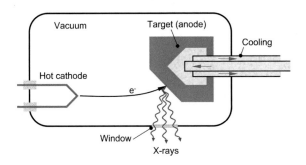

FIG. 19

Schema of an X-ray tube with a side window (many other geometries are used too).

Monochromation allows the X-rays generated from the L_{II} to K transition only ($K_{\alpha 1}$ radiation) to be selected which results in higher resolution and prevents peak splitting at large scattering angles. Typical metal targets are copper, molybdenum and, in recent technology, a gallium liquid metal jet [36]. Fig. 20 shows the X-ray emission spectra obtained from a device like that in Fig. 19 operated with a copper anode and an electron impact energy of 35 keV. Also shown is the emission from a tungsten target which is Bremsstrahlung in this energy range, appropriate for use in the Laue camera described earlier.

Synchrotron radiation was once just an unwanted side effect in particle accelerators and storage rings developed for high energy physics. However, since 1960 or so, synchrotrons were first adapted from high energy physics machines or later built in many places around the world, specifically to produce and use the intense electromagnetic radiation (usually X-rays, but generically known as light) they emit. A synchrotron consists of a particle accelerator (usually for electrons) designed to produce a beam in the GeV energy range (>99.999999% of the speed of light), an injection system which deposits pulses of particles into a storage ring, and the storage ring itself in which particle pulses circulate continuously in orbits maintained by powerful magnetic fields. The storage ring is an approximately toroidal tube, often 100s m in diameter in which a very high vacuum is maintained so that the pulses do not loose particles through gas scattering. Every time the path of an electron is bent by a magnetic field, the (relativistic) electron radiates X-rays in its direction of travel—as if it were wearing an intense head torch flicked on and off by the field. Several tangential tubes beam lines) penetrate the ring, placed so that the pulses

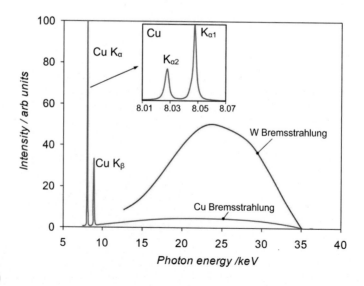

FIG. 20

Emission spectra from copper and tungsten bombarded with 35 keV electrons. Copper data after [37, 38], tungsten data after [39].

of light from orbiting electron bunches travel down them. A synchrotron may have 30 or more beam lines and allows that number of independent experiments to be conducted simultaneously. The pulses leaving the ring are very short and, for this reason, they constitute a broadband source of radiation. The radiation is monochromated using diffraction from a (usually) silicon crystal and may be tuned to a selected energy (wavelength) by rotating the crystal appropriately. After further beam conditioning optics it arrives in the experimental hutch ready to use.

XFELs were a development from optical devices called undulators which are placed in synchrotron storage rings to further intensify and cohere the radiation. The XFEL takes electron bunches down a linear accelerator which may be several kilometers long which accelerates them to (typically) 10–20 GeV (>99.9999999% of light speed). They then enter a stage around 0.1 km long in which they are forced to veer from side to side by a chain of intense magnets of alternating polarities (the undulator). The electrons are steered into a dump and intense femtosecond X-ray pulses continue to the experimental halls. One potential heritage application for XFEL-based crystallography would be the examination of ancient biomolecules, viruses, bacteria and so on, although we are not aware that such work has been attempted at the time of writing. Because XFEL pulses destroy the sample, data are acquired using a series of identical micro or nanoscopic samples fed serially into the beam, but diffraction can be carried out on particles as small as a single virus [40].

5.4 Zero, one, and two dimensional detectors

The original XRD X-ray detectors were sheets of photographic film. This meant that the full range of diffraction measurements from the acquisition of Laue patterns from single crystals to powder patterns using monochromatic radiation could be measured using sheets or strips of film placed around the sample, and the general behavior of diffraction in three dimensions was well understood early on in the history of XRD. Film is still used today even after 120 years of development as it is cheap and flexible in its application, and patterns recorded on film are easily compared by eye. However, it has largely been supplanted by electronic detectors whose output can be read directly into computer memory.

A review of current detector technology will likely become out of date as soon as it is written firstly because research and development for industrial, medical and consumer imaging and position sensing is a field of rapid development, much of which is directly or indirectly transferable to X-ray detection and secondly, at the other end of the scale, highly specialized inventions for particle detection in high energy physics almost always find their way into less esoteric applications [41]. At the time of writing, the most common detectors are based on semiconductor devices which are compact and can be fabricated singly or in arrays [42]. The objective of such a device is to convert an X-ray entering the semiconductor material into an electrical pulse. Pulses can then be amplified and counted by an electrical circuit and the count read as a function of angle, time, position on the detector face etc. and stored. Detectors can broadly be classified as 0D (point) detectors, 1D (a line of point detectors) and 2D (a rectangular array of point detectors). For further terminology see, for example, [19].

A 2D device is often known as an X-ray camera, for obvious reasons. For example, Fig. 21 shows powder diffraction data from an impure cuprite patina removed from a copper coupon. The patina had been formed by the conversion of a nantokite (CuCl) corrosion layer in a weak sodium sesquicarbonate solution [43], but the nantokite had already partly decomposed to paratacamite ($Cu_2(OH)_3Cl$) in the air. Compared to Fig. 17, the Cuprite 221 and 310 reflections are too weak to be immediately visible. The outermost rings are the 311 and 222 reflections, not shown in Fig. 17.

A 2D image like Fig. 21 can contain a huge amount of extra information. For example, in this case, the rings are textured and the range of crystalline grain sizes in the sample can be estimated from this [9, 18].

Modern X-ray cameras based on hybrid photon counting technology (HPC) can detect X-rays efficiently between 2 and 25 keV and each element (pixel) can count up to 10^7 photons per second against a noise count of zero [41]. Whilst such devices were originally the province of synchrotron and XFEL beam lines, they have been appearing on laboratory instruments for some time now.

5.5 Plotting, comparing and analyzing data

Fig. 22 is a 1D diffraction pattern obtained by integrating around the rings in the 2D pattern of Fig. 21. Most diffraction data from heritage applications is presented like this, usually because it comes from a scanning diffractometer. However, we have plotted the data using a square root intensity scale which has the advantage that it matches the dynamic range of XRD rather well and allows both small and large

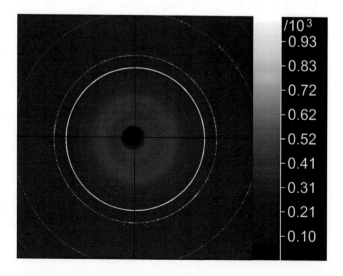

FIG. 21

False color (ironbow scale) powder diffraction image taken on beamline 2.3 at SRS Daresbury, UK, using 0.87 Å X-rays. The sample was slightly impure cuprite scraped off a patina on a copper coupon.

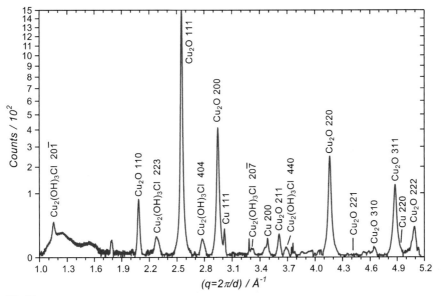

FIG. 22

The 1D pattern obtained by circular integration of the image in Fig. 21. The intensity is plotted on a square root scale to bring up small reflections without the noise amplification and distortion which results from a logarithmic scale. The peak position is given versus q the angular wavenumber. This is independent of the wavelength of the source. Data processing and display was done using esaProject (EVA Surface Analysis, UK).

reflections to be displayed without the noise amplification and distortion that often accompany logarithmic plots. Although it is hard to pick out in the form of a distributed ring (Fig. 21), the cuprite 310 reflection is now visible, although the 221 reflection is still lost in the noise. The precursor compound (nantokite) has been completely converted to cuprite, although some paratacamite remains. The copper comes from the substrate coupon from which the material was scraped. Data of this kind is often plotted as intensity versus 2θ. This has the disadvantage that a 2θ scale is wavelength dependent, so data taken on diffractometers with different wavelength sources are hard to compare without the digital data and a conversion program. When using a 2θ scale in a publication it is essential to give the wavelength or energy of the X-rays used, and this is not always done. An alternative is to use Bragg's law and plot versus d. This is a very nonlinear scale which compresses the small d (large 2θ) reflections to the left of the plot. A solution is to plot intensity versus angular wavenumber q where

$$q = \frac{2\pi}{d} = \frac{4\pi \sin\theta}{\lambda} \tag{5}$$

This scale is the same for any wavelength so that data measured on an instrument with an Mo or Cu Kα source, or a synchrotron will give identical peak positions. The physical significance of q is that it is the angular wavenumber for a wavelength of d and it arises from using the vector form of the Laue equations.

Modern powder diffractometers usually come with powerful software often linked to international data bases for crystal structure, for example [44–47]. Minerals, which are often just metallic corrosion products, are particularly well studied [45]. It has therefore become relatively straightforward to identify the mixtures of crystalline substances responsible for a pattern from a heritage sample. However, automated pattern matching still needs some assistance, and amongst the most important prior knowledge required to select sensible matches is an atomic analysis using a spectroscopy technique with a similar volumetric sampling range to XRD. EDS in an SEM, or PIXE are especially useful in this context. Other parameters, such as the density of the material may also be useful although this may be impossible to measure for a heritage material.

XRD is not a spectroscopic tool. Its data come from the arrangement of atoms in space, or more precisely the 3D electron density distribution which accompanies this. At least for a single substance the peaks, rings, or spots in the pattern can be used to deduce this distribution, reveal the departures from perfect crystallinity which arise from defects or stress, and even allow the amplitudes of an atom's oscillation about its mean position as a function of temperature to be measured. Originally, this was an intense trial and error procedure. Since 1969, the deduction of crystal structure has become increasingly automated through the use of Rietveld refinement [48] which uses a mixture of peak fitting and simulation to find a spatial arrangement of atoms which would give the measured powder pattern. The simulation is complete in the sense that it gives the identity of the atoms in the basis, its structure and its arrangement in space. High quality data with an excellent signal to noise ratio are required for this purpose. Acquisition often takes many hours, even on a synchrotron.

5.6 Information content for heritage

XRD not only provides chemical and structural analysis of crystalline materials at the atomic and molecular levels, but also the characterization of crystalline compounds from simple elemental and ionic solids to complex macromolecules and mixtures. It therefore finds many applications in heritage science. With metals and other materials its ability to give quantitative information on alloys, crystal deformation and imperfection can provide evidence for manufacturing techniques and use and thus reveal something of the way of life of the cultures examined. XRD analysis can be indispensable for conservation practice in distinguishing the original parts of an object from later additions or restoration interventions. Also, since cultural heritage is prone to deteriorate due to ageing and the effects of climate change and pollution, characterizing degradation mechanisms helps to prevent or at least decrease the chance that cultural treasures are lost forever. Table 2 summarizes a fraction of the analytical information which can be obtained and its relevance to heritage science.

Table 2 Types of information and typical materials.

XRD analysis	Obtained information	Cultural heritage materials
Qualitative analysis	Phase identification of crystalline compounds	Pigments, building materials, metals, coatings, corrosion products, glass, ceramics, etc.
Quantitative analysis	Provenance, production, degradation kinetics, forgery	Metals, metal corrosion, pigments, ceramics, stone, etc.
In situ real-time qualitative & quantitative analysis	Degradation kinetics, corrosion rates, change of crystalline phases, deposition of coatings, ceramic firing studies	Metals, metal corrosion, metal protection, ceramic simulants
Crystal structure analysis	Production process, material alteration, polymorphs	Different phases of pigments in paints and corrosion products on surfaces
Physical microstructural analysis	Size distribution of crystals, physical information, stratigraphy	Metals, stone, glass
Texture / strain analysis	Manufacturing process, physical and mechanical properties	Rocks, ceramics, metals, metallography, polymers

6 Applications—Laboratory instrumentation

6.1 Wiener Neustadt treasure trove

The Wiener Neustadt Treasure Trove forms part of an archaeological discovery reported to the Austrian Federal Department for the Protection of Monuments in 2010. It was analyzed by X-ray based techniques including powder XRD (Fig. 23). The treasure trove contained about 130 gilded silver objects and was probably hidden towards the end of the 14th century. The scientific questions dealt with the circumstances of the find and the type of landfill as well as the specific composition of the hoard, which are of central importance for the interpretation of the find. In the course of the conservation of the Wr. Neustadt treasure trove, the corrosion products on the objects were mechanically removed and then used as samples to identify and characterize the chemical composition of these degradation products. Different powder samples were analyzed by XRD in order to obtain accurate information of the present crystalline corrosion products of the object samples. Analyses of all samples revealed that the corrosion products consist mainly of copper sulfide, carbonate, hydroxide, sulfate, oxide, nitrate, and chloride, as well as mixed phases. Besides, simple and complex crystalline corrosion products such as rouaite, malachite, atacamite, chlorargyrite, cuprite, and stromeyerite were identified by XRD analysis. Additionally, identified crystalline phases like quartz, calcite, dolomite and phyllosilicates could correlate with soil contamination. These results gave chemical information about the soil in which the objects were buried [49].

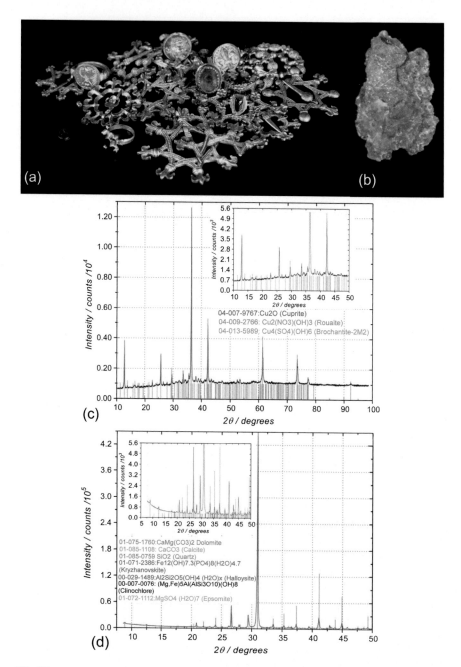

FIG. 23

(A) Wiener Neustadt treasure trove, photograph courtesy of Atelier Olschonsky. (B) Typical flake of corrosion crust removed from s shield ring plate. (C) XRD pattern and phase assignment of corrosion products shown in (B). The nine-digit number before the stoichiometry of the phase indicates the database entry of the ICDD reference pattern. The red-brown corrosion phase was assigned to a copper oxide (cuprite). For the green and green-blue corrosion products of the investigated sample, a copper nitrate hydroxide phase Rouait, a dimorph of gerhardtite and broncherite, a copper-hydroxide-sulphate, was detected. Inset: Detail of the first 50° in 2θ. (D) XRD pattern from a soil contamination in which the objects were buried These crystalline phases were identified as quartz, calcite, dolomite and phyllosilicates. Inset: detail of low intensity peaks.

6.2 Macroscopic X-ray powder diffraction mapping (MA-XRPD)

Once the province of the synchrotron, macroscopic X-ray powder diffraction mapping (MA-XRPD) has recently been developed and characterized for laboratory X-ray sources by Vanmeert et al. [50]. This method is able to map specific crystalline phases over areas of several thousand square millimeters with a lateral resolution in the range 100–400 μm. Quantitative data obtained from an illuminated parchment sheet from the 15th and 16th century revealed three lead white pigments with different hydrocerussite-cerussite composition in certain pictorial elements. This quantitative information could be linked to the macroscopic spatial distribution of the three different lead white pigments throughout the cartouches. The blue paint used for the decorative scrollwork consists of azurite (copper carbonate). Impurities in azurite such as barite or quartz can play an important role in provenance studies. Quantification analysis of impurities in the blue azurite pigment in the investigated manuscript resulted in two different types of azurite: one rich in barite and one rich in quartz.

6.3 Composite materials, e.g., enameled metal

Historic objects composed of soda glass/copper alloy (e.g., enamels) in museum collections are often exposed to formaldehyde and formic acid emitted from wooden storage cabinets, adhesives, etc. Dinnebier et al. used high-resolution X-ray powder diffraction (XRPD) to determine the previously unknown crystal structure of corrosion products and the refinement process on historic art object (Historic clasp, collection of the Rosgartenmuseum Konstanz, Germany). Full quantitative phase analysis from a corroded sample contained $Cu_4Na_4O(HCOO)_8(OH)_2 \cdot 4H_2O$ as the main phase and $Cu_2(OH)_3(HCOO)$ and Cu_2O as minor phases. This degradation phenomenon has recently been characterized as "glass induced metal corrosion" [51].

6.4 Synthetic corrosion protocols

The production of synthetic corrosion layers is important for several purposes, one of which is the testing of metals conservation methods [43]. However, it is necessary to establish that a protocol results in the corrosion product expected and that the physical properties of the product are sufficiently like those found in nature for the simulant to be useful. For example, we studied a protocol supposed by conservators to produce atacamite (an isomer of copper (II) hydroxychloride) and adapted from a recipe for producing a decorative patina on copper, no. 3.151 in Hughes and Rowe [52]. A solution of 15.07 g of $(NH_4)_2CO_3 \cdot NH_3$ (Fluka) and 10.02 g of NH_4Cl (Aldrich) in 100 mL of deionized water was prepared and a clean polished copper coupon was wetted with drops of reagent twice a day for up to 5 days and then left in the air for a further 5 days to complete the process. There was no rinsing stage.

Fig. 24A shows surface powder XRD of the product which contains no obvious atacamite. In fact, very few of the sharp reflections have been identified except

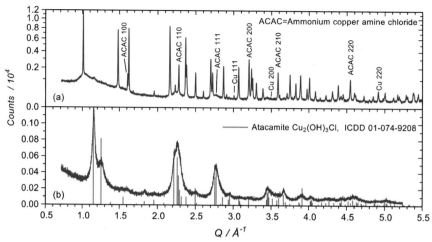

FIG. 24

(A) Surface powder XRD pattern from patina produced by Hughes and Rowe protocol 3.151 [] on a square root intensity scale. Most of the sharp reflections have not been identified but are believed to be copper-ammonium complexes. (B) The patina in (a) after adding a rinsing stage. Only atacamite remains, but the crystal size is small (~0.9 nm).

Data courtesy D. Walker and P.A. Thomas, University of Warwick.

(tentatively) ammonium copper amine chloride. Nevertheless, the product was scraped off the copper substrate, rinsed in deionized water and dried. The rinsing solution was also allowed to dry out and recrystallize at room temperature. Fig. 24B shows the pattern from the rinsed product which is similar to the background in Fig. 24A. Atacamite is now the only substance detected, but the reflections are broad, probably indicating a very small crystal size (~0.9 nm). The recrystallized solution contained only residual atacamite and ammonium chloride reflections. We presume the sharp peaks in Fig. 24A to be due to copper-ammonium compounds which hydrolyzed in the rinse. Further work [53] using Cu-K edge XANES shows that the small crystallite size results in an over optimistic result when this synthetic patina is used to test soaking protocols for chloride removal: The (insoluble) nanocrystals are lifted off the surface by water or dilute sodium sesquicarbonate solution, and become suspended in the liquid by Brownian motion. So, XRD establishes that rinsing is a required step to isolate the atacamite, but XRD and XANES show that with the crystal size which results and the friability of the layer make it an unrealistic test.

6.5 Pigments: Inorganic analysis in painting cross-sections

Paintings consist of superimposed layers of inorganic and organic materials (pigments and binders). A knowledge of stratigraphy of these complex inhomogeneous layers is fundamental to understanding the artist's painting technique and for future conservation interventions. The analysis of cross-sections by analytical techniques such as XRD, XRF, IR- and Raman Spectroscopy provides insights on the sources of the pigments and binders, their preparation, chemical composition and physical and chemical properties.

Concerning traditional artists' materials, natural ultramarine, derived from the mineral lazurite (lapis lazuli), was by far the most valuable blue pigment for centuries. Famous painters were especially fond of ultramarine, due to its distinctive blue color and its durability [54–56]. It was not only used for blues, but also in mixtures with yellow lake or lead-tin yellow to obtain a bright green color. The mineral lapis lazuli is a rock of complex composition out of mineralized limestone containing lazurite. Lazurite, which is a blue cubic mineral $[(Na,Ca)_8 (AlSiO_4)_6(SO_4,S, Cl)_2$ [57, 58], is responsible for the intense blue hue. Moreover, lapis lazuli can contain other minerals such as sodalite $[Na_8(Al_6Si_6O_{24})(Cl_2)]$, nosean $[Na_8(Al_6Si_6 O_{24}) (SO_4) \cdot H_2O]$ and hauyne $[(Ca,Na)_{4-8}(Al_6Si_6O_{24})(SO_4,S,Cl)_{1-2}]$. Other mineral impurities are also common, including pyrite (FeS_2), calcite $(CaCO_3)$, diopside $(CaMgSi_2O_6)$, forsterite (Mg_2SiO_4), wollastonite $(CaSiO_3)$, etc. Synthetic ultramarine, which was introduced in 1828, is very similar to the natural form, except that it is much finer and more uniform in particle size. Furthermore, it does not contain the mineral impurities often associated with natural ultramarine.

In the example given here, different ultramarine pigments from the artists' material collection of the Institute of Science and Technology in Art (ISTA) at the Academy of Fine Arts Vienna and dated to 19th and 20th centuries, were analyzed by XRD and complementary analytical methods to create a pigment database [59–61] including material information and spectral data. To identify the crystalline phases of the blue paint layer in a cross-section (Fig. 25), the pigment was analyzed by laboratory XRD and the diffraction patterns obtained were compared to the institute's database. The XRD patterns clearly show that the main blue color component is sodalite $[Na_8(Al_6 Si_6 O_{24})]$. Additionally, crystalline phases of SiO_2 could be identified. Due to the purity and also the uniform particle size the ultramarine used in this painting was identified as being synthetic.

7 Applications—Synchrotron XRD

7.1 Real time monitoring

Amongst other attributes, synchrotron data can often be acquired very fast because of the beam intensity and also the availability of 1D and 2D parallel detectors. So it becomes possible to study evolving systems, provided one takes care that the high photon flux does not modify the chemistry. To some extent, X-ray damage can be guarded against by using beams which are hundreds of microns in footprint dimensions, rather than microfocus beams, and by shuttering the beam line so that the sample is only exposed to X-rays during acquisition. In another way, the high flux is an advantage because reasonable signals can be measured through a window into a

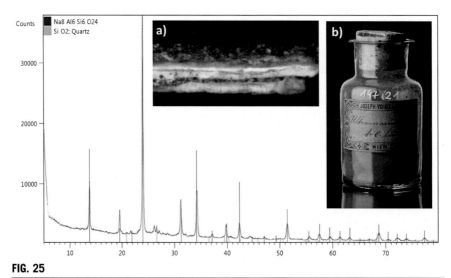

FIG. 25

XRD diffraction pattern of a blue paint layer (A) compared to XRD patterns of ultramarine pigments of the Pigment Database of the ISTA (B). XRD analysis was performed with a Siemens D5000 Diffractometer, a 2-theta goniometer using Cu K-α radiation.

controlled environment cell, and through liquids over the sample. Examples include thermal cycling in a controlled atmosphere to study the firing of glazes [62], exposure to controlled atmospheres of deteriogens [63] and coating and corrosion studies [64]. For example, a sealed cell such as the electrochemical/environmental cell (eCell) shown in Fig. 26 [65, 66] enables us to perform studies of reactions in liquids with or without parallel electrochemistry, or exposure to gases, e.g., anthropogenic sources of corrosion.

7.1.1 Corrosion of silver in gaseous mixtures

The atmospheric corrosion of silver is not well understood. Fundamental research into the chemistry occurring on surfaces exposed to the environment is needed so that material degradation can be controlled. The most common corrosion products (the sulfides) are blueish to black, but we expect to see silver with a lustrous surface often with intricate chasing. Thus, for example, silver museum artifacts, have, been repeatedly cleaned and polished over 100s or even 1000s of years, leading to serious loss of material, surface detail and other damage [67].

In these experiments [63] we examined, in real time, the early stages of the corrosion of silver exposed to the anthropogenic gases H_2S and O_3 at various levels of relative humidity (RH). A particular goal was to look for synergistic effects in gas mixtures to closer approach the real environment. Experiments were carried out in the eCell fed by a gas mixing unit designed and built at the Academy of Fine Arts in Vienna [68]. Synthetic air containing 500 ppb O_3 and/or 500 ppb or 10 ppm H_2S at 50% or 90% RH was blown through the eCell which contained a high purity silver

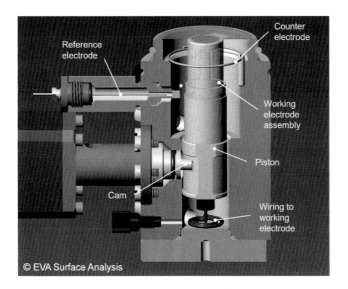

FIG. 26

Section through the eCell. The cell provides a self contained environment for a 12 mm diameter sample which may be the working electrode of an electrochemical system. The environment may be gaseous or liquid. The cell material is impervious to most chemicals.

Figure courtesy EVA Surface Analysis, UK.

coupon (99.9% Ag). In experiments lasting 12–24 h SR-XRD patterns were recorded every 10 min using the Mar CCD 165 camera at beamline 28 (XMaS) at the ESRF with an acquisition time per image of 20 s. Images were reduced to one dimensional patterns and trends in peak area over time were extracted using the esaProject package [69].

Overall, the effects were neither linear with time, nor monotonic and the patterns reveal different behavior for different compounds coexisting on the same surface, changes in rates of corrosion over time, and a strong dependence of the rate of corrosion on the gas mixture. For example, the growth of silver oxides in O_3 at 50% RH was shown to proceed via cubic Ag_2O for 3 h or so, then AgO-related reflections appear followed by more complex oxides after 10 h (polymorphic AgO, Ag_2O_3 and Ag_3O_4). The total coverage is still increasing rapidly after 24 h. Conversely, when O_3 is replaced by H_2S at the same level (Fig. 27), a fresh coupon shows little evidence of corrosion on the same timescale. However, if O_3 is added to the H_2S + 50% RH mix after 12 h, a very rapid increase in coverage of both sulfides and oxides is observed. We conclude that at concentrations close to those found in city centers, synergistic effects between O_3 and H_2S lead to greatly accelerated rates of corrosion.

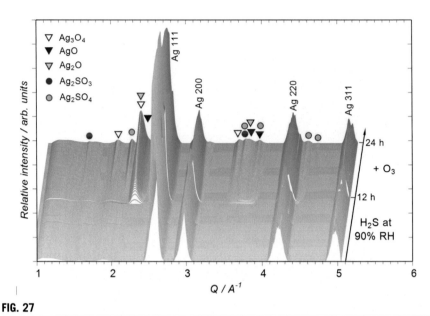

FIG. 27

Waterfall plot of extracted and normalized SR-XRD patterns for two-part experiment with initial exposure to synthetic air containing 10 ppm H_2S and 90% RH with the addition of O_3 after 12 h. Note the rapid appearance of corrosion peaks upon the addition of O_3.

7.1.2 Coating growth—lead monocarboxylates

European lead heritage ranges from simple ornaments to vast artistic and engineering masterpieces such as pipe organs—all in danger of decay and loss through corrosion. Indeed, lead organ pipes are amongst the most seriously jeopardized artefacts: their sound is critically dependent on the material, as well as their shape, size, and condition. Corrosion usually starts at the foot of the pipe and proceeds invisibly from the inside towards its mouth, resulting in changes to the tonal quality due to distortion of the mouth geometry. Cracks and holes then develop, at which point the pipe becomes mute and requires repair or replacement [70]. Lead is seriously affected by the presence of organic acids (formic, acetic, etc.) in the environment, and both original and restored wood (e.g., in the wind chest) are a source of these.

Coating lead artifacts to protect them is fraught with difficulty since badly researched coatings and treatments can lead to irreversible changes in appearance and function, and even more rapid degradation. Only after meticulous investigation can a coating be applied with relative confidence. Coating the lead with its mono-carboxylate $[CH_3(CH_2)_nCOO]_2Pb$ where $n > 8$ is one possible approach [71].

For n around 8 or 10, the coating is simple to apply by soaking in an aqueous solution [71] of the appropriate sodium carboxylate, e.g., sodium decanoate which is nontoxic. The result is a dark coating, similar to aged lead in appearance, providing it is not made too thick. In this initial study, we set out to measure the growth

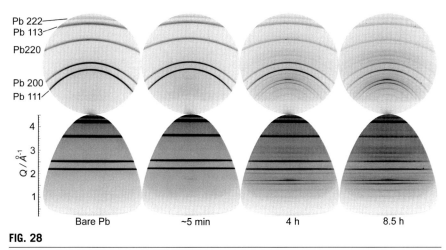

FIG. 28

XRD images (upper row) measured during growth and esaProject transforms (lower row) from the developing coating. The lead rings are indexed; other rings are due to the lead monocarboxylate.

characteristics of such layers using time resolved XRD, electrochemical impedance spectroscopy (EIS), scanning electron microscopy (SEM) and mass gain measurements, as a prelude to testing its performance in a corrosive environment.

On beamline 28 (XMaS) at the ESRF we made simultaneous EIS and XRD measurements whilst the lead was soaking in the eCell, (Fig. 26). Fig. 28 shows a time sequence of diffraction images from a Mar CCD 165 camera and their mapping to Q-γ space (γ is the out of plane scattering angle) from a growing lead carboxylate coating. In detail, the lead rings are spotty whereas the carboxylate rings are smooth, which is indicative of a dense interlocking microcrystalline structure, ideal in a potential protective coating.

Fig. 29 shows the change in peak height extracted from a sequence of CCD images. That the lead is covered over is clearly evident and the data correlate well with the mass gain measurements. The simultaneous EIS measurements are indicative of the development of a high integrity layer free from pinholes (not shown here).

The combination of XRD and EIS shows that the carboxylate coating is promising and has the basic characteristics required for a protective coating on the lead. Improved coatings with $n > 12$ can be deposited from ethanolic solution of the carboxylic acid or a mixed aqueous/ethanolic solution of sodium carboxylate [64, 72, 73]. The longer chain length leads to increased water resistance but from a pure ethanolic solution, the crystalline layer grows not during soaking but as the liquid dries.

7.1.3 Corrosion study—Nantokite conversion

Like many other metals, copper corrodes when it comes into contact with an aggressive environment, such as the sea or the atmosphere. However, heritage copper and its alloys are often preferred in the corroded state: The colors are aesthetically

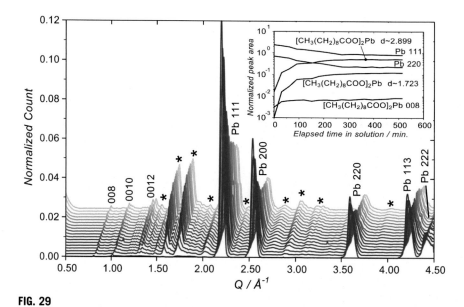

FIG. 29

Partly indexed (00x) and other peaks from growing lead decanoate $CH_3(CH_2)_8COO]_2Pb$ (*) showing the simultaneous decrease in the Pb intensity. Inset shows the growth/decay in peak area versus time for representative peaks.

pleasing, the right patina protects the surface, and the presence of corrosion products provides evidence of time past and passing, thereby adding extra value to the object. However, active corrosion (corrosion which leads to the deterioration of the underlying metal) will result in the destruction of the object and is undesirable.

Corroded archaeological cuprous artifacts recovered from wet saline environments, are particularly susceptible to further corrosion. They are typically saturated with chlorides whose attack on the metal will be greatly accelerated upon air exposure. To alleviate this problem, objects are stored in tap water or sodium sesquicarbonate solution for periods of up to several years [74–76]. This is believed both to remove potentially damaging soluble chlorides from the microstructure of the metal and to change various copper chlorides into more benign compounds such as cuprite. During treatment, the chloride concentration in solution can be measured. When this exceeds a predetermined value, the solution is replaced. This procedure is repeated until the chloride concentration in solution falls below some threshold. The disadvantage of this method is the fact that it is indirect: the conservator has no idea what is happening with the metal surface. Hence, a different monitoring method is desirable. A method which has been proposed for monitoring the progress of the treatment and determining its end-point in-situ is the measurement of the corrosion potential Ecorr, the open circuit potential of the electrochemical cell in which, effectively, the artifact is immersed.

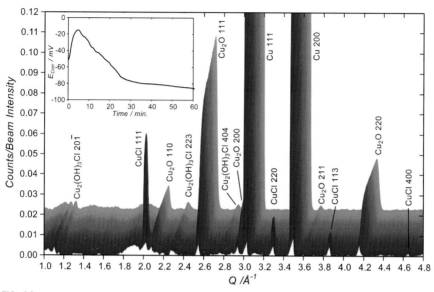

FIG. 30

A time sequence of XRD patterns displayed as a waterfall plot showing the conversion of nantokite to cuprite in 1% wt/v sodium sesquicarbonate solution. In the course of an hour, the nantokite disappears. Paratacamite, on the other hand, is unaffected on this timescale. Inset: Behavior of Ecorr measured in parallel. There is no correlation, and repeating the experiment gives a different result for Ecorr versus time.

To evaluate whether Ecorr measurements may have a role to play we immersed artificially corroded copper coupons in a 1% (wt/v) sodium sesquicarbonate solution. The possible use of corrosion potential measurements as a monitoring method was evaluated by recording the corrosion potential of the immersed coupons as a function of time while analyzing the surface of the coupon simultaneously using SR-XRD.

The study was carried out using the eCell at stations 2.3 and MPW 6.2 of the SRS at the Daresbury Laboratory in Cheshire, UK [43] and beamlines 28 (XMaS) and 26 (DUBBLE) at the ESRF in Grenoble, France [53, 77].

Fig. 30 shows a waterfall plot of an hour long sequence of SR-XRD patterns from the RAPID II detector at SRS 6.2 of a copper sample initially covered with synthetic nantokite and some cuprite and also paratacamite (formed through hydrolyzation of the nantokite in natural humidity). The sequence shows a rapid conversion of nantokite to cuprite either via [74]

$$2CuCl + H_2O \rightarrow Cu_2O + 2Cl^- + 2H^+ \tag{6}$$

or possibly through [78]

$$2CuCl_2^- + 2OH^- \rightarrow Cu_2O + 4Cl^- + H_2O \tag{7}$$

The inset shows the parallel variation of Ecorr. Repeated measurements over hours to days showed no useful correlation between Ecorr and the nantokite conversion but a variation in the speed with which the conversion occurred, possibly depending on the thickness of the layer. The removal of the hydroxychloride is clearly far slower (if it occurs at all).

7.2 Structural analysis—Pigments

In-depth knowledge of the crystal structure of heritage materials is sometimes the key to understand the physicochemical properties and reactivity of the phase. The identification of different phases of pigments in paints and frescoes is required to localize the nature of the color mechanisms, the state of conservation and their pre-dicted variation (stability) with time. Highly advanced laboratory and synchrotron-based diffraction techniques are increasingly used to identify pigments, distribution in paint layers [79, 80], which even allows access to information from subsurface layers and the micro-spatial distribution of the pigment [81, 82] and effects of alter-ation. In complex cases, however, the definition is of the pigment and full under-standing of its physico-chemical behavior requires knowledge of atomic structure, and this can be a challenging task.

Gonzalez et al. have investigated the famous Chrome yellow pigment, as used by Van Gogh, and an example of an artists' pigment with multigrade composition. Van Gogh used different grades of chrome yellows which have different color and com-position, resulting from different synthesis protocols. The reconstruction of the orig-inal recipes of Winsor and Newton 19th revealed that this company produced mainly three pigment types: a lemon/pale type ($PbCr_{1-x}S_xO_4$, with $0 < x < 0.5$), a middle type ($PbCrO_4$) and orange/red type ($PbCrO_4 \cdot PbO$); these different types are obtained in acid, neutral and basic conditions [34]. The $PbCrO_4$ usually crystallizes in mono-clinic form while the $PbCr_{1-x}SxO_4$ can exist in both monoclinic and orthorhombic forms. These different grades were differentiated by both SR-μXRD and MA-XRPD to identify these different compositions in both paint fragments and paintings [83]. The information obtained is essential not only to know which pigments were used by the artists but also since these different grades exhibit different photostability; nota-bly, orthorhombic $PbCr_{1-x}S_xO_4$, rich in SO_4^{2-} ($x \geq 0.4$) is prone to darkening while the monoclinic $PbCrO_4$ is stable [84]. The XRD mapping revealed that van Gogh used crystallographically different chrome yellows, in different ratios and with dif-ferent pigments to paint the flower heart and the petals of the sunflowers. This mul-timodal study revealed evidence of degradation of some chrome yellows in the Van Gogh museum's version of the Sunflowers.

Another structure analysis study using synchrotron X-ray and neutron diffraction dealing with one of the oldest **blue organic pigments, namely Indigo** reveals that the molecule was absorbed in the zeolitic channels of the palygorskite clay and that there is a strong interaction between the organic molecule and the hosting framework. How-ever, it is known that the palygorskite mineral is composed of at least two polymorphs, an orthorhombic and a monoclinic one causing the structure to have extensive poly-somatic defects and therefore, sepiolite contaminants are often present [85].

7.3 Microstructural analysis—Geology to cosmetics

A recent study by Hansford et al. using energy-dispersive X-ray diffraction (ED-XRD) in back-reflection geometry was implemented on beamline B18 at the Diamond Light Source synchrotron in Oxfordshire, UK [86]. This set-up allows analysis without any sample preparation and is therefore nondestructive. XRD analysis of geological samples, fossils and archaeological samples was performed and besides phase identification, the derivation of unit-cell parameters and extraction of microstructural information was possible. Generally, XRD analysis of clay and phyllosilicate minerals is delicate and needs a variety of sample preparation steps, including glycolation and dehydration by heating. By using back-reflection EDXRD for an unprepared clay containing sample, the authors could impressively show the identification of a specific polytype of muscovite in a mica schist sample [86].

Microstructural analysis can be obtained from the profile shape of the diffraction maxima of polycrystalline material and provides information on the size distribution of crystals and the presence of lattice strains [87].

Microstructural effects in a sample can be identified from the diffraction peak broadening. These peak broadening effects may be inadvertently caused through sample preparation [39]. An example of EDXRD analysis of a **limestone rock sample** [86] in comparison with a powder pellet made from a portion of this stone shows an apparent peak broadening of the diffraction peaks obtained from the pellet, indicating the introduction of crystallite size or lattice strain effects during milling of the rock sample.

Another prominent example of microstructural analysis obtained from the diffraction peak broadening of heritage materials was the analysis of the Egyptian cosmetics performed by high resolution synchrotron XRD [88, 89]. The investigations showed that the large grain size and the significant particle strain of Egyptian kohl could be related to hand grinding of galena particles, whereas the small crystal size and lack of strain of Egyptian white pigments (namely laurionite, hydrocerussite) could be due to chemical precipitation [90].

8 Conclusion

XRD, like many aesthetic aspects of heritage artifacts is characterized by symmetries and patterns. XRD is a structural tool, not a spectroscopy—although at the beginning of the 20th century, what we now call patterns were known by the originators of the technique as spectra [2, 91]. However, the term X-ray spectrum is now reserved for techniques like XRF and XAS. XRD tells us about the precise location of atoms in space if the sample has long range order. It tells us about the spatial extent over which long range order persists, and about departures from ordering. Indeed, it tells us whether a material is crystalline at all. Even for an amorphous material or a gas it can say something about the average atomic spacing.

Laboratory and synchrotron instrumentation is continuously improving in speed, sensitivity of analysis, and in resolution through the development of X-ray sources of increased power and pixelated detectors with low (or even no) intrinsic noise and high saturation levels. It is important, though, that the user tests the nondestructivity of the technique, especially when using high power sources.

The vast range of applications for both laboratory and synchrotron XRD of heritage objects makes our sample of these just the tip of the iceberg.

Acknowledgements

Figures 1–14, 15b and 16–20 are ©2021 EVA Surface Analysis and used with permission.

References

[1] J.B. Jones, E.R. Segnit, The nature of opal I. Nomenclature and constituent phases, J. Geol. Soc. Aust. 18 (1971) 57–68.

[2] R.W. James, in: Sir Lawrence Bragg (Ed.), The Optical Principles of the Diffraction of X-rays: The Crystalline State, vol. II, G. Bell and Sons Ltd, London, 1948 (Chapters IX and X).

[3] M. Schreiner, B. Frühmann, D. Jembrih-Simbürger, R. Linke, X-rays in art and archaeology—an overview, Adv. X-ray Anal. 17 (2004) 1–17.

[4] M. Schreiner (Ed.), Naturwissenschaften in der Kunst., Beitrag der Naturwissenschaften zur Erforschung und Erhaltung unseres kulturellen Erbes, Böhlau Verlag, Köln, 1995.

[5] W. Röntgen, Ueber eine neue Art von Strahlen: Vorläufige Mitteilung, Sitzungsberichten der Würzburger Physik.-medic, Gesellschaft Würzburg, 1896.

[6] W. Friedrich, P. Knipping, M. Laue, Interferenz-Erscheinungen bei Röntgenstrahlen, Sitzungsberichte der Königlich Bayerischen, Akademie der Wissenschaften (1912) 303–322.

[7] P.P. Ewald, 50 Years of X-Ray Diffraction, IUCr Publications 1962, 1999 (Chapter 4) https://www.iucr.org/publ/50yearsofxraydiffraction. (Accessed 30 April 2021).

[8] J.C. Maxwell, A dynamical theory of the electromagnetic field, Phil. Trans. Roy. Soc. 155 (1865) 459–512.

[9] B.E. Warren, X-Ray Diffraction, Dover Publications Inc., New York, 1990.

[10] A. Einstein, Über einen die Erzeugung und Verwandlung des Lichtes betreffenden heuristischen Gesichtspunkt, Ann. Phys. 17 (1905) 132–148.

[11] P.A.M. Dirac, Relativistic quantum mechanics, Proc. R. Soc. A 136 (1932) 453–464.

[12] R.P. Feynman, Space-time approach to non-relativistic quantum mechanics, Rev. Mod. Phys. 20 (1948) 367–403.

[13] J. Schwinger, Quantum electrodynamics. I. A covariant formulation, Phys. Rev. 74 (1948) 1439–1461.

[14] S. Tomonaga, On infinite field reactions in quantum field theory, Phys. Rev. 74 (1948) 224–225.

[15] H. Karabiyik, A practical scheme for ab initio determination of a crystal structure based on the Dirac equation, Theor. Chem. Acc. 118 (2007) 785–790.

[16] F. Kika, J. Matějková-Pišková, S. Jiwajinda, P. Dechkrong, M. Shiojiri, Photonic crystal structure and coloration of wing scales of butterflies exhibiting selective wavelength iridescence, Materials 5 (2012) 754–771.

[17] A. Bravais, Mémoire sur les systèmes formeés par des points distribués regulièerement sur un plan or dans d'espace, J. de L'Ecole Polytech. 19 (1850) 1–128.

[18] H. Lipson, H. Steeple, Interpretation of X-Ray Powder Diffraction Patterns, Macmillan, London, New York, 1970.

[19] B.B. He, Two-Dimensional X-Ray Diffraction, second ed., John Wiley & Sons, Hoboken, NJ, 2018.

[20] F. Hoffmann, Introduction to Crystallography, Springer, Switzerland, 2020, doi:https://doi.org/10.1007/978-3-030-35110-6 (Accessed 04 May 2021).

[21] R.D. Shannon, Revised effective ionic radii and systematic studies of interatomic distances in halides and chalcogenides, Acta Crystallogr. A32 (1976) 751–767.

[22] H.W. Kroto, J.R. Heath, S.C. O'Brien, R.F. Curl, R.E. Smalley, C_{60}: Buckminsterfullerine, Nature 318 (1985) 162–163.

[23] W.I.F. David, R.M. Lbberson, J.C. Matthewman, K. Prassides, T.J.S. Dennis, J.P. Hare, H.W. Kroto, R. Taylor, D.R.M. Walton, Crystal structure and bonding of ordered C_{60}, Nature 353 (1991) 147–149.

[24] R.P. Feynman, QED—The Strange Theory of Light and Matter, Princeton University Press, Princeton, 2014.

[25] D.C. Phillips, William Lawrence Bragg. https://royalsocietypublishing.org/doi/pdf/10.1098/rsbm.1979.0003. (Accessed 4 May 2021).

[26] W.L. Bragg, The specular reflection of X-rays, Nature 90 (1912) 410.

[27] M. Laue, Eine quantitative Prüfung der Theorie für die Interferenz-Erscheinungen bei Röntgenstrahlen, Sitzungsberich. Kgl. Bayer. Akad. der Wiss. (1912) 363–373. reprinted in Ann. Phys. 41 (1913) 989–1002.

[28] W.H. Miller, A Treatise on Crystallography, The Pitt Press, Cambridge, 1839.

[29] European Synchrotron Radiation Facility (ESRF) www.esrf.fr (Accessed 04 May 2021).

[30] Diamond Light Source, www.diamond.ac.uk (Accessed 04 May 2021).

[31] European XFEL, www.xfel.eu, (Accessed 04 May 2021).

[32] A.B. Christian, J.J. Neumeier, R.L. Cone, Economical Laue X-ray diffraction using photographic film and orientation of single crystals, Rev. Sci. Instrum. 91 (2020), https://doi.org/10.1063/1.5139611, 051401.

[33] V. Simova, P. Bezdicka, J. Hradilova, D. Hradil, T. Grygar, X-ray powder microdiffraction for routine analysis of paintings, Powder Diffr. 20 (2005) 224–229, https://doi.org/10.1154/1.1938983.

[34] D. Benedetti, S. Valetti, E. Bontempi, C. Piccioli, L.E. Depero, Study of ancient mortars from the Roman Villa of Pollio Felice in Sorrento (Naples), Appl. Phys. A 79 (2004) 341–345.

[35] W.D. Coolidge, A powerful Röntgen ray tube with a pure electron discharge, Phys. Rev. II (1913) 409–430.

[36] M. Otendal, T. Tuohimaa, A. Vogt, H.M. Hertz, A 9 keV electron-impact liquid-galliumjet X-ray source, Rev. Sci. Instrum. 79 (2008), https://doi.org/10.1063/1.2833838, 016102.

[37] https://www.deanza.edu/faculty/lunaeduardo/documents/CharacteristicXraysofCopper.pdf (Accessed 20 March 2021).

[38] M. Deutsch, et al., X-ray spectrometry of copper: new results on an old subject, J. Res. Natl. Inst. Stand. Technol. 109 (2004) 75–98.

[39] C.T. Ulrey, An experimental investigation of the energy in the continuous X-ray spectra of certain elements, Phys. Rev. 11 (1918) 401–410.

[40] J.C.H. Spence, XFELs for structure and dynamics in biology, IUCrJ 4 (2017) 322–339,- https://doi.org/10.1107/S2052252517005760.

[41] A. Förster, S. Brandstetter, C. Schulze-Briese, Transforming X-ray detection with hybrid photon counting detectors, Phil. Trans. R. Soc. A377 (2019) 20180241, https://doi.org/10.1098/rsta.2018.0241.

[42] Y. Amemiya, U.W. Arndt, B. Buras, J. Chikawa, L. Gerward, J.I. Langford, W. Parrish, P.M. de Wolff, Detectors for X-rays, in: E. Prince (Ed.), International Tables for Crystallography Volume C: Mathematical, Physical and Chemical Tables. International Tables for Crystallography Volume C, Kluwer, Dordrecht, 2004, https://doi.org/10.1107/97809553602060000604.

[43] K. Leyssens, A. Adriaens, M. Dowsett, B. Schotte, I. Oloff, E. Pantos, A.M.T. Bell, S. Thompson, Simultaneous in-situ time resolved SR-XRD and corrosion potential analyses to monitor the corrosion on copper, Electrochem. Commun. 7 (2005) 1265–1270.

[44] ICDD data bases: https://www.icdd.com (Accessed 04 May 2021).

[45] A.V. Chichagov, D.A. Varlamov, R.A. Dilanyan, T.N. Dokina, N.A. Drozhzhina, O.L. Samokhvalova, T.V. Ushakovskaya, MINCRYST: a crystallographic database for minerals, local and network (WWW) versions, Crystallogr. Rep. 46 (2001) 876–879. http://database.iem.ac.ru/mincryst/. (Accessed 4 May 2021).

[46] C.R. Groom, I.J. Bruno, M.P. Lightfoot, S.C. Ward, The Cambridge structural database, Acta Crystallogr. B72 (2016) 171–179, https://doi.org/10.1107/S2052520616003954. https://www.ccdc.cam.ac.uk/solutions/csd-core/components/csd/. (Accessed 4 May 2021).

[47] http://www.crystallography.net/ (Accessed 04 May 2021).

[48] H.M. Rietveld, A profile refinement method for nuclear and magnetic structures, J. Appl. Crystallogr. 2 (1969) 65–71.

[49] R. Wiesinger, K. Hradil, I. Martina, M. Schreiner, Chemische Analysen der Korrosionsprodukte von Objekten des Wiener Neustädter Schatzfundes, in: N. Hofer (Ed.), Der Schatzfund von Wiener Neustadt, Berger & Söhne, 2014, pp. 72–79.

[50] F. Vanmeert, W. De Nolf, J. Dik, K. Janssens, Macroscopic X-ray powder diffraction scanning: possibilities for quantitative and depth-selective parchment analysis, Anal. Chem. 90 (2018) 6445–6452, https://doi.org/10.1021/acs.analchem.8b00241.

[51] R.E. Dinnebier, A. Fischer, G. Eggert, T. Runčevski, N. Wahlberg, X-ray powder diffraction in conservation science: towards routine crystal structure determination of corrosion products on heritage art objects, J. Vis. Exp. 112 (2016), https://doi.org/10.3791/54109, e54109.

[52] R. Hughes, M. Rowe, The Colouring, Bronzing and Patination of Metals, Thames and Hudson, London, 1991.

[53] A. Adriaens, M. Dowsett, G. Jones, K. Leyssens, S. Nikitenko, An in-situ X-ray absorption spectroelectrochemisty study of the response of artificial chloride corrosion layers on copper to remedial treatment, J. Anal. Atom. Spectrom. 24 (2009) 62–68, https://doi.org/10.1039/b814181a.

[54] N. Costaras, A study of the materials and techniques of Johannes Vermeer, in: J.M. Gaskell (Ed.), Vermeer Studies, Studies in the History of Art 55, National Gallery of Art/ Yale University Press, Washington/New Haven and London, 1998, pp. 145–168.

[55] A. Van Suchtelen, Q. Buvelot, M. Goverde, A. Vandivere, L. van der Vinde, E. Buijsen, C. Pottasch, S. Meloni, Genre paintings in the Mauritshuis, Waanders Publishers, The Hague/Zwolle: Mauritshuis, 2016.

[56] E.M. Gifford, L.D. Glinsman, Collective style and personal manner: materials and techniques of high-life genre painting, an essay, in: A.E. Waiboer, A.K. Wheelock Jr., B. Ducos, P. Bakker, P. Buvelot, E.J. Sluijter, M.E. Wieseman (Eds.), Vermeer and the Masters of Genre Painting: Inspiration and Rivalry, Yale University Press, 2017, pp. 65–83.

[57] J. Plesters, Ultramarine blue, natural and artificial, in: A. Roy (Ed.), Artists' Pigments: A Handbook of their History and Characteristics Volume 2, National Gallery of Art, Washington, DC, 1993, pp. 37–65.

[58] J. Goettlicher, A. Kotelnikov, N. Suk, A. Kovalski, T. Vitova, R. Steininger, Sulfur K X-ray absorption near edge structure spectroscopy on the photochrome sodalite variety hackmanite, Z. Kristallogr. Cryst. Mater. 228 (2013) 157–171.

[59] V. Desnica, K. Furić, M. Schreiner, Multianalytical characterisation of a variety of ultramarine pigments, e-PS 1 (2004) 15–21.

[60] B. Hochleitner, V. Desnica, M. Mantler, M. Schreiner, Historical pigments: a collection analyzed with X-ray diffraction analysis and X-ray fluorescence analysis in order to create a database, Spectrochim. Acta B 58 (2003) 641–649.

[61] D. Jembrih-Simbuerger, E. Wenger, M. Schreiner, A database including material information and spectral data, in: Conference: Archäometrie und Denkmalpflege, Vienna, Austria, 2019.

[62] T. Pradell, J. Molera, E. Pantos, A.D. Smith, C.M. Martin, A. Labrador, Temperature resolved reproduction of medieval luster, Appl. Phys. A90 (2008) 81–88.

[63] R. Wiesinger, R. Grayburn, M. Dowsett, P. Sabbe, P. Thompson, M. Adriaens, M. Schreiner, In situ time-lapse synchrotron radiation X-ray diffraction of silver corrosion, J. Anal. Atom. Spectrom. 30 (2015) 694–701.

[64] R. Grayburn, M. Dowsett, M. De Keersmaecker, E. Westenbrink, J.A. Covington, J.B. Crawford, M. Hand, D. Walker, P.A. Thomas, D. Banerjee, A. Adriaens, Time-lapse synchrotron X-ray diffraction to monitor conservation coatings for heritage lead in atmospheres polluted with oak-emitted volatile organic compounds, Corros. Sci. 82 (2014) 280–289.

[65] M. Dowsett, A. Adriaens, Cell for simultaneous synchrotron radiation X-ray and electrochemical corrosion measurements on cultural heritage metals and other materials, Anal. Chem. 78 (2006) 3360–3365.

[66] A. Adriaens, M. Dowsett, The coordinated use of synchrotron spectro-electrochemistry for corrosion studies on heritage metals, Acc. Chem. Res. 43 (2010) 927–935.

[67] http://www.vam.ac.uk/content/articles/c/cleaning-metals-basic-guidelines/ (Accessed 05 May 2021).

[68] R. Wiesinger, M. Schreiner, C. Kleber, Investigations of the interactions of CO_2, O_3 and UV light with silver surfaces by in-situ IRRAS/QCM and ex situ TOF-SIMS, Appl. Surf. Sci. 256 (2010) 2735–2741.

[69] M.G. Dowsett, esaProject 2020 Manual, EVA Surface Analysis, 2020.

[70] C.M. Ortel, A. Richards, Music and materials: art and science of organ pipe metal, MRS Bull. 42 (2017) 55–61.

[71] E. Rocca, C. Rapin, F. Mirambet, Inhibition treatment of the corrosion of lead artefacts in atmospheric conditions and by acetic acid vapour: use of sodium decanoate, Corr. Sci. 46 (2004) 653–665.

[72] R.A. Grayburn, M. Dowsett, M. DeKeersmaecker, D. Banerjee, S. Brown, A. Adriaens, Towards a new method for coating heritage lead, Herit. Sci. 2 (2014) 14, https://doi.org/10.1186/2050-7445-2-14.

[73] A.A.M. Elbeshary, M. De Keersmaecker, K. Verbeken, A. Adriaens, Saturated long linear aliphatic chain sodium monocarboxylates for the corrosion inhibition of lead objects—an initiative towards the conservation of our lead cultural heritage, J. Solid State Electrochem. 21 (2017) 693–704, https://doi.org/10.1007/s10008-016-3402-5.

[74] W.A. Oddy, M.J. Hughes, The stabilization of 'active' bronze and iron antiquities by the use of sodium sesquicarbonate, Stud. Conserv. 15 (1970) 183–189.

[75] I.D. MacLeod, Conservation of corroded copper alloys: a comparison of new and traditional methods for removing chloride ions, Stud. Conserv. 32 (1987) 25–40.

[76] A.M. Pollard, R.G. Thomas, P.A. Williams, Mineralogical changes arising from the aqueous sodium carbonate solutions for the treatment of archaeological copper objects, Stud. Conserv. 35 (1990) 148–152.

[77] A. Adriaens, M. Dowsett, K. Leyssens, B. Van Gasse, Insights into the electrolytic stabilization with weak polarization as treatment for archaeological copper objects, Anal. Bioanal. Chem. 377 (2007) 861–868.

[78] G. Kear, B.D. Barker, F.C. Walsh, Electrochemical corrosion of unalloyed copper in chloride media—a critical review, Corros. Sci. 46 (2004) 109–135.

[79] A. Duran, M.C. Jiminez de Haro, J.L. Perez-Rodriguez, M.L. Franquelo, L.K. Herrera, A. Justo, Determination of pigments and binders in Pompeian wall paintings using synchrotron radiation—high-resolution X-ray powder diffraction and conventional spectroscopy—chromatography, Archaeometry 52 (2010) 286–307, https://doi.org/10.1111/j.1475-4754.2009.00478.x.

[80] L.K. Herrera, M. Cotte, M.C. Jimenez de Haro, A. Duran, A. Justo, J.L. Perez-Rodriguez, Characterization of iron oxide-based pigments by synchrotron-based micro X-ray diffraction, Appl. Clay Sci. 42 (2008) 57–62, https://doi.org/10.1016/j.clay.2008.01.021.

[81] M. Cotte, J. Susini, J. Dik, K. Janssens, Synchrotron-based x-ray absorption spectroscopy for art conservation: looking back and looking forward, Acc. Chem. Res. 43 (2010) 705–714, https://doi.org/10.1021/ar900199m.

[82] M. Cotte, E. Welcomme, V.A. Solé, M. Salomé, M. Menu, P. Walter, J. Susini, Synchrotron-based X-ray spectromicroscopy used for the study of an atypical micrometric pigment in 16th century paintings, Anal. Chem. 79 (2007) 6988–6994, https://doi.org/10.1021/ac0708386.

[83] V. Gonzalez, M. Cotte, F. Vanmeert, W. de Nolf, K. Janssens, X-ray diffraction mapping for cultural heritage science: a review of experimental configurations and applications, Chem. Eur. J. 26 (2019) 1703–1719, https://doi.org/10.1002/chem.201903284.

[84] L. Monico, G. Van der Snickt, K. Janssens, W. De Nolf, C. Miliani, J. Dik, M. Radepont, E. Hendriks, M. Geldof, M. Cotte, Degradation process of lead chromate in paintings by Vincent van Gogh studied by means of synchrotron X-ray spectromicroscopy and related methods. 2. Original paint layer samples, Anal. Chem. 83 (2011) 1224–1231, https://doi.org/10.1021/ac1025122.

[85] G. Ferraris, E. Makovicky, S. Merlino, Crystallography of Modular Materials, Oxford Scholarship, 2008, https://doi.org/10.1093/acprof:oso/9780199545698.001.0001.

[86] G.M. Hansford, S.M.R. Turner, P. Degryse, A.J. Shortland, High-resolution X-ray diffraction with no sample preparation, Acta. Crystallogr. A Found. Adv. 73 (2017) 293–311, https://doi.org/10.1107/S2053273317008592.

[87] R.L. Snyder, J. Fiala, H.J. Bunge, Defect and Microstructure Analysis by Diffraction, Oxford University Press, 2000.

[88] P. Walter, P. Martinetto, G. Tsoucaris, R. Brniaux, M.A. Lefebvre, G. Richard, J. Talabot, E. Dooryhee, Making make-up in ancient Egypt, Nature 397 (1999) 483–484.

[89] G. Tsoucaris, P. Martinetto, P. Walter, J.-L. Lévêque, Chimie et maquillage dans l'antiquité, Ann. Pharm. Fr. 59 (2001) 415–422. APF-11-2001-59-6-003-4509-101019-ART8.

[90] P. Martinetto, M. Anne, E. Dooryhée, G. Tsoucaris, P. Walter, A synchrotron X-ray diffraction study of Egyptian cosmetics, in: D. Creagh, D. Bradley (Eds.), Radiation in Art and Archaeometry, Elsevier Science, Amsterdam, 2000, pp. 297–316.

[91] W.H. Bragg, W.L. Bragg, X-Rays and Crystal Structure, G. Bell and Sons Ltd., London, 1915.

Laser-induced breakdown spectroscopy in cultural heritage science

7

Rosalba Gaudiuso

Université du Québec à Montréal, Montréal, QC, Canada

1 Introduction

Laser-Induced Breakdown Spectroscopy (LIBS), also known as Laser-Induced Plasma Spectroscopy (LIPS), is the optical emission spectroscopy of Laser-Induced Plasmas (LIPs). By focusing an intense laser pulse on a sample, a plasma can be formed by vaporizing (ablating), ionizing, and exciting a tiny portion of the material. The high-intensity radiation emitted by the Laser-Induced Plasma (LIP) constitutes the LIBS signal [1–4]. In a book offering an overview about many techniques for the scientific analysis of cultural heritage, the first question one needs to answer is: why LIBS? The answer is that, despite not being yet well-established in the field, LIBS has much of what is required of a technique for the analysis of tangible cultural heritage. The variety of materials falling under this definition is intrinsically immense, as it includes movable, immovable, and underwater cultural heritage. A "one-size-fits-all" approach is therefore unfeasible, and the "right" technique depends on the material under investigation and on the kind of information to be retrieved. Due to the historical significance of heritage samples, an obvious requirement is that the analysis should maximize the amount of information about the investigated artifacts, at the same time minimizing the damage. In the subdivision between invasive/non-invasive and destructive/nondestructive techniques, LIBS can be considered micro-invasive or micro-destructive. It is noninvasive, in that the objects can be analyzed without sampling or even any sample preparation procedure. On the other hand, the laser does vaporize a portion of sample to produce the analytical signal, in fact destroying the ablated material and producing an ablation crater. Nonetheless, the amount of vaporized material is in the order of 10–10^2 nanograms per laser pulse, and the damage can be virtually invisible to the naked eye (diameter of few µm up to few hundred µm), as illustrated in Fig. 1.

Moreover, LIBS makes up for its micro-destructivity with a large amount of spectroscopic information and several practical advantages, especially useful in heritage science. Some well-known features of LIBS as an analytical technique include [5]: speed (e.g., hundreds of spectra can be acquired in seconds to minutes);

209

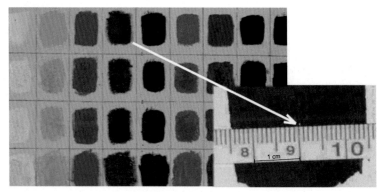

FIG. 1

Close-up of crater induced by focusing 20 laser shots on an acryl painting layer containing cadmium red pigment. Laser energy = 30 mJ; crater diameter < 300 μm.

Adapted from J. Marczak, A. Koss, P. Targowski, M. Góra, M. Strzelec, A. Sarzyński, W. Skrzeczanowski, R. Ostrowski, A. Rycyk, Characterization of laser cleaning of artworks, Sensors 8 (2008) 6507–6548.

multielemental analysis, including both light and heavy elements; simultaneous analysis of major, minor, and trace components; good sensitivity (generally on the order of ppm); possibility of calibration-free quantitative analysis. Moreover, LIBS is a point analysis technique, therefore it can be used for spatially-resolved measurements and elemental maps of highly inhomogeneous samples. Spatial resolution is not limited to the surface: one of the most important features of LIBS is in fact that it enables depth profiles up to several hundred microns underneath the object surface, without any sampling procedure. Moreover, the experimental setup is flexible, compact, and field-deployable, thus enabling in situ analysis, also in remote and standoff configurations. Finally, another valuable advantage is the possibility of analyzing submerged objects, which can be of particular interest for underwater archaeology. Based on this brief introduction, it appears clear that LIBS can help address several issues in heritage science and has the potential to find its way in heritage science laboratories, in particular when requirements for complete nondestructivity are not absolute.

This chapter starts with an overview about the principles of laser-induced plasmas, LIBS as an analytical technique and instrumentation (Section 2), and goes on by reviewing the relevant literature, with a focus on the three most representative applications of LIBS in heritage science: depth-profiles (Section 3); in situ and standoff measurements (Section 4); underwater LIBS (Section 5). Finally, Section 6 briefly reviews relevant papers that do not fall under any of these categories. The chapter covers publications from the last decade, while less recent literature was previously reviewed by this author [6] and others [7–9]. Several recent reviews on this and related topics have been published, and interested readers may refer to them for additional points of view [10–13].

2 **Principles of LIBS**

2.1 **Laser-induced plasmas**

Laser-Induced Plasmas (LIP) can be formed by focusing a laser pulse of appropriate irradiance on a portion of matter, thus vaporizing, atomizing, and ionizing the material at the irradiated spot. (Irradiance is defined as the power deposited per unit area, and measured in W/cm^2.) The laser-induced breakdown, i.e., the plasma formation, is a threshold phenomenon, and the irradiance threshold value depends on properties of the laser (such as pulse amplitude, wavelength) and the irradiated material (such as state of aggregation; optical, thermal, and electrical properties; porosity and surface morphology). These properties also determine the actual physical processes involved in the breakdown process. As a general rule, the breakdown is dominated by thermal phenomena (melting, vaporization, boiling) for long pulses, long wavelengths, and metallic targets, while it is mostly electrostatic (Coulomb explosion), with limited thermal contribution, in the case of short pulses, short wavelength, and dielectric targets. In all cases, the formed plasma is not stationary, but expands at supersonic speed, having overall persistence time in the order of a few to tens of microseconds (in typical work conditions), after which it extinguishes [14]. Excited ions and atoms in the LIP can de-excite by emitting radiation, which forms the LIBS signal and exhibits time-dependent features, as shown in Fig. 2A. The temporal evolution of LIBS spectra shown in Fig. 2 is related to the elementary processes in LIPs, that will

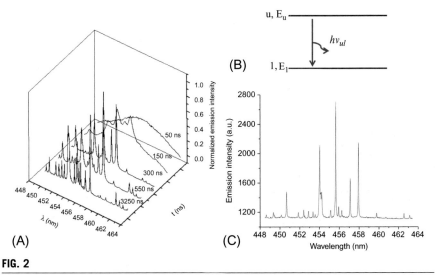

(A)   (B)   (C)

FIG. 2

(A) Time evolution of nanosecond laser-induced plasma of pure metallic Ti in air at atmospheric pressure as background environment; (B) schematic representation of de-excitation from an upper energy level "u" to a lower energy level "l" with emission of one photon with energy $E = h\nu_{ul}$; (C) LIBS spectrum of Ti at delay 3250 ns after the laser pulse.

be qualitatively overviewed here (interested readers may refer to [1–4] for a detailed description).

In their early stages, LIPs have high electron density ($\sim 10^{19}\,cm^{-3}$) and high temperature ($\sim 10^4\,K$), which progressively decrease with time due to expansion and recombination phenomena. Collisions between free electrons and heavy species (atoms and ions) are responsible for establishing excitation/de-excitation and ionization/recombination equilibria. In LIPs, electron collisions are the main route for these processes, but radiative routes are also possible. Spontaneous emission is the de-excitation phenomenon responsible for the existence of LIBS signals: in this process, excited atoms (or ions) relax to a lower energy state by emitting photons of specific energy, which appear as a peak at a given wavelength in LIBS spectra (Fig. 2B and C). In radiative recombination, instead, excess energy resulting from ion-electron recombination is released as photons with nonquantized energy, which appear as continuum radiation in LIBS spectra. Continuum dominates LIBS spectra at the beginning of the plasma evolution, due to the high initial electron density, and gradually decreases with time, while emission lines arise, due to radiative de-excitation of ions and atoms (Fig. 2A).

Tens of microseconds after the plasma formation (depending on experimental conditions), LIPs are virtually extinguished and no emission lines are visible (not shown in Fig. 2). In late stages of plasma evolution, molecular signals can often be acquired, usually due to chemical reactions within the plasma (e.g., formation of C_2 fragments during ablation of organic materials) or between the plasma species and the background environment (e.g., formation of CN fragments when ablating organic materials in air), as shown in Fig. 3.

Molecular analysis is a relatively small niche in the LIBS community, due to the nature and high temperatures of LIPs. In the breakdown event, chemical bonds in the irradiated samples are broken and the ablated portion is atomized and ionized.

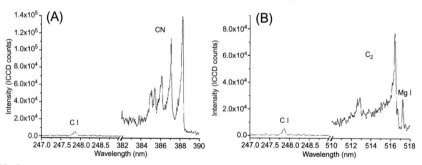

FIG. 3

LIBS spectra of *E. coli* bacterium showing 247.86 nm carbon atomic line and (A) CN molecular band around 387 nm, (B) C_2 molecular band around 516 nm.

Credit: Adapted from M. Baudelet, L. Guyon, J. Yu, J.-P. Wolf, T. Amodeo, E. Fréjafon, P. Laloi, Femtosecond time-resolved laser-induced breakdown spectroscopy for detection and identification of bacteria: a comparison to the nanosecond regime, J. Appl. Phys. 99 (2006) 084701, https://doi.org/10.1063/1.2187107.

Nonetheless, several authors have investigated the relation of atomic signals to the molecular structure of the sample, and the evolution of molecular signals ([15] and references therein). In its standard version, though, LIBS is mostly considered an atomic spectroscopy technique, as far as analytical applications are concerned. In particular, though early and late stages are useful for fundamental studies [16–19], analytical LIBS is usually conducted in the temporal window where emission lines of atoms and ions are well resolved and not affected by continuum (e.g., from 550 ns on in Fig. 2A). In the absence of spectral interference, each emission peak corresponds to a single transition, which can be identified and assigned to a given ion or atom with atomic spectra databases (e.g., [20,21]).

In addition, LIBS can be used for quantitative analysis, since the emission intensity of a peak corresponding to a transition from an upper energy level "u" to a lower energy level "l" is proportional to the number of emitters undergoing the transition:

$$I_{ul} = \frac{1}{4\pi} G N_u A_{ul} h\nu_{ul} \tag{1}$$

Here, I_{ul} is the peak intensity; N_u is the number density of emitters in the upper state; A_{ul} is the Einstein coefficient of spontaneous emission; h is Planck's constant; ν_{ul} is the frequency of the emitted radiation; and G is an experimental factor related to the employed setup and configuration that can be determined using radiometric calibration sources. This is not usually necessary, and relative intensities expressed in "arbitrary units" are commonly used instead.

Using LIBS for quantitative analysis requires some important conditions to be met, i.e., (1) stoichiometric ablation, (2) optically thin plasma, and (3) local thermodynamic equilibrium (LTE). Stoichiometric ablation means that the plasma has the same elemental composition of the irradiated target, that is, that all the species contained in the target are carried into the plasma phase with the same efficiency. For multielement targets, evidence of preferential ablation (or fractionation) has been reported, but it is generally agreed upon that this risk is negligible when operating at laser irradiance well above the breakdown threshold, which is the usual working conditions in LIBS [22]. The condition of an optically thin plasma may be more challenging to meet. In optically thin plasmas, all the emitted photons are able to escape the plasma itself and reach the detector, so to contribute to the acquired signal I_{ul}. LIPs often suffer from self-absorption, that is, the emitted radiation can be partially reabsorbed in the plasma, which alters the proportionality between I_{ul} and N_u and makes the self-absorbed transition unsuitable for quantitative analysis. In cases of strong self-absorption, the line profiles can appear severely affected, and show a typical self-inversion "dip," but in most cases self-absorption is not so obvious, and specifically designed experimental or computational approaches [23–26] may be needed to evaluate and correct it. A careful selection of transitions is anyway always suggested to minimize errors due to self-absorption, especially if the element under investigation has high concentration. In particular, it is a good practice to avoid transitions involving the ground state and low-lying energy levels, as well as transitions with high relative intensity, as they are the most affected by self-absorption [27].

Finally, the use of LIBS as an analytical technique relies on LIPs being in LTE. As mentioned, LIPs are expanding physical systems that emit radiation throughout their finite persistence time. In LTE, energy losses through emission of radiation are negligible with respect to the energy exchanged in the equilibria involving material particles, which are instantaneously and locally established despite the transient nature of the LIPs. In this condition, the number density of emitters in excited levels is given by the Boltzmann distribution:

$$N_u = \frac{g_u N_0}{Z} e^{-\frac{E_u}{kT}} \tag{2}$$

Here, g_u is the degeneracy of the excited level with energy E_u, N_0 is the total number density of the species, Z is the partition function, k is the Boltzmann's constant, and T is the plasma temperature.

By acquiring the experimental emission intensities I_{ul} and combining Eqs. (1), (2), it is possible to determine the plasma temperature, using the so-called Boltzmann plots [28]. Temperature and electron density (N_e) are important parameters for plasma characterization. The latter can be determined from the spectral broadening of LIBS peaks. Several line broadening phenomena cause emission peaks to have a spectral profile instead of a single sharp wavelength corresponding to the emission of a photon with energy equal to the jump between levels "u" and "l". In LIBS plasmas, the main contribution to line broadening is the Stark effect, which is due to interactions between emitters and charged particles (in particular electrons). The Stark effect causes both the peak to be broadened and its central wavelength to be shifted, in an extent that is proportional to N_e. For this reason, the Stark effect can be used to determine N_e, by comparing the experimental values of line broadening (and shift) with database and literature values of the same peaks in known conditions of temperature and electron density [29].

2.2 LIBS as an analytical technique

Eq. (1) shows that, in LTE, the emission intensity of LIBS peaks is proportional to the number density of emitters in the upper level "u," and, through the Boltzmann distribution (Eq. 2) to the total number density of the given species, N_0, i.e., to its concentration. Therefore if the conditions of stoichiometric ablation and optically thin plasma are met, LIBS can be used for quantitative analysis.

Like with other analytical techniques, quantitative measurements in LIBS require calibration to determine the numerical relationship between the emission intensity of a given species and its concentration. This is done by drawing calibration lines using samples with known concentration of the analytes of interest. An important point is that the matrix of the calibration samples must match that of the unknown ones, because of matrix effects. These are due to the fact that physical–chemical properties of materials affect the way they couple with the laser, which, in turn, can cause species with similar concentration but contained in different matrices to have very different emission intensity. Moreover, to account for experimental fluctuations such as laser energy instability and sample surface inhomogeneity, it is customary to normalize the emission intensity when drawing calibration lines [30]. A radically different approach to quantitative analysis has been proposed for LIBS, the Calibration-Free (CF) method, which does not require matrix-matched reference samples and calibrations.

CF methods strictly rely on LTE and require determining the plasma parameters (electron density and plasma temperature) and the emission intensity of all the species in the target. Using these data, it is possible to find the total number density of each species, and then their percentage with the following normalization procedure (often referred to as "closure equation") [31]:

$$wt\%_i = \frac{N_{0,i}M_i}{\sum_i N_{0,i}M_i} \tag{3}$$

Here, $wt\%_i$ is the mass percentage of the ith species; $N_{0,i}$ is its total number density; and M_i its molar mass.

CF methods can be very useful, or even the only possible choice, when acquiring or preparing matrix-matched standards is unfeasible, and several variants of the original approach have appeared in the literature [32,33].

Finally, in the 2000s a trend for the use of chemometric methods emerged, which has gained considerable momentum in the last decade. Being a fast and rather straightforward technique, LIBS lends itself to the acquisition of many spectra for large sets of samples, each spectrum containing numerous transitions. Such abundance of spectroscopic information can make manual processing for quantitative or qualitative analysis unfeasible or impractical, especially when information is needed in real time and in situ. Multivariate statistical methods can therefore be used for tasks such as classification, prediction/regression, clustering, and data reduction [34,35].

2.3 Instrumentation

A typical LIBS setup, schematically shown in Fig. 4, usually comprises: one or more laser sources (the most common being nanosecond Nd:YAG lasers, which have fundamental harmonic at 1064 nm and higher harmonics at 532, 355, 266, and 213 nm); one spectrometer to separate radiation in its components; and a detector interfaced with a computer for spectra acquisition and visualization. In Fig. 4, the plasma is generated in an interaction chamber, which enables measurements in vacuum and in background gases different from air. Operating in noble gas atmosphere (typically He or Ar) or in reduced pressure can increase the plasma persistence time and emission intensity [36,37], but LIBS can be, and in fact usually is, carried out in air at atmospheric pressure. The target motion control reported in Fig. 4 is used to offer the incoming laser beam a fresh surface to ablate, instead of digging the sample at the same location during the spectra acquisition. Translating the sample with respect to the laser beam can also be used for spatially resolved analysis, for example

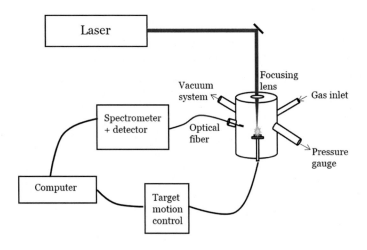

FIG. 4

Block diagram of typical LIBS setup.

in the case of inhomogeneous samples (e.g., inclusions in rocks [38]) or for chemical mapping [39].

Due to the transient nature of LIPs, plasma generation and radiation acquisition must be synchronized, and an appropriate selection of time parameters is important. As schematically shown in Fig. 5, two time parameters have to be controlled in a classical LIBS experiment, i.e., the delay after the end of the laser pulse, at which the acquisition starts (t_d, delay time), and the duration of the acquisition itself (t_g, gate width or integration time).

Broadly speaking, two temporal acquisition modes are possible in LIBS: time-integrated and time-resolved. In the first, the gate width is comparable to the plasma

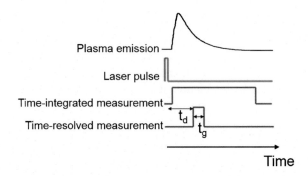

FIG. 5

Schematic representation of the temporal profile of LIP generation and evolution and of the time parameters in LIBS measurements: t_d, delay time after the laser pulse, and t_g, gate width or integration time.

persistence time, so that all the emitted radiation emitted throughout the LIP evolution is acquired, starting from the set initial delay time; in the second, a short gate width is used, which captures only a portion of the plasma emission profile. Nongated detectors, which have integration time in the order of milliseconds (thus much longer than the plasma persistence time) have become increasingly popular, also thanks to their affordability and reduced dimensions. The "blocks" of LIBS setups can be compacted into mobile systems, as several groups have demonstrated with in-house prototypes (see Section 4). Nowadays, commercial portable instruments are also available, including man-portable and handheld instruments. The main issue for these is that the trade-off of miniaturization is a lower instrumental performance, especially in terms of laser energy and spectral resolution. LIBS with portable systems, both research prototypes and commercial instruments, was reviewed and critically evaluated in [40,41].

3 Depth profiles

Fig. 6 illustrates the simple principle of depth profile analysis by LIBS. Since every laser shot ablates a minute portion of the sample, focusing more shots at a given location will locally remove subsequent layers, and the plasma emission will reflect changes in the elemental composition.

The depth of analysis usually ranges between few microns to some hundreds of microns, and the crater dimensions depend on the number and energy of laser shots, and, consequently, on the ablation rate (Fig. 7).

With appropriate experimental precautions, such as tightly focusing the laser beam, using as low laser energy as possible and accurately choosing the laser wavelength so to reduce the thermal damage, it is possible to produce an almost invisible or negligible crater, compatible with curatorial needs. A special application of LIBS depth profiles is the online monitoring of laser cleaning procedures. Spectra of the locally removed layers can be acquired during the cleaning and therefore offer a

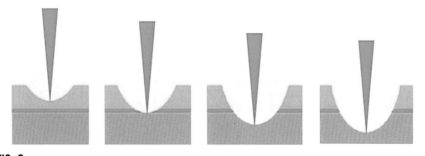

FIG. 6

Schematics of LIBS depth-profile analysis. The *green* wedge exemplifies an incoming laser beam focused at the surface of the multilayered sample.

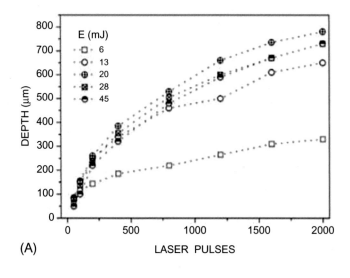

(A)

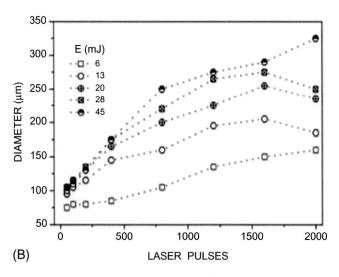

(B)

FIG. 7

Depth (A) and diameter (B) of laser-induced craters during depth profiles of Cu-based alloys at different laser energies and as functions of the number of laser shots focused on the sample. Laser source: 8 ns Nd:YAG laser, 1064 nm.

Credit: J. Agresti, I. Cacciari, S. Siano, Optimized laser ablation elemental depth profiling of bronzes using 3D microscopy. in: 18th Italian National Conference on Photonic Technologies (Fotonica 2016), Rome, Italy, 6–8 June, 2016, pp. 1–4. https://doi.org/10.1049/cp.2016.0958.

simple way to evaluate the effectiveness of the procedure, prevent damaging the object by overcleaning and assess the surface conditions after the treatment. Moreover, with portable instruments, conservation treatments and their assessment can be performed at the same time, on site and without sampling, which represents a considerable advantage for built heritage and immovable objects, or during field campaigns [42,43].

The possibility of probing hundreds of microns without invasive procedures is an important advantage of LIBS, in particular for heavily corroded or weathered objects. On the other hand, the formation of deep craters can be accompanied by phenomena such as melting and re-deposition of the material inside the crater, changes in focusing conditions, and plasma confinement in the cavity. These, in turn, can affect plasma evolution and LIBS spectra, in particular with thermal ablation, i.e., with long pulses (nanoseconds and longer) and long wavelength (IR, VIS) [22,44–50].

Despite these possible issues, the applications of LIBS depth-resolved analysis in heritage science are numerous, and include assessment of conservation state (e.g., corrosion, weathering, pollution); authentication and provenance studies; and monitoring of conservation treatments. An overview of these applications is given in the following.

3.1 Depth profiles of cultural heritage materials

One of the most active groups in the field, both for fundamental investigations and studies about various heritage materials, is that of Siano and coworkers, from Florence (Italy). In [51,52], they combined portable LIBS and 3D optical microscopy for characterization of laser-induced craters and depth calibration. In [52], they validated their combined approach with three historical materials (i.e., marble baluster with a black crust; paleo-Christian wall painting with a calcareous concretion; corroded Roman coin), while in [51] they conducted a parametric study to optimize deep depth profiles of bronzes. In [53], they used LIBS and terahertz (THz) spectroscopy to assess the potential of the latter for characterization of mineralized layers on archeological bronze surfaces. The authors concluded that, due to the high surface roughness of heavily corroded artifacts, THz spectroscopy was unsuitable for precise microstratigraphy, but nonetheless useful to estimate the presence of a metallic bulk underneath the corrosion layer. The same group also proposed elemental depth profiles of ancient earthenware artifacts [54], archeological copper-based artifacts [55], coins [56], and bronze Egyptian figurines [57,58] for authentication studies, since ancient manufacturing techniques and burial conditions are associated to phenomena of surface enrichment/depletion of some elements (e.g., Sn enrichment in bronze patinas [59]), which can be used to discriminate ancient objects from modern counterfeits.

Several groups conducted LIBS depth-resolved analysis of metallic objects. Osorio et al. [60] analyzed a Japanese jug of unknown matrix from the mid-20th century, and established that it was made of Pb and covered by a thin layer of Cu (estimated thickness: 27 μm). An investigation of 19th-century Spanish silver canopies was carried out in [61], which assessed the conservation state of the samples (some pieces showed evidence of previous restoration and others of corrosion layers that required conservation interventions) and revealed that silver was a thin layer deposited over a Cu-Zn alloy with an electrodeposition technique (Ruolz method). Orlić Bachler et al. [62] studied Roman bronze coins coated in silver (*follis*). These authors employed two chemometric methods to classify the samples and evaluate their similarities throughout the depth profile analysis, as well as to correlate changes in the elemental composition with historical events influencing the amount of silver used for the coinage. Alberghina et al. coupled XRF and LIBS (with a portable instrument, but in the laboratory) for stratigraphic analysis of various materials. These included [63] artificially corroded mock bronze samples, buried in the archeological site of Tharros (Sardinia, Italy) to imitate the corrosion process of ancient Roman alloys; and fragments of mural paintings detached from historical buildings in Sicily (Italy) [64]. Being intrinsically multilayered, paintings are among the most common targets in depth profile studies, often to characterize the number and thickness of layers and the employed pigments. Such as are the works by: Caneve et al. [65] who determined elemental intensity ratios (Ca, Mg and C, over Fe) in a wall painting fragment from Pompeii; Palladini and coworkers, who carried out multitechnique investigations of mural painting fragments from the Sabina area, near Rome (Italy), in the context of a study about the dynamics of Roman settlement in the area [66,67]; Borba et al. [68], who used principal component analysis (PCA) to discriminate the external layers, due to impurities and weathering processes, from the painted layer and underlying rock in two Brazilian prehistoric paintings; Lofrumento et al. [69], who showed that black pigments found in prehistoric rock paintings from the Hararghe region (Ethiopia) did not contain any iron or manganese, and confirmed the differences between pictorial techniques used in Ethiopia (only carbonaceous materials for black) and in Eritrea (manganese hydro-oxides). Kaszweska et al. [70] developed a novel approach for UV-LIBS/Optical Coherence Tomography (OCT) integrated in a hybrid mobile instrument for pigment analysis and online monitoring of laser-induced crater depths on a modern painting of unknown dating. Other studies used mock samples, either to improve the depth resolution by characterizing the effect of different laser parameters [71,72], or to develop signal normalization methods for in situ depth profile classification of painted samples using portable LIBS instrumentation [73] (see Section 4). Depth profiles of ancient bones and teeth can provide valuable anthropological information, such as dietary habits and mobility, since these calcified tissues are very stable and can retain their trace element content over long periods of time. For this purpose, Alvira et al. monitored the Mg/Ca and Sr/Ca ratios in dentin and enamel of Neolithic, middle age, and modern *Homo sapiens* teeth, both with nanosecond [74] and femtosecond [75] pulses. Kasem et al. used Sr/Ca depth profiles to estimate diagenetic effects responsible from alterations of the burial environment [76] and infer dietary habits of ancient Egyptians of different dynasties in relation to modern Egyptians. In [77], they compared three different laser wavelengths, and observed that UV provided the best depth resolution and signal stability. Rusak et al. [78] observed that depth intensity ratios of archeological sheep and cattle bones (in particular, Ca/F) correlated well with the preservation state, previously assessed by independent measurements, and could thus be used to estimate it. Other LIBS depth profile studies, often carried out with the support of completely nondestructive techniques, include investigations about: an archeological Egyptian cartonnage (panels used to cover mummified bodies) [79]; classification of ceramics and pottery (Islamic and Byzantine ceramics from East Plain Cilicia (Turkey) [80]; Egyptian pottery shards and ceramic glaze [81,82]); mock wall painting samples, prepared with pigments and binding media used in the Paleolithic and analyzed in ambient conditions simulating those of prehistoric caves [83]; authentication of archeological Ushabtis (Egyptian funerary figurines) from modern fakes [84]; characterization of pollution encrustations from the Santa Maria La Blanca church in Seville (Spain) [85]. Most studies discussed so far used trends with depth of intensities of given elements and/or their ratios as the diagnostic criterion to identify the layers and estimate their thickness, without determining the elemental composition. Different approaches to depth-resolved quantitative analysis, though, have been proposed by several groups. In [49] Agresti et al. developed a method for depth-dependent calibration in deep depth profiles (several hundreds of microns), based on calibration surfaces taking into account the phenomena occurring during laser ablation at increasing depth. Lazic et al. [86] used a correction procedure for the variation of plasma parameters during depth profiles in standoff mode (see Section 4). Gaudiuso et al. proposed a variant of the calibration-free method (inverse method [87,88]) and validated it with a set of copper-based archeological objects [89] and with one Sasanian silver coin covered with an ancient Hg layer [90]. Finally, some authors used LIBS to monitor the penetration depth, and consequently the effectiveness, of cleaning or antifouling treatments of cultural heritage materials. Carmona-Quiroga et al. [91] estimated the penetration depth of two antigraffiti coatings (a fluoroalkyl siloxane and a Zr propoxide) in several construction materials and showed that LIBS could be used with the F-containing product, but not for the Zr-containing one, since no Zr transition was detectable. The suitability of silver and titanium oxide nanoparticles as biocides to prevent the biodegradation of stone monuments was instead tested in [92,93], using limestone samples from three Spanish quarries.

3.2 Monitoring of laser cleaning

Laser cleaning is a well-established technique for removing pollution, impurities, bio-growths, old conservation treatments, and overall extraneous or unwanted layers from cultural heritage objects [94]. Being able to provide online feedback about the composition of removed layers, LIBS appears as an obvious choice for monitoring of laser cleaning procedures. In particular, it can help to identify the causes of degradation (and plan consequent restoration interventions) and to self-regulate the process so to preserve the object. The years 1990s and 2000s have seen a considerable growth of laser cleaning/LIBS studies [6,42], which instead have been comparatively less investigated in the last decade. The more recent works aim at two goals, in particular: optimizing the procedure, in terms of laser wavelength and pulse amplitude, to improve depth resolution and selectivity of the cleaning; and testing portable and handheld LIBS instrumentation for in situ laser cleaning.

Kono et al. [95] used ultrashort laser pulses (170 fs) to selectively remove corroded layers from the gold thread in a very delicate material, cloth of gold, without affecting the silk thread (see Fig. 8).

The authors monitored the cleaning process with LIBS and used the emission intensity ratio of Ag (from the gold thread) to C (from the dirt) as a parameter to stop the cleaning. Moreover, they demonstrated through SEM images that laser cleaning with a ns laser, though effective, caused partial melting of the gold thread, due to the thermal component of laser ablation with long pulses. Nanosecond lasers, though, are currently still more user-friendly, rugged and cheap than their ultrashort counterpart, and therefore more easily adaptable to in situ applications. Despite ultrashort LIBS studies in heritage science are still relatively few [75,95], the need to reduce the thermal stress of materials is strongly felt, especially for depth-resolved analyses. To meet this need, the use of UV lasers has been gradually gaining momentum [70–72,77,96–98], though the first and second harmonic of Nd:YAG lasers remain the most common LIBS laser sources, both in with benchtop and portable

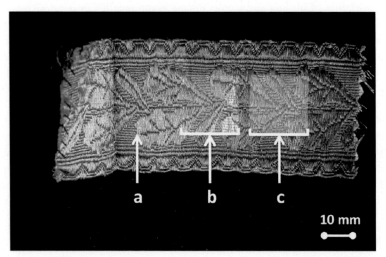

FIG. 8

Gold braid, showing different levels of laser treatment. Due to the storage conditions, the *left side* of the object retained its original color and was used as control (area "a"). The *right side*, instead, was exposed to dirt and tarnish, and therefore needed cleaning. The areas marked as "b" and "c" were treated with the laser, respectively at the optimal laser irradiance, which cleaned the sample without damaging it, and at a deliberately high laser irradiance, to show the discoloration effects.

Credit: M. Kono, K.G.H. Baldwin, A. Wain, A.V. Rode, Treating the untreatable in art and heritage materials: ultrafast, laser cleaning of "cloth-of-gold", Langmuir 31 (4) (2015) 1596–1604.

instruments, for prospective in situ studies [43]. Roberts et al. demonstrated that the LIBS signals of phosphorus acquired during laser removal of stone encrustations from fossil bones could be efficiently used to discriminate between the two materials during excavation campaigns [99]. Senesi et al. analyzed black weathering encrustations on a limestone fragment, sampled from the Castello Svevo in Bari (Italy), with benchtop [100,101], mobile [102] and handheld [103] instruments. In [102] they also associated surface spatial analysis, generating 3D maps and "virtual thin sections" based on changes in the emission intensity of the main elements, in depth and throughout the surface. Instead of acquiring LIBS spectra during laser cleaning, Khedr et al. [104] proposed to use optical imaging of the LIP, i.e., to not resolve the radiation in its spectral components. The authors observed that, while cleaning marble from dark encrustations, the emission intensity gradually decreased, which they attributed to the different optical properties of the dirt layer and the underlying white marble, and to the lower absorptivity of the latter.

4 **Onsite, remote, and standoff LIBS**

The possibility of micro-destructive field analyses is doubtless one of the most attractive features of LIBS in heritage science. Transferring objects of historical and cultural interest to scientific laboratories for chemical analysis is not always feasible. Some objects are intrinsically immovable, such as built heritage, or require special arrangements for the transfer and for storage in their new location, such as wall paintings or historical shipwrecks. For others, transfer to different locations may not be a suitable option due to issues such as fragility, conservation state, museum policies, sociopolitical implications, storage needs. In such cases, the only possibility to carry out accurate physical-chemical characterization of cultural heritage objects is performing onsite investigations with portable instrumentation. Some techniques are available for this purpose, such as XRF and Raman spectroscopy, which have the intrinsic advantage of being entirely nondestructive. Nonetheless, these techniques have some drawbacks that can make the provided data insufficient for an accurate chemical analysis. Some known caveats are: limited sensitivity to the analysis of light elements (XRF), impossibility of depth-resolved analysis (Raman, unless using a nonportable confocal microscope; XRF, beyond the outermost layers), strong interference from fluorescence signals, that can prevent unambiguous identification or, in some cases, even the detection, of spectral fingerprints (Raman).

To overcome these issues, LIBS can in fact be a very efficient choice, in particular for deep depth profiles and for the simultaneous detection of heavy and light elements. On the other hand, LIBS is not typically the technique of choice for molecular analysis, and, though only micro-destructive, is as such a less desirable option for cultural heritage objects. For these reasons, LIBS, XRF, and Raman spectroscopy should rather be considered complementary than mutually exclusive, and several studies have exploited this complementarity to obtain a comprehensive characterization of the investigated objects. Of particular interest is the possibility of combining various laser spectroscopy in the same portable instrument. A network of European groups is currently working to develop a prototype coupling Raman, LIBS, and Laser-Induced Fluorescence (LIF) in a single apparatus [105] and to address the many instrumental challenges related to the choice of laser sources, collection optics, and experimental configurations ensuring the best compromise for the three techniques [106,107].

The LIBS analysis of distant objects, usually several meters from the operator and the experimental apparatus, may be regarded to as a special in situ case, since it requires not only field-deployable instrumentation, but also optical elements for long-distance delivery of the laser radiation and collection of the emitted radiation. Two configurations are possible for LIBS of distant objects: remote and standoff, schematically shown in Fig. 9. In remote measurements, an optical fiber is used to deliver the laser beam to the target and the emitted plasma radiation to the spectroscopic system (Fig. 9A). In standoff measurements (Fig. 9C), instead, laser beam and emitted radiation are transmitted in open path. Standoff LIBS has stringent requirements in terms of high quality and energy of laser beam, and necessitates specific optical elements, such as telescopes, to make up for the attenuation of radiation traveling through the atmosphere for several meters.

Moreover, further effects must be considered in open path, such as atmospheric conditions that can affect the index of refraction of the air traveled by the light, as well as the presence of particulates and aerosols along the optical path [40]. As for compact portable systems, as mentioned in Section 2.3, the main issue is their usually lower performance with respect to benchtop instruments, which can negatively affect the sensitivity of measurements.

Considering advantages and challenges of in situ LIBS, and the fact the commercial availability of man-portable systems is relatively recent, it is not surprising that, to date, papers about actual field measurements are still few (Section 4.2), while laboratory studies aiming to validate portable instrumentation, as well as standoff configurations, are more numerous (Section 4.1). Technological progress and more widespread availability of affordable portable instruments on one hand, and scientific advancements on the other hand, are likely going to further propel the field in this direction.

4.1 Laboratory feasibility studies

Works about standoff LIBS of heritage materials are rather few. A group from Brno (Czech Republic) analyzed samples of archeological interest at distances in the range 5–15 m. In [109,110], they studied mineralized biological tissues with LIBS and LA-ICP-MS, while in [108,111], they developed supervised chemometric models for standoff and in situ classification of archeological and paleontological samples.

Lazic et al. performed depth profile analysis of multilayered heritage materials [86,112] at standoff distances ∼10 m. In [86], they applied a previously proposed [113] correction procedure based on the total plasma emission and ionic to atomic line intensity ratios from selected elements, aimed at reducing shot-to-shot intensity

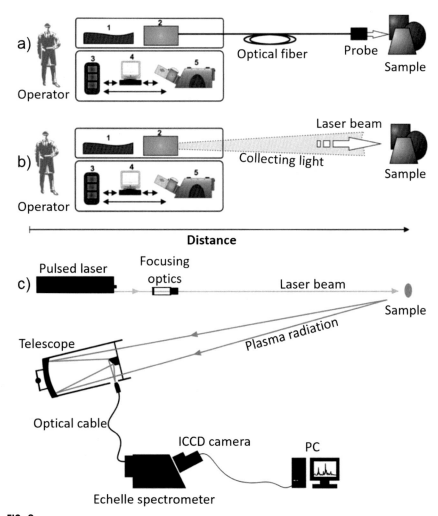

FIG. 9

Schematic diagram of remote (A) and standoff (B) LIBS setup: (1) laser, (2) optical module, (3) laser power supply, (4) personal computer, and (5) detector and spectrograph [5]. (C) Example of detailed schematics of open path stand-off LIBS [108].

Credit: (A) and (B) F.J. Fortes, J. Moros, P. Lucena, L.M. Cabalín, J. J. Laserna, Laser-induced breakdown spectroscopy, Anal. Chem. 85 (2013) 640–669. (C) G. Vítková, K. Novotný, L. Prokeš, A. Hrdlička, J. Kaiser, J. Novotný, R. Malina, D. Prochazka, Fast identification of biominerals by means of stand-off laser-induced breakdown spectroscopy using linear discriminant analysis and artificial neural networks, Spectrochim. Acta Part B At. Spectrosc. 73 (2012) 1–6.

fluctuations and account for variations of plasma parameters caused by the formation of craters (Section 3.1). In [112], they compared standoff LIBS (also in depth-profile mode) with XRF and Particle-Induced X-ray Emission, PIXE, and elucidated their pros and cons for specific issues posed by the analyzed materials.

Laboratory studies with portable instrumentation are more numerous, and, like the standoff ones, they are often multianalytical investigations aiming to highlight advantages, drawbacks and complementarity of LIBS and other techniques.

Concretions found in prehistoric caves in the Vézère valley (France) were characterized in [114,115] with portable and benchtop LIBS. With the latter, spatially-resolved elemental maps were obtained (Fig. 10), that were useful to differentiate the concretions and clarify their growth processes.

The sensitivity of LIBS for light element analysis was exploited in [116,117]. In the first cited work, the authors analyzed low temperature glazes of the Qing Dynasty [116] focusing on boron, which was used by craftsmen to reduce the glaze melting temperature and control the thermal expansion coefficient. Three kinds of enamel were showed to contain boron, which was instead absent in a fourth kind, historically considered more difficult to manufacture and therefore more precious. The second study [117] used benchtop and portable LIBS to investigate hydrated compounds in marble samples. Due to overlapping spectral fingerprints, the Raman spectra of these samples were difficult to interpret, which justified the LIBS investigation to quantify the water content by determining the amount of H with a series of standard samples. In [118], researchers from the Heraklion (Greece) group investigated wall painting pigments used in Crete (Greece) from the Bronze Age to the Roman and Byzantine periods for a dual purpose investigation: studying the continuity of painting materials and techniques for wall paintings in Crete; and testing the onsite feasibility of combined LIBS/Raman. The feasibility study was followed by field measurements [119] (Section 4.2), and highlighted a well-known limit of Raman spectroscopy, i.e., the interference from fluorescence. They tested a different technique to overcome this issue for pigment analysis in [120], by developing a hybrid LIBS/diffuse reflectance spectroscopy mobile instrument. Following a previous work of theirs [121], where they combined LIBS and reflectance spectroscopy to identify pigments in illuminated manuscripts, the authors successfully tested their portable hybrid instrument with pigments in powder form and with painted Minoan pottery fragments.

The Applied Laser Spectroscopy group in Pisa (Italy) contributed to develop and commercialize a mobile DP-LIBS instrument [122], which they and other groups employed for several studies. These include: dating a black ceramic vase of uncertain origin that, upon multitechnique analysis, proved much older (6th century BCE) than its attributed age (CE 1st century) [123]; characterizing organic and inorganic fractions in reference madder lakes containing different metal cations [124]; supporting thermogravimetric and pyrolytic studies about the suitability of Aquazol (i.e., poly (2-ethyl-2-oxazoline)) as a binder for reversible retouching paints [125]; a provenance study of raw materials used in limestone Nuragic statues from Sardinia (Italy) [126]; spatially resolved analysis of Roman [127] and Norman mortars [128,129] with chemometric and CF methods. For the highly heterogeneous samples used in [127,128], the authors acquired elemental maps with high spatial resolution and validated a computational method for image analysis (SOM, Self-Organizing Maps), which provided supporting information to petrographic and mineralogical investigations. An unusual application was described in [130], where the authors used

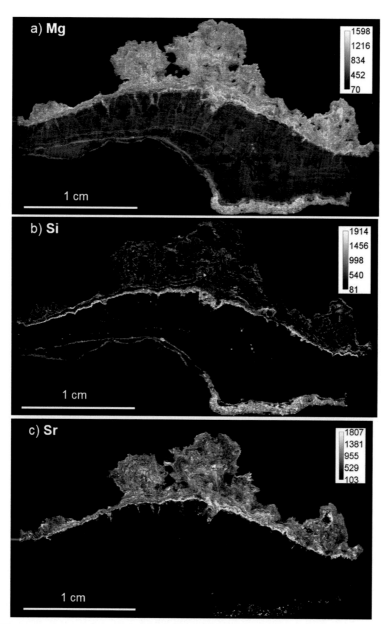

FIG. 10

LIBS elemental images of the two concretions (coralloid) analyzed in [115] for three elements: (A) Mg (using the transition at 285.21 nm), (B) Si (288.16 nm), (C) Sr (407.77 nm).

Credit: L. Bassel, V. Motto-Ros, F. Trichard, F. Pelascini, F. Ammari, R. Chapoulie, C. Ferrier, D. Lacanette, B. Bousquet, Laser-induced breakdown spectroscopy for elemental characterization of calcitic alterations on cave walls, Environ. Sci. Pollut. Res. 24 (2017) 2197–2204. https://doi.org/10.1007/s11356-016-7468-5.

surface-enhanced LIBS (SENLIBS), to detect trace elements in inorganic dyeing mordant in textiles. In SENLIBS, micro-drops of liquid samples are deposited and dried on nonporous solid substrates prior to laser irradiation [131]. This has proved very effective to overcome the inherent issues of LIBS analysis of liquid samples (such as splashing, generation of aerosols, and energy losses due to interaction of laser and plasma with the liquid [14]). In [130], they extracted dyes from reference and historic dyed textiles, and analyzed their inorganic fraction (mordant) with SEN-LIBS. The same mobile instrument was used by Agrosì and coworkers to discriminate synthetic emeralds, based on trace elements content (Cr, V, Fe) [132] and to distinguish different varieties of red beryl, using Cs and Li signals [133]. In [133], the authors minimized the damage induced on gemstones by focusing the laser through a microscope objective.

Several chemometric studies with portable instrumentation have been prompted by the need to obtain real-time data in situations such as archeological excavations or classification of museum collections, which would represent a great benefit for cultural heritage professionals and a major asset of LIBS.

Duchêne et al. [134] analyzed model painting specimens with benchtop LIBS, and historical and contemporary wall paintings with portable LIBS. They compared two approaches to distinguish pigments and painting techniques, i.e., direct investigation of spectral intensities and chemometric models. The chemometric investigation highlighted the main issues affecting the robustness of models and their in situ applicability. In situ investigations, being performed in less controlled experimental conditions than in the laboratory, can suffer from significant shot-to-shot laser fluence fluctuations, due, for example, to chemical or textural inhomogeneity of samples, commonly found in heritage materials because of surface alterations (corrosion, patinas, or actual layers such as in painted samples) or irregular shape of the analyzed object.

A numerical strategy to deal with such fluctuations, in particular during stratigraphic analysis of painted and metallic samples, was proposed by Syvilay et al. [73,135]. In these studies, the authors used a standard normal variate (SNV) method for LIBS and Raman spectra normalization and demonstrated its potential to account for laser fluence fluctuations [73] and improve the identification of organic and inorganic layers and interfaces [136], at the same time warning against the risk of potentially losing quantitative information due to heavy normalization.

The same authors also carried out a provenance study of obsidian samples from four Western Mediterranean islands [135], to investigate migration routes and interactions of early humans that used obsidian tools. To this end, the authors built a QDA (quadratic discriminant analysis) model, which discriminated the four source-islands, though with a relatively high percentage of misclassified samples (6.8%). This showed that the relatively low sensitivity of portable instruments can in fact degrade the classification accuracy if highly discriminating elements are below the limit of detection (LOD). Pagnotta et al. [137] built a neural network (NN) model to distinguish similar objects, otherwise difficult to discriminate. The authors trained the NN with modern brass standards and successfully tested it with historical specimens, i.e., two almost identical fragments of a medieval brass statue. In [138], instead of a single NN, the authors used an ensemble of NNs to predict the composition of unknown samples after training the models with known standards and averaging the results obtained by all NNs.

Grégoire et al. [139] suggested using molecular signals in LIBS spectra for in situ identification of polymers used for restoration of wall paintings, both by directly comparing the emission intensities and with chemometric methods. They demonstrated that the observed molecular signals depended on the molecular composition of samples, and in particular that the C2 fragment band allowed discriminating polymers with very similar elemental composition but different structure (e.g., polypropylene and polystyrene).

Further laboratory studies with portable instruments include: semiquantitative investigation of Islamic copper artifacts from Umm Qais (Jordan) [140]; characterization of construction materials (limestones, mortars, and bricks) from Kom El-Dikka, Alexandria (Egypt) [141]; online monitoring of laser cleaning procedures and depth profile analyses [60,63,64,70,100–103] (discussed in Sections 3.1 and 3.2).

4.2 In situ and stand-off field measurements

As mentioned, measurements with field-deployed LIBS instruments are relatively few, but represent case studies of great interest, both for the artistic value of the analyzed artifacts, and for the contributions of such studies to the development of LIBS and its penetration in the heritage science community. Pardini et al. used the mobile instrument designed and commercialized by the Pisa group [142] for a numismatic study of Roman silver *denarii*. The authors selected the most appropriate technique based on a preliminary investigation of a small set of samples with μ-XRF and LIBS, and, due to the high correlation between the two techniques, completed the analysis of the whole collection using only the intrinsically nondestructive μ-XRF. Using their noncommercial prototype, Siano and Agresti [143] conducted an archaeometallurgical investigation of several copper-based artifacts by one of the masters of Italian renaissance, Donatello, including the famous *David* (Fig. 11).

By applying their previously developed depth-dependent calibration method ([49], Section 3.1), they obtained quantitative compositional analysis of the artworks and contributed useful information to studies about the evolution of alloy compositions used in Early Renaissance Italian workshops. The Heraklion group carried out several in situ investigations of objects and built heritage. In [119], they used their noncommercial portable system to study decorative pigments in stone sculptures produced in Crete during the Venetian and Ottoman dominations with LIBS and Raman at indoors and outdoors locations in Heraklion. Fig. 12 shows details about the experimental configurations used for small/movable pieces (Fig. 12A and B) and for large/unmovable pieces and walls (Fig. 12C). The combined molecular and elemental information provided by the two techniques unequivocally identified the pigments in most cases, and a rough ranking of elements based on their intensity

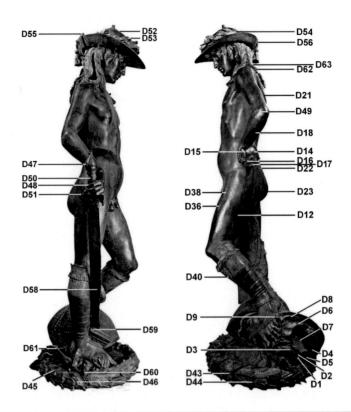

FIG. 11

Map of the locations of LIBS analysis of Donatello's *David* [143].

Credit: S. Siano, J. Agresti, Archaeometallurgical characterisation of Donatello's Florentine copper alloy masterpieces using portable laser-induced plasma spectroscopy and traditional techniques, Stud. Conserv. 60 (1) (2015) 106–119.

(i.e., strong, medium, weak) provided a semiquantitative indication of their relative amounts.

In [144], they used the same LIBS instrument to analyze various materials from the archeological site of Tiryns (Greece), including ceramics, metals (lead and copper alloys objects), pigments, glass and faience fragments, and mineral ore fragments. In [145], they described a monitoring protocol for two Heraklion monuments (Palace of Knossos; Venetian coastal fortress of Koules), within the frame of the European Project "HERACLES" (HEritage Resilience Against CLimate Events on Site). This project targets the effects of climate change on cultural heritage, and this specific aim focuses on characterizing and monitoring the degradation products on the two monuments. The protocol developed in [145] was largely based on LIBS and Raman spectroscopy and combined in situ and ex situ analyses, as well as information such as meteorological data, for a holistic approach to the safeguard of built heritage. In situ LIBS and Raman were also used by Veneranda et al. to

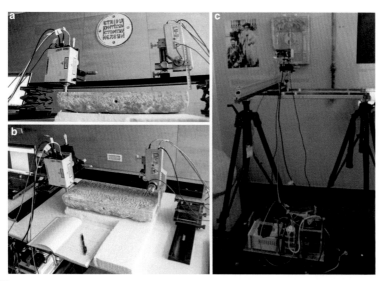

FIG. 12

Details of the field campaign carried out by the Heraklion groups at the Historical
Museum of Crete. (A) and (B) LIBS and Raman instruments are mounted on a rail during
the analysis of a stone inscription. (C) Raman instrument mounted on tripods during the
analysis of a stone relief.

*Credit: Z.E. Papliaka, A. Philippidis, P. Siozos, M. Vakondiou, K. Melessanaki, D. Anglos, A multi-technique
approach, based on mobile/portable laser instruments, for the in situ pigment characterization of stone sculptures
on the island of Crete dating from venetian and ottoman period, Herit. Sci. 4 (15) (2016). 1–18.*

investigate the materials and different causes and degradation patterns of two murals
in Ariadne's House, in Pompeii [146], found in the same location but having different
exposure and, consequently, conservation state. As for standoff measurements, to the
best of the author's knowledge, at the time when this chapter was written there was
only one study describing a field study of this application [147]. In this paper, Gaona
et al. analyzed selected areas on the façade of the Cathedral of Malaga (Spain) with a
field deployable standoff LIBS system for in situ long distance analysis (Fig. 13).

The operating distance was 35 m, and the elemental composition both of the
original construction materials (sandstone, different kinds of marble, ornamental
metallic objects) and of the deposited pollutants were characterized in open path
mode, with no scaffolding or sampling. The authors demonstrated that they could
correctly identify the different kinds of materials found in the main façade and discriminate black, rose, and white marbles based on Mg/Ca and Sr/Ca intensity ratios,
thus obtaining a sort of spectral fingerprint for material sorting and provenance studies. As for pollutants, they carried out depth-profile analyses and found evidence of
natural (sand, marine aerosols) and anthropogenic (road traffic) contamination, also
qualitatively estimating the thickness of pollutants' encrustations.

FIG. 13

Schematic representation of standoff LIBS measurements of the façade of the Cathedral of Malaga, Spain [147].

Credit: I. Gaona, P. Lucena, J. Moros, F.J. Fortes, S. Guirado, J. Serrano, J.J. Laserna, Evaluating the use of standoff LIBS in architectural heritage: surveying the Cathedral of Málaga, J. Anal. At. Spectrom. 28 (2013) 810.

5 Underwater LIBS

To date, LIBS is the only technique that can provide chemical analysis of bulk liquids and of submerged objects, by inducing a breakdown directly inside the liquid bulk. Several groups are active in underwater LIBS, for applications such as: analysis of seawater, sediments, and deep-ocean hydrothermal vents; production of nanomaterials through pulsed laser ablation in liquids (PLAL); and underwater archaeology. This section will illustrate the fundamentals of LIBS of submerged objects and then focus on underwater archaeology applications. Interested readers may find further resources about LIBS in liquids in [14,148,149] and references therein.

Broadly speaking, the generation of LIPs under water is similar to its counterpart in gaseous environment. In a liquid background, though, the much higher density and lower compressibility of the surrounding medium causes the LIP to be strongly confined and quenched, and to extinguish within a short time after the generation (hundreds of nanoseconds in water at atmospheric pressure). In this situation, few emission lines are visible, and, throughout its short persistence time, LIP spectra are featureless. The fast release of LIP internal energy, though, causes an interesting phenomenon, i.e., it generates a cavitation bubble in the liquid bulk by vaporizing the water layer surrounding the hot extinguishing plasma. Cavitation bubbles have much

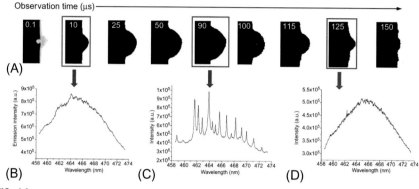

FIG. 14

(A) shadowgraphs showing the temporal evolution of a cavitation bubble formed by irradiating a Ti target under water at atmospheric pressure with nanosecond Nd:YAG laser (532 nm). The numbers reported in each frame indicate the observation time after the laser pulse (the LIP is visible in the first frame, as a light area indicating the brightness, at 100 ns delay). DP-LIBS spectra obtained by focusing a second laser pulse in the cavitation bubble at precise interpulse delays are shown in (B)–(D), corresponding to the interpulse delays and stages in the cavitation bubble evolution framed in (A).

Adapted from M. Dell'Aglio, R. Gaudiuso, O. De Pascale, A. De Giacomo, Mechanisms and processes of pulsed laser ablation in liquids during nanoparticle production, Appl. Surf. Sci. 348 (2015) 4–9. Credit: M. Dell'Aglio, R. Gaudiuso, O. De Pascale, A. De Giacomo, Mechanisms and processes of pulsed laser ablation in liquids during nanoparticle production, Appl. Surf. Sci. 348 (2015) 4–9.

longer persistence than the LIP (in the order of 10^2 μs in water at atmospheric pressure), and a dynamic behavior, outlined in Fig. 14A.

The cavitation bubble expands against water till its internal pressure drops below the hydrostatic pressure, which prompts the subsequent contraction and collapse of the bubble itself. An efficient approach to underwater LIBS is shooting a second laser pulse in the cavitation bubble at its maximum expansion. In this situation, the plasma is formed inside the bubble and thus expands in a gaseous environment, with similar evolution and spectra to those observed in air (Fig. 14C). Many studies have successfully employed double pulse (DP)-LIBS for laboratory analysis of submerged objects, as reviewed in [14,148,149], but this method has two limitations: the fact that it can introduce some instrumental complication, as it requires two collinear laser sources (or extracting more than one pulse from a single source); and the dramatic decrease of the cavitation bubble persistence time at high hydrostatic pressures, that is, in the conditions relevant to deep sea investigations and underwater archaeology. Moreover, as hydrostatic pressure increases, the water vapor pressure inside the bubble at its maximum expansion also increases; the emission intensity of the plasma produced in the bubble decreases; and the spectra become gradually featureless, like in bulk liquids [150,151]. Fig. 15 shows how the spectra of a titanium target in water become increasingly continuum-like as pressure increases.

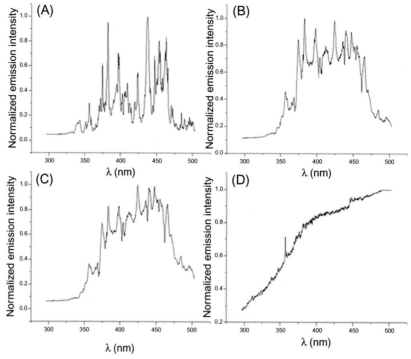

FIG. 15

DP-LIBS spectra of Ti under water at increasing hydrostatic pressure: (A) $P=1$ bar, interpulse delay $=100\,\mu s$; (B) $P=57$ bar, interpulse delay $=2.9\,\mu s$; (C) $P=70$ bar, interpulse delay $=2.9\,\mu s$; (D) $P=148$ bar, interpulse delay $=2\,\mu s$ [151].

Credit: A. De Giacomo, A. De Bonis, M. Dell'Aglio, O. De Pascale, R. Gaudiuso, S. Orlando, A. Santagata, G. S. Senesi, F. Taccogna, R. Teghil, Laser ablation of graphite in water in a range of pressure from 1 to 146 atm using single and double pulse techniques for the production of carbon nanostructures, J. Phys. Chem. C 115 (2011) 5123–5130.

Two alternative approaches can overcome these limitations: in the first, a gas flow is directed at the submerged target, so to locally displace water and generate a favorable gas—sample interface during the formation and evolution of the LIP [152]; the second uses hundred nanosecond-long laser pulses to generate LIPs with different mechanisms and more favorable evolution than with shorter pulses [153]. Laserna and coworkers, from Malaga (Spain), adopted the first approach to build a prototype for underwater archeological explorations in the Mediterranean Sea. Their instrument comprises a hand-held probe, connected to the main unit through an umbilical cable, containing an optical fiber (to deliver the laser beam to the target and collect the plasma emission) and a coaxial pressurized air flow [154]. In [154], the authors carried out a feasibility trial in the Mediterranean Sea, where the instrument was mounted on a research vessel and operated by a diver up to depths of 30 m (corresponding to hydrostatic pressure of 4 bar). The spectra of three bronze samples were generated at the seafloor and showed well-resolved features, useful for sample identification. Nonetheless, the damage threshold of the optical fiber required using low laser energy and thus restricted the applicability only to materials with low breakdown threshold (such as metals). Adopting a multipulse (MP) approach solved this issue. In [155], the same group demonstrated that delivering a train of pulses generated by the same laser source provided high-intensity LIBS spectra without using a single high-energy laser pulse, which would have damaged the optical fiber, and thus allowed analyzing samples with higher breakdown threshold, such as organic materials or concrete. They tested their upgraded instrument for both SP and MP excitation in an underwater archeological site (in the Bay of Cadiz, at the wreck of the ship Bucentaure, 17 m depth) and were able to acquire spectra from both metallic and nonmetallic materials. In [156], they analyzed (under water, but in the laboratory) a set of objects collected from several archeological shipwrecks and built a chemometric model (LDA, Linear Discriminant Analysis) for onsite classification of underwater archeological remains with their underwater LIBS probe. They tested their approach with two shipwrecks in the Bay of Cadiz, and successfully assigned all the samples in the correct group (bronze alloy; other metals; ceramic; marble). The Malaga group also explored the possibility of an open-path standoff configuration for underwater LIBS, by focusing a laser beam through a seawater bulk for recognition of distant submerged objects. The challenges of this approach are many. In addition to the mentioned issues related to LIPs generation under water, in standoff configurations laser radiation and plasma emission must be delivered through the water bulk, and thus are unavoidably attenuated. To address these issue, in [157] they built a scaled-down laboratory setup to study the effect on LIBS spectra of several parameters relevant to open-path operation (distance; water temperature; and interpulse delay).

The long nanosecond pulse approach was proposed for oceanographic explorations by Japanese researchers from University of Tokyo and University of Kyoto. In [153], Sakka et al. demonstrated that irradiating submerged targets with long nanosecond pulses (i.e., 150 ns pulse duration, instead of the usual 5–10 ns of most LIBS laser sources) provided well-resolved and long-lasting SP-LIBS spectra. This prompted several subsequent works, both fundamental, aiming to clarify the basic mechanisms of this radically different behavior and assess the analytical performance of this method, and applied, leading to the construction and successful deployment, at depths in the order of thousands of meters, of a remotely operated vehicle mounting a long-pulse LIBS probe [158–165].

6 Other applications

The common thread of studies included in this section is that they employed LIBS as a traditional analytical technique for the surface analysis of cultural heritage materials or mock samples, using benchtop instrumentation.

A trend observed in several laboratory studies is that LIBS is often combined with LA-ICP-MS, with the latter being used as a baseline to validate LIBS results. This is the case of the works by: Gaudiuso et al. [166] and Rossi et al. [167] who used calibration-free LIBS methods respectively for the analysis of archeological copper-based brooches from the National Archaeological Museum of Egnatia (Italy) and gemstones; Syvylay et al. [168] who characterized Pb tiles from the roof of the Beauvais cathedral (France); and Rai et al. who proposed using LIBS signals of rare earth elements (REEs) in archeological pottery and bricks from Kausambi (India) for provenance studies [169]. Some authors performed tandem LIBS/MS measurements of the material ablated by the same laser shot, so to obtain two kinds of complementary information for each ablation event: Kokkinaki et al. [170] analyzed painting materials with LIBS and LA-TOF-MS, both in tandem and sequentially, but concluded that the separate experiments provided better results; Syta et al. [171] used a commercial LIBS/LA-ICP-MS instrument for elemental imaging of cross sections of medieval Nubian wall paintings, and successfully identified the blue pigments (Cu-containing Egyptian blue and Na-containing lapis lazuli) and their distribution. Another joint study investigated a less common sample for LIBS, i.e., paper. Two forensic studies, of potential interest for cultural heritage, reported high accuracy for classification of paper and inks with LIBS, combined with other laser-based techniques, and chemometric models. Trejos et al. [172] used LIBS and LA-ICP-MS and were able to discriminate, with almost 100% accuracy, paper produced by different manufacturers, and by the same manufacturer at different times, as well as gel and ballpoint inks from different producers; Hoehse et al. coupled LIBS and Raman and showed that the combination of LIBS and Raman provided better classification of different kinds of inks on paper than the individual techniques [173]. A minimally destructive experimental configuration was described by Metzinger et al. who achieved remarkably high classification accuracy using single-shot micro-LIBS to discriminate clean paper and printed sheets obtained with digital printers (96.3% for paper kind; 83.3% for print types) [174]. Historical textiles are also unusual in LIBS studies. To the best of the author's knowledge, only three papers in the recent LIBS literature have studied these kinds of material, two of which were already discussed in Section 3.1 in the frame of depth profile studies [79,95]. In the third, a fragment of an Egyptian textile from King Tutankhamun's tomb was analyzed [175], to qualitatively identify the elemental composition of the colors. As shown throughout this chapter, ceramics and pottery are, on the other hand, rather common samples. These often represent the largest fractions of objects excavated during archeological campaigns, and therefore call for fast classification and analysis methods. To this end, Qi et al. [176] used a random forest (RF) machine learning algorithm to classify a set of ceramic specimens, both modern and archeological samples from 5 different Chinese dynasties, obtaining an average classification accuracy of 94% and validating their method for micro-destructive and fast authentication of archeological ceramics. A more classical quantitative approach was instead described in [177] for archeological Roman ceramics, and validated with Atomic Absorption Spectroscopy (AAS). The authors fabricated reference samples in a KBr matrix, using Zn as internal standard, and prepared the archeological samples for LIBS in the same way, by pulverizing a minute fraction of ancient ceramics. Singh et al. [178] identified main and minor elements in Neolithic and Historical potsherds from Northeastern India, and found good qualitative agreement between LIBS and energy dispersive X-ray (EDX) analysis. Provenance studies of various lithic materials have also been reported. In [179,180], the spectral signatures of marbles from different parts of Pakistan were identified, and showed some differences in the relative intensities of main and minor elements. In [181], the authors applied chemometric methods to LIBS and XRF spectra of limestone, sandstones, and calcarenite samples from quarries and historical buildings in Andalusia (Spain), and classified materials of unknown origin sampled from historical buildings, in good agreement with petrographical description. A small scale chemometric study was carried out in [182] with 5 ancient Indian coins of different kinds. Finally, several experimental approaches exist to increase the LIBS signal and improve its sensitivity (as reviewed in [183]). Here only two are briefly discussed, due to their promise for enhancing the spectral intensity at the same time limiting the sample damage, and their consequent usefulness in cultural heritage science. In [184], Tang et al. showed a resonant LIBS method [185], based on splitting a tunable laser beam and using the two separate portions to ablate and to selectively re-excite spectral transitions at specific wavelengths (Fig. 16A). This approach can efficiently increase the LIBS signal while working with low laser energies, but it may not be easily employed in field measurements and it requires previous knowledge on the element of interest and its transitions, and it may therefore be useful when high sensitivity is needed for selected elements. On the other hand, De Giacomo et al. developed a practical sample treatment method based on depositing colloidal solutions of metallic nanoparticles on the target prior to laser irradiation (nanoparticle-enhanced LIBS, NELIBS, Fig. 16B) to

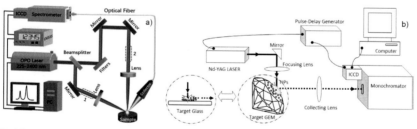

FIG. 16

(A) Schematic representation of two experimental approaches for signal enhancement in LIBS: double-pulse resonant LIBS [184] and (B) nanoparticle-enhanced LIBS for transparent samples [188].

Credit: Z. Tang, R. Zhou, Z. Hao, S. Ma, W. Zhang, K. Liu, X. Li, X. Zeng, Y. Lu, Micro-destructive analysis with high sensitivity using double-pulse resonant laser-induced breakdown spectroscopy, J. Anal. At. Spectrom. 34 (2019) 1198; C. Koral, M. Dell'Aglio, R. Gaudiuso, R. Alrifai, M. Torelli, A. De Giacomo, Nanoparticle-enhanced laser induced breakdown spectroscopy for the noninvasive analysis of transparent samples and gemstones. Talanta. 182 (2018) 253–258.

enhance signals from all the elements contained in the target [186,187]. In [188], they successfully applied NELIBS to the analysis of transparent and fragile samples (gemstones and glasses), without inducing any cracks or any visible crater.

7 Conclusions

Laser-induced breakdown spectroscopy (LIBS) has reached its maturity and is ready to be integrated in specialized laboratories for several analytical applications. Cultural heritage science is one of the fields that can benefit the most from the practical advantages offered by this technique, which include micro-destructivity and portability. This chapter reviewed the recent LIBS literature in heritage science, with a focus on three applications of particular interest for the analysis of cultural heritage objects: noninvasive depth-profiles; in situ investigations and analysis of distant objects; and underwater archaeology.

Some conclusions may be drawn on the foreseeable future of the technique in heritage science.

The complementarity of LIBS to well-established techniques like Raman spectroscopy and XRF (e.g., for the detection of light elements and deep depth profiles) holds great potential. It is easy to imagine the three techniques being routinely used together, each contributing a different piece to the puzzle of a comprehensive investigation of heritage objects, and many of the reviewed papers in fact demonstrated this positive synergy. Commercial availability of affordable and compact LIBS instruments for laboratory analysis and, even more, for field campaigns is key, especially considering that the aforementioned "sister techniques" are portable. The development of robust chemometric models is likely to play a role, in particular for provenance and authentication studies, while their application in field campaigns appears still hindered by the relatively low sensitivity of currently available portable instruments. Validation of experimental approaches for signal enhancement that limit the sample damage and are easy to implement in field campaigns will also represent a major asset in the future of the technique.

References

[1] L.J. Radziemski, D.A. Cremers (Eds.), Laser-Induced Plasma and Applications, Marcel Dekker, New York, NY, USA, 1989.

[2] L.J. Radziemski, D.A. Cremers, Handbook of Laser-Induced Breakdown Spectroscopy, second ed., John Wiley & Sons, Ltd., 2013.

[3] A.W. Miziolek, V. Palleschi, I. Schechter (Eds.), Laser Induced Breakdown Spectroscopy (LIBS): Fundamentals and Applications, Cambridge University Press, 2006.

[4] S. Musazzi, U. Perrini (Eds.), Laser-Induced Breakdown Spectroscopy. Theory and Applications, Springer Series in Optical Sciences, vol. 182, Springer, 2014.

[5] F.J. Fortes, J. Moros, P. Lucena, L.M. Cabalín, J.J. Laserna, Laser-induced breakdown spectroscopy, Anal. Chem. 85 (2013) 640–669.

[6] R. Gaudiuso, M. Dell'Aglio, O. De Pascale, G.S. Senesi, A. De Giacomo, Laser induced plasma spectroscopy for elemental analysis in environmental, cultural heritage and space applications: a review of methods and results, Sensors 10 (2010) 7434–7468.

[7] A. Giakoumaki, K. Melessanaki, D. Anglos, Laser-induced breakdown spectroscopy (LIBS) in archaeological science – applications and prospects, Anal. Bioanal. Chem. 387 (2007) 749–760.

[8] A. Nevin, G. Spoto, D. Anglos, Laser spectroscopies for elemental and molecular analysis in art and archaeology, Appl. Phys. A Mater. Sci. Process. 106 (2012) 339–361.

[9] D. Anglos, J.C. Miller, Cultural heritage applications of LIBS, in: A.W. Miziolek, V. Palleschi, I. Schechter (Eds.), Laser Induced Breakdown Spectroscopy (LIBS): Fundamentals and Applications, Cambridge University Press, 2006 (Chapter 9).

[10] D. Anglos, V. Detalle, Cultural heritage applications of LIBS, in: S. Musazzi, U. Perrini (Eds.), Laser-Induced Breakdown Spectroscopy. Theory and Applications, Springer Series in Optical Sciences, vol. 182, Springer, 2014.

[11] V. Spizzichino, R. Fantoni, Laser induced breakdown spectroscopy in archeometry: a review of its application and future perspectives, Spectrochim. Acta Part B At. Spectrosc. 99 (2014) 201–209.

[12] A. Botto, B. Campanella, S. Legnaioli, M. Lezzerini, G. Lorenzetti, S. Pagnotta, F. Poggialini, V. Palleschi, Applications of laser-induced breakdown spectroscopy in cultural heritage and archaeology: a critical review, J. Anal. At. Spectrom. 34 (2019) 81–103.

[13] F. Ruan, T. Zhang, H. Li, Laser-induced breakdown spectroscopy in archeological science: a review of its application and future perspectives, Appl. Spectrosc. Rev. 54 (2019) 573–601.

[14] A. De Giacomo, M. Dell'Aglio, R. Gaudiuso, S. Amoruso, O. De Pascale, Effects of the background environment on formation, evolution and emission spectra of laser-induced plasmas, Spectrochim. Acta Part B At. Spectrosc. 78 (2012) 1–19.

[15] J.M. Anzano, C. Bello-Gálvez, R.J. Lasheras, Identification of polymers by means of LIBS, in: S. Musazzi, U. Perrini (Eds.), Laser-Induced Breakdown Spectroscopy, Theory and Applications, Springer Series in Optical Sciences, vol. 182, Springer, 2014, pp. 421–438.

[16] A. De Giacomo, R. Gaudiuso, M. Dell'Aglio, A. Santagata, The role of continuum radiation in laser induced plasma spectroscopy, Spectrochim. Acta Part B At. Spectrosc. 65 (2010) 385–394.

[17] J. Lam, V. Motto-Ros, D. Misiak, C. Dujardin, G. Ledoux, D. Amans, Investigation of local thermodynamic equilibrium in laser-induced plasmas: measurements of rotational and excitation temperatures at long time scales, Spectrochim. Acta Part B At. Spectrosc. 101 (2014) 86–92.

[18] C.G. Parigger, Atomic and molecular emissions in laser-induced breakdown spectroscopy, Spectrochim. Acta Part B At. Spectrosc. 79–80 (2013) 4–16.

[19] C.G. Parigger, A.C. Woods, M.J. Witte, L.D. Swafford, D.M. Surmick, Measurement and analysis of atomic hydrogen and diatomic molecular AlO, C2, CN, and TiO spectra following laser-induced optical breakdown, J. Vis. Exp. 84 (2014), https://doi.org/10.3791/51250.

[20] National Institute of Standards and Technology (NIST), Atomic Spectra Database, 2019. https://www.nist.gov/pml/atomic-spectra-database. (Updated December 11, 2019).

[21] R.L. Kurucz, B. Bell, Atomic Line Data Kurucz CD-ROM No. 23, 1995. https://www.cfa.harvard.edu/amp/ampdata/kurucz23/sekur.html.

[22] R.E. Russo, X. Mao, J.J. Gonzalez, V. Zorba, J. Yoo, Laser ablation in analytical chemistry, Anal. Chem. 85 (2013) 6162–6177.

[23] H.-Y. Moon, K.K. Herrera, N. Omenetto, B.W. Smith, J.D. Winefordner, On the usefulness of a duplicating mirror to evaluate self-absorption effects in laser induced breakdown spectroscopy, Spectrochim. Acta Part B At. Spectrosc. 64 (2009) 702–713.

[24] D. Bulajic, M. Corsi, G. Cristoforetti, S. Legnaioli, V. Palleschi, A. Salvetti, E. Tognoni, A procedure for correcting self-absorption in calibration free-laser induced breakdown spectroscopy, Spectrochim. Acta Part B At. Spectrosc. 57 (2002) 339–353.

[25] A.N. Kadachi, M.A. Al-Eshaikh, K. Ahmad, Self-absorption correction: an effective approach for precise quantitative analysis with laser induced breakdown spectroscopy, Laser Phys. 28 (2018), 095701.

[26] T. Li, Z. Hou, Y. Fu, J. Yu, W. Gu, Z. Wang, Correction of self-absorption effect in calibration-free laser-induced breakdown spectroscopy (CF-LIBS) with blackbody radiation reference, Anal. Chim. Acta 1058 (2019) 39–47.

[27] A. De Giacomo, M. Dell'Aglio, O. De Pascale, S. Longo, M. Capitelli, Laser induced breakdown spectroscopy on meteorites, Spectrochim. Acta Part B At. Spectrosc. 62 (2007) 1606–1611.

[28] C. Aragón, J.A. Aguilera, Characterization of laser induced plasmas by optical emission spectroscopy: a review of experiments and methods, Spectrochim. Acta Part B At. Spectrosc. 63 (2008) 893–916.

[29] N. Konjevic, Plasma broadening and shifting of non-hydrogenic spectral lines: present status and applications, Phys. Rep. 316 (1999) 339–401.

[30] N.B. Zorov, A.A. Gorbatenko, T.A. Labutin, A.M. Popov, A review of normalization techniques in analytical atomic spectrometry with laser sampling: from single to multivariate correction, Spectrochim. Acta Part B At. Spectrosc. 65 (2010) 642–657.

[31] A. Ciucci, M. Corsi, V. Palleschi, S. Rastelli, A. Salvetti, E. Tognoni, New procedure for quantitative elemental analysis by laser-induced plasma spectroscopy, Appl. Spectrosc. 53 (1999) 960–964.

[32] E. Tognoni, G. Cristoforetti, S. Legnaioli, V. Palleschi, Calibration-free laser-induced breakdown spectroscopy: state of the art, Spectrochim. Acta Part B At. Spectrosc. 65 (2010) 1–14.

[33] E. Grifoni, S. Legnaioli, G. Lorenzetti, S. Pagnotta, F. Poggialini, V. Palleschi, From calibration-free to fundamental parameters analysis: a comparison of three recently proposed approaches, Spectrochim. Acta Part B At. Spectrosc. 124 (2016) 40–46.

[34] L.J. Radziemski, D.A. Cremers, Chemometric analysis in LIBS, in: J.L. Gottfried (Ed.), Handbook of Laser-Induced Breakdown Spectroscopy, Second Edition, John Wiley & Sons, Ltd., 2013, pp. 223–255.

[35] A. Safi, B. Campanella, E. Grifoni, S. Legnaioli, G. Lorenzetti, S. Pagnotta, F. Poggialini, L. Ripoll-Seguer, M. Hidalgo, V. Palleschi, Multivariate calibration in laser-induced breakdown spectroscopy quantitative analysis: the dangers of a 'black box' approach and how to avoid them, Spectrochim. Acta Part B At. Spectrosc. 144 (2018) 46–54.

[36] A. De Giacomo, M. Dell'Aglio, R. Gaudiuso, S. Amoruso, O. De Pascale, Effects of the background environment on formation, evolution and emission spectra of laser-induced plasmas, Spectrochim. Acta Part B At. Spectrosc. 78 (2012) 1–19.

[37] A.J. Effenberger Jr., J.R. Scott, Effect of atmospheric conditions on LIBS spectra, Sensors 10 (2010) 4907–4925.

[38] M. Dell'Aglio, A. De Giacomo, R. Gaudiuso, O. De Pascale, S. Longo, Laser induced breakdown spectroscopy of meteorites as a probe of the early solar system, Spectrochim. Acta Part B At. Spectrosc. 101 (2014) 68–75.

[39] B. Busser, S. Moncayo, J.-L. Coll, L. Sancey, V. Motto-Ros, Elemental imaging using laser-induced breakdown spectroscopy: a new and promising approach for biological and medical applications, Coord. Chem. Rev. 358 (2018) 70–79.

[40] F.J. Fortes, J.J. Laserna, The development of fieldable laser-induced breakdown spectrometer: no limits on the horizon, Spectrochim. Acta Part B At. Spectrosc. 65 (2010) 975–990.

[41] J. Rakovský, P. Čermák, O. Musset, P. Veis, A review of the development of portable laser induced breakdown spectroscopy and its applications, Spectrochim. Acta Part B At. Spectrosc. 101 (2014) 269–287.

[42] J. Marczak, A. Koss, P. Targowski, M. Góra, M. Strzelec, A. Sarzyński, W. Skrzeczanowski, R. Ostrowski, A. Rycyk, Characterization of laser cleaning of artworks, Sensors 8 (2008) 6507–6548.

[43] S. Siano, J. Agresti, I. Cacciari, D. Ciofini, M. Mascalchi, I. Osticioli, A.A. Mencaglia, Laser cleaning in conservation of stone, metal, and painted artifacts: state of the art and new insights on the use of the Nd:YAG lasers, Appl. Phys. A Mater. Sci. Process. 106 (2012) 419–446.

[44] M. Corsi, G. Cristoforetti, M. Hidalgo, D. Iriarte, S. Legnaioli, V. Palleschi, A. Salvetti, E. Tognoni, Effect of laser-induced crater depth in laser-induced breakdown spectroscopy emission features, Appl. Spectrosc. 59 (2005) 853–860.

[45] X. Zeng, S.S. Mao, C. Liu, X. Mao, R. Greif, R.E. Russo, Plasma diagnostics during laser ablation in a cavity, Spectrochim. Acta Part B At. Spectrosc. 58 (2003) 867–877.

[46] X. Zeng, X. Mao, S.S. Mao, J.H. Yoo, R. Greif, R.E. Russo, Laser–plasma interactions in fused silica cavities, J. Appl. Phys. 95 (2004) 816–822.

[47] X. Zeng, X. Mao, S.S. Mao, S. Wen, R. Greif, R.E. Russo, Laser-induced shockwave propagation from ablation in a cavity, Appl. Phys. Lett. 88 (2006), 061502.

[48] J. Picard, J.-B. Sirven, J.-L. Lacour, O. Musset, D. Cardona, J.-C. Hubinois, P. Mauchien, Characterization of laser ablation of copper in the irradiance regime of laser-induced breakdown spectroscopy analysis, Spectrochim. Acta Part B At. Spectrosc. 101 (2014) 164–170.

[49] J. Agresti, S. Siano, Depth-dependent calibration for quantitative elemental depth profiling of copper alloys using laser-induced plasma spectroscopy, Appl. Phys. A Mater. Sci. Process. 117 (2014) 217–221.

[50] M. Abdelhamid, S. Grassini, E. Angelini, G.M. Ingo, M.A. Harith, Depth profiling of coated metallic artifacts adopting laser-induced breakdown spectrometry, Spectrochim. Acta Part B At. Spectrosc. 65 (2010) 695–701.

[51] J. Agresti, I. Cacciari, S. Siano, Optimized laser ablation elemental depth profiling of bronzes using 3D microscopy, in: 18th Italian National Conference on Photonic Technologies (Fotonica 2016), Rome, Italy, 6–8 June, 2016, pp. 1–4, https://doi.org/10.1049/cp.2016.0958.

[52] S. Siano, I. Cacciari, A. Mencaglia, J. Agresti, Spatially calibrated elemental depth profiling using LIPS and 3D digital microscopy, Eur. Phys. J. Plus 126 (120) (2011), https://doi.org/10.1140/epjp/i2011-11120-y.

[53] I. Cacciari, J. Agresti, S. Siano, Combined THz and LIPS analysis of corroded archaeological bronzes, Microchem. J. 126 (2016) 76–82.

[54] I. Osticioli, J. Agresti, C. Fornacelli, I. Turbanti Memmi, S. Siano, Potential role of LIPS elemental depth profiling in authentication studies of unglazed earthenware artifacts, J. Anal. At. Spectrom. 27 (2012) 827–833.

[55] J. Agresti, I. Osticioli, A.A. Mencaglia, S. Siano, Laser-induced plasma spectroscopy depth profile analysis: A contribution to authentication, in: Proceedings. SPIE 8790, Optics for Arts, Architecture, and Archaeology IV, 87900I, SPIE Optical Metrology, Munich, Germany, May, 2013, https://doi.org/10.1117/12.2020701.

[56] L. Bartoli, J. Agresti, M. Mascalchi, A. Mencaglia, I. Cacciari, S. Siano, Combined elemental and microstructural analysis of genuine and fake copper-alloy coins, Quantum Electron. 41 (2011) 663668.

[57] J. Agresti, I. Osticioli, M.C. Guidotti, G. Capriotti, N. Kardjilov, A. Scherillo, S. Siano, Combined neutron and laser techniques for technological and compositional investigations of hollow bronze figurines, J. Anal. At. Spectrom. 30 (2015) 713–720.

[58] J. Agresti, I. Osticioli, M.C. Guidotti, N. Kardjilov, S. Siano, Non-invasive archaeometallurgical approach to the investigations of bronze figurines using neutron, laser, and X-ray techniques, Microchem. J. 124 (2016) 765–774.

[59] L. Robbiola, J.-M. Blengino, C. Fiaud, Morphology and mechanisms of formation of natural patinas on archaeological cu-Sn alloys, Corros. Sci. 40 (1998) 2080–2111.

[60] L. Moreira Osorio, L.V. Ponce Cabrera, M.A. Arronte García, T. Flores Reyes, I. Ravelo, Portable LIBS system for determining the composition of multilayer structures on objects of cultural value, J. Phys. Conf. Ser. 274 (2011), 012093.

[61] M.A. Gómez-Morón, P. Ortiz, R. Ortiz, J.M. Martín, M.P. Mateo, G. Nicolás, Laser-induced breakdown spectroscopy study of silversmith pieces: the case of a Spanish canopy of the nineteenth century, Appl. Phys. A Mater. Sci. Process. 122 (2016) 548.

[62] M. Orlić Bachler, M. Bišćanb, Z. Kregar, I. Jelovica Badovinac, J. Dobrinić, S. Milošević, Analysis of antique bronze coins by laser induced breakdown spectroscopy and multivariate analysis, Spectrochim. Acta Part B At. Spectrosc. 123 (2016) 163–170.

[63] M.F. Alberghina, R. Barraco, M. Brai, T. Schillaci, L. Tranchina, Integrated analytical methodologies for the study of corrosion processes in archaeological bronzes, Spectrochim. Acta Part B At. Spectrosc. 66 (2011) 129–137.

[64] M.F. Alberghina, R. Barraco, M. Brai, L. Tranchina, LIBS and XRF analysis for a stratigraphic study of pictorial multilayer surfaces, Period. Mineral. 84 (2015) 569–589.

[65] L. Caneve, A. Diamanti, F. Grimaldi, G. Palleschi, V. Spizzichino, F. Valentini, Analysis of fresco by laser induced breakdown spectroscopy, Spectrochim. Acta B 65 (2010) 702–706.

[66] F. Toschi, A. Paladini, F. Colosi, P. Cafarelli, V. Valentini, M. Falconieri, S. Gagliardi, P. Santoro, A multi-technique approach for the characterization of Roman mural paintings, Appl. Surf. Sci. 284 (2013) 291–296.

[67] A. Paladini, F. Toschi, F. Colosi, G. Rubino, P. Santoro, Stratigraphic investigation of wall painting fragments from Roman villas of the Sabina area, Appl. Phys. A Mater. Sci. Process. 118 (2015) 131–138.

[68] F.d.S.L. Borba, J. Cortez, V.K. Asfora, C. Pasquini, M.F. Pimentel, A.-M. Pessis, H.J. Khoury, Multivariate treatment of LIBS data of prehistoric paintings, J. Braz. Chem. Soc. 23 (2012) 958–965.

[69] C. Lofrumento, M. Ricci, L. Bachechi, D. De Feo, E.M. Castellucci, The first spectroscopic analysis of Ethiopian prehistoric rock painting, J. Raman Spectrosc. 43 (2012) 809–816.

[70] E.A. Kaszewska, M. Sylwestrzak, J. Marczak, W. Skrzeczanowski, M. Iwanick, E. Szmit-Naud, D. Anglos, P. Targowski, Depth-resolved multilayer pigment identification in paintings: combined use of laser-induced breakdown spectroscopy (LIBS) and optical coherence tomography (OCT), Appl. Spectrosc. 67 (2013) 960–972.

[71] E. Pospíšilová, K. Novotný, P. Pořízka, D. Hradil, J. Hradilová, J. Kaiser, V. Kanický, Depth-resolved analysis of historical painting model samples by means of laser-induced breakdown spectroscopy and handheld X-ray fluorescence, Spectrochim. Acta Part B At. Spectrosc. 147 (2018) 100–108.

[72] X. Bai, D. Syvilay, N. Wilkie-Chancellier, A. Texier, L. Martinez, S. Serfaty, D. Martos-Levif, V. Detalle, Influence of ns-laser wavelength in laser-induced breakdown spectroscopy for discrimination of painting techniques, Spectrochim. Acta Part B At. Spectrosc. 134 (2017) 81–90.

[73] D. Syvilay, N. Wilkie-Chancellier, B. Trichereau, A. Texier, L. Martinez, S. Serfaty, V. Detalle, Evaluation of the standard normal variate method for laser-induced breakdown spectroscopy data treatment applied to the discrimination of painting layers, Spectrochim. Acta Part B At. Spectrosc. 114 (2015) 38–45.

[74] F.C. Alvira, F.R. Rozzi, G.M. Bilmes, Laser-induced breakdown spectroscopy microanalysis of trace elements in Homo sapiens teeth, Appl. Spectrosc. 64 (2010) 313–319.

[75] F.C. Alvira, F.V. Ramirez Rozzi, G.A. Torchia, L. Roso, G.M. Bilmes, A new method for relative Sr determination in human teeth enamel, J. Anthropol Sci 89 (2011) 153–160.

[76] M.A. Kasem, R.E. Russo, M.A. Harith, Influence of biological degradation and environmental effects on the interpretation of archeological bone samples with laser-induced breakdown spectroscopy, J. Anal. At. Spectrom. 26 (2011) 1733–1739.

[77] M.A. Kasem, J.J. Gonzalez, R.E. Russo, M.A. Harith, Effect of the wavelength on laser induced breakdown spectrometric analysis of archaeological bone, Spectrochim. Acta Part B At. Spectrosc. 101 (2014) 26–31.

[78] D.A. Rusak, R.M. Marsico, B.L. Taroli, Using laser-induced breakdown spectroscopy to assess preservation quality of archaeological bones by measurement of calcium-to-fluorine ratios, Appl. Spectrosc. 65 (2011) 1193–1196.

[79] O.A. Nassef, H.E. Ahmed, M.A. Harith, Surface and stratigraphic elemental analysis of an ancient Egyptian cartonnage using laser-induced breakdown spectroscopy (LIBS), Anal. Methods 8 (2016) 7096–7106.

[80] B.G. Oztoprak, M.A. Sinmaz, F. Tülek, Composition analysis of medieval ceramics by laser-induced breakdown spectroscopy (LIBS), Appl. Phys. A Mater. Sci. Process. 122 (5) (2016) 557.

[81] A. Khedr, M.A. Harith, In-depth micro-spectrochemical analysis of archaeological Egyptian pottery shards, Appl. Phys. A Mater. Sci. Process. 113 (2013) 835–842.

[82] A. Khedr, O. Abdel-Kareem, S.H. Elnabi, M.A. Harith, Compositional analysis of ceramic glaze by laser induced breakdown spectroscopy and energy dispersive X-ray, AIP Conf. Proc. 1380 (2011) 87, https://doi.org/10.1063/1.3631815.

[83] S. Touron, B. Trichereau, D. Syvilay, In-depth analyses of Paleolithic pigments in cave climatic conditions, Proc. SPIE 10331 (2017), https://doi.org/10.1117/12.2269965. Optics for Arts, Architecture, and Archaeology VI, 103310M, 11 July 2017.

[84] A. Khedr, H. Sadek, O.A. Nassef, M. Abdelhamid, M.A. Harith, Discrimination between the authentic and fake Egyptian funerary figurines "Ushabtis" via laser-

induced breakdown spectroscopy, J. Cult. Herit. 40 (2019) 25–33, https://doi.org/10.1016/j.culher.2019.05.006.

[85] P. Ortiz, M.A. Vázquez, R. Ortiz, J.M. Martin, T. Ctvrtnickova, M.P. Mateo, G. Nicolas, Investigation of environmental pollution effects on stone monuments in the case of Santa Maria La Blanca, Seville (Spain), Appl. Phys. A Mater. Sci. Process. 100 (2010) 965–973.

[86] V. Lazic, A. Trujillo-Vazquez, H. Sobral, C. Márquez, A. Palucci, M. Ciaffi, M. Pistilli, Corrections for variable plasma parameters in laser induced breakdown spectroscopy: application on archeological samples, Spectrochim. Acta Part B At. Spectrosc. 122 (2016) 103–113.

[87] R. Gaudiuso, M. Dell'Aglio, O. De Pascale, A. Santagata, A. De Giacomo, Laser-induced plasma analysis of copper alloys based on local thermodynamic equilibrium considerations: an alternative approach and archaeometric applications, Spectrochim. Acta Part B At. Spectrosc. 74–75 (2012) 38–45.

[88] A. De Giacomo, M. Dell'Aglio, O. De Pascale, R. Gaudiuso, A. Santagata, G. Senesi, M. Rossi, M.R. Ghiara, F. Capitelli, A laser induced breakdown spectroscopy application based on local thermodynamic equilibrium assumption for the elemental analysis of alexandrite gemstone and copper-based alloys, Chem. Phys. 398 (2012) 233–238.

[89] R. Gaudiuso, Calibration-free inverse method for depth-profile analysis with laser-induced breakdown spectroscopy, Spectrochim. Acta Part B At. Spectrosc. 123 (2016) 105–113.

[90] R. Gaudiuso, K. Uhlir, M. Griesser, Micro-invasive depth profile analysis by laser-induced breakdown spectroscopy (LIBS): the case of mercury layers on Sasanian coins, J. Anal. At. Spectrom. 34 (2019) 2261–2272.

[91] P.M. Carmona-Quiroga, S. Martínez-Ramírez, S. Sánchez-Cortés, M. Oujja, M. Castillejo, M.T. Blanco-Varela, Effectiveness of antigraffiti treatments in connection with penetration depth determined by different techniques, J. Cult. Herit. 11 (2010) 297–303.

[92] M. Mateo, J. Becerra, A.P. Zaderenko, P. Ortiz, G. Nicolás, Laser-induced breakdown spectroscopy applied to the evaluation of penetration depth of bactericidal treatments based on silver nanoparticles in limestones, Spectrochim. Acta Part B At. Spectrosc. 152 (2019) 44–51.

[93] J. Becerra, M. Mateo, P. Ortiz, G. Nicolás, A.P. Zaderenko, Evaluation of the applicability of nano-biocide treatments on limestones used in cultural heritage, J. Cult. Herit. 38 (2019) 126–135.

[94] S. Siano, Principles of laser cleaning in conservation, in: M. Schreiner, M. Strlic, R. Salimbeni (Eds.), Handbook on the Use of Lasers in Conservation and Conservation Science, COST Office, Brussels, Belgium, 2008.

[95] M. Kono, K.G.H. Baldwin, A. Wain, A.V. Rode, Treating the untreatable in art and heritage materials: ultrafast, laser cleaning of "cloth-of-gold", Langmuir 31 (4) (2015) 1596–1604.

[96] V. Atanassova, P. Penkova, I. Kostadinov, S. Karatodorov, G.V. Avdeev, Laser removal of chlorine from historical metallic objects, in: Proc. SPIE 11047, 20th International Conference and School on Quantum Electronics: Laser Physics and Applications, 29 January, Vol. 110470C, 2019, https://doi.org/10.1117/12.2516813.

[97] I. Apostol, V. Damian, F. Garoi, I. Iordache, M. Bojan, D. Apostol, A. Armaselu, P.J. Morais, D. Postolache, I. Darida, Controlled removal of overpainting and painting layers under the action of UV laser radiation, Opt. Spectrosc. 111 (2011) 287–292.

[98] A. Staicu, I. Apostol, A. Pascu, I. Urzica, M.L. Pascu, V. Damian, Minimal invasive control of paintings cleaning by LIBS, Opt. Laser Technol. 77 (2016) 187–192.

[99] D.E. Roberts, A. du Plessis, J. Steyn, L.R. Botha, S. Pityana, L.R. Berger, An investigation of laser induced breakdown spectroscopy for use as a control in the laser removal of rock from fossils found at the Malapa hominin site, South Africa, Spectrochim. Acta Part B At. Spectrosc. 73 (2012) 48–54.

[100] G.S. Senesi, I. Carrara, G. Nicolodelli, D.M.B.P. Milori, O. De Pascale, Laser cleaning and laser-induced breakdown spectroscopy applied in removing and characterizing black crusts from limestones of Castello Svevo, Bari, Italy: a case study, Microchem. J. 124 (2016) 296–305.

[101] G.S. Senesi, G. Nicolodelli, D.M.B.P. Milori, O. De Pascale, Depth profile investigations of surface modifications of limestone artifacts by laser-induced breakdown spectroscopy, Environ. Earth Sci. 76 (16) (2017) 565.

[102] G.S. Senesi, D. Manzini, O. De Pascale, Application of a laser-induced breakdown spectroscopy handheld instrument to the diagnostic analysis of stone monuments, Appl. Geochem. 96 (2018) 87–91.

[103] G.S. Senesi, B. Campanella, E. Grifoni, S. Legnaioli, G. Lorenzetti, S. Pagnotta, F. Poggialini, V. Palleschi, O. De Pascale, Elemental and mineralogical imaging of a weathered limestone rock by double-pulse micro-laser-induced breakdown spectroscopy, Spectrochim. Acta B 143 (2018) 91–97.

[104] A. Khedr, V. Papadakis, P. Pouli, D. Anglos, M.A. Harith, The potential use of plume imaging for real-time monitoring of laser ablation cleaning of stonework, Appl. Phys. B Lasers Opt. 105 (2011) 485–492.

[105] V. Detalle, X. Bai, E. Bourguignon, M. Menu, I. Pallot-Frossard, LIBS-LIF-Raman: A new tool for the future E-RIHS, in: Proc. SPIE Optical Metrology 2017, Munich, Germany 10331, Optics for Arts, Architecture, and Archaeology VI, 11 July, Vol. 103310N, 2017, https://doi.org/10.1117/12.2272027.

[106] D. Syvilay, X.S. Bai, N. Wilkie-Chancellier, A. Texier, L. Martinez, S. Serfaty, V. Detalle, Laser-induced emission, fluorescence and Raman hybrid setup: a versatile instrument to analyze materials from cultural heritage, Spectrochim. Acta Part B At. Spectrosc. 140 (2018) 44–53.

[107] A. Martínez-Hernández, M. Oujja, M. Sanza, E. Carrasco, V. Detalle, M. Castillejo, Analysis of heritage stones and model wall paintings by pulsed laser excitation of Raman, laser-induced fluorescence and laser-induced breakdown spectroscopy signals with a hybrid system, J. Cult. Herit. 32 (2018) 1–8.

[108] G. Vítková, K. Novotný, L. Prokeš, A. Hrdlička, J. Kaiser, J. Novotný, R. Malina, D. Prochazka, Fast identification of biominerals by means of stand-off laser-induced breakdown spectroscopy using linear discriminant analysis and artificial neural networks, Spectrochim. Acta Part B At. Spectrosc. 73 (2012) 1–6.

[109] A. Hrdlička, L. Prokeš, A. Staňková, K. Novotný, A. Vitešníková, V. Kanický, V. Otruba, J. Kaiser, J. Novotný, R. Malina, K. Páleníková, Development of a remote laser-induced breakdown spectroscopy system for investigation of calcified tissue samples, Appl. Opt. 49 (2010) C16.

[110] M. Galiová, J. Kaiser, F.J. Fortes, K. Novotný, R. Malina, L. Prokeš, A. Hrdlička, T. Vaculovič, M.N. Fišáková, J. Svoboda, V. Kanický, J.J. Laserna, Multielemental analysis of prehistoric animal teeth by laser-induced breakdown spectroscopy and laser

ablation inductively coupled plasma mass spectrometry, Appl. Opt. 49 (13) (2010) C191–C199.

[111] G. Vítková, L. Prokeš, K. Novotný, P. Pořízka, J. Novotný, D. Všianský, L. Čelko, J. Kaiser, Comparative study on fast classification of brick samples by combination of principal component analysis and linear discriminant analysis using stand-off and table-top laser-induced breakdown spectroscopy, Spectrochim. Acta Part B At. Spectrosc. 101 (2014) 191–199.

[112] V. Lazic, M. Vadrucci, R. Fantoni, M. Chiari, A. Mazzinghi, A. Gorghinian, Applications of laser-induced breakdown spectroscopy for cultural heritage: a comparison with X-ray fluorescence and particle induced X-ray emission techniques, Spectrochim. Acta Part B At. Spectrosc. 149 (2018) 1–14.

[113] V. Lazic, R. Fantoni, F. Colao, A. Santagata, A. Morone, V. Spizzichino, Quantitative laser induced breakdown spectroscopy analysis of ancient marbles and corrections for the variability of plasma parameters and of ablation rate, J. Anal. At. Spectrom. 19 (2004) 429–436.

[114] R. Chapoulie, L. Bassel, G. Mauran, C. Ferrier, A. Queffelec, D. Lacanette, P. Malaurent, B. Bousquet, V. Motto-Ros, F. Trichard, F. Pelascini, V. Rodriguez, Photons and electrons for the study of a white veil covering some walls in prehistoric caves, Acta Imeko 6 (3) (2017) 82–86.

[115] L. Bassel, V. Motto-Ros, F. Trichard, F. Pelascini, F. Ammari, R. Chapoulie, C. Ferrier, D. Lacanette, B. Bousquet, Laser-induced breakdown spectroscopy for elemental characterization of calcitic alterations on cave walls, Environ. Sci. Pollut. Res. 24 (2017) 2197–2204, https://doi.org/10.1007/s11356-016-7468-5.

[116] L. Qu, X. Zhang, H. Duan, R. Zhang, G. Li, Y. Lei, The application of LIBS and other techniques on Chinese low temperature glaze, MRS Adv. Mater. Issues Art Archaeol. XI 2 (39–40) (2017) 2081–2094, https://doi.org/10.1557/adv.2017.85.

[117] J. Aramendia, L. Gómez-Nubla, S. Fdez-Ortiz de Vallejuelo, K. Castro, G. Arana, J. Manuel Madariaga, The combination of Raman imaging and LIBS for quantification of original and degradation materials in cultural heritage, J. Raman Spectrosc. 50 (2019) 193–201.

[118] P. Westlake, P. Siozos, A. Philippidis, C. Apostolaki, B. Derham, A. Terlixi, V. Perdikatsis, R. Jones, D. Anglos, Studying pigments on painted plaster in Minoan, Roman and early byzantine Crete. A multi-analytical technique approach, Anal. Bioanal. Chem. 402 (2012) 1413–1432.

[119] Z.E. Papliaka, A. Philippidis, P. Siozos, M. Vakondiou, K. Melessanaki, D. Anglos, A multi-technique approach, based on mobile/portable laser instruments, for the in situ pigment characterization of stone sculptures on the island of Crete dating from venetian and ottoman period, Herit. Sci. 4 (15) (2016) 1–18.

[120] P. Siozos, A. Philippidis, D. Anglos, Portable laser-induced breakdown spectroscopy/diffuse reflectance hybrid spectrometer for analysis of inorganic pigments, Spectrochim. Acta Part B At. Spectrosc. 137 (2017) 93–100.

[121] K. Melessanaki, V. Papadakis, C. Balas, D. Anglos, Laser induced breakdown spectroscopy and hyper-spectral imaging analysis of pigments on an illuminated manuscript, Spectrochim. Acta Part B At. Spectrosc. 56 (2001) 2337–2346.

[122] A. Bertolini, G. Carelli, F. Francesconi, M. Francesconi, L. Marchesini, P. Marsili, F. Sorrentino, G. Cristoforetti, S. Legnaioli, V. Palleschi, L. Pardini, A. Salvetti, Modì: a new mobile instrument for in situ double-pulse LIBS analysis, Anal. Bioanal. Chem. 385 (2006) 240–247.

[123] S. Legnaioli, F. Anabitarte Garcia, A. Andreotti, E. Bramanti, D. Díaz Pace, S. Formola, G. Lorenzetti, M. Martini, L. Pardini, E. Ribechini, E. Sibilia, R. Spiniello, V. Palleschi, Multi-technique study of a ceramic archaeological artifact and its content, Spectrochim. Acta A Mol. Biomol. Spectrosc. 100 (2013) 144–148.

[124] B. Campanella, E. Grifoni, M. Hidalgo, S. Legnaioli, G. Lorenzetti, S. Pagnotta, F. Poggialini, L. Ripoll-Seguer, V. Palleschi, Multi-technique characterization of madder lakes: a comparison between non- and micro-destructive methods, J. Cult. Herit. 33 (2018) 208–212.

[125] J. La Nasa, F. Di Marco, L. Bernazzani, C. Duce, A. Spepi, V. Ubaldi, I. Degano, S. Orsini, S. Legnaioli, M.R. Tiné, D. De Luca, F. Modugno, Aquazol as a binder for retouching paints. An evaluation through analytical pyrolysis and thermal analysis, Polym. Degrad. Stabil. 144 (2017) 508–519.

[126] S. Columbu, S. Carboni, S. Pagnotta, M. Lezzerini, S. Raneri, S. Legnaioli, V. Palleschi, A. Usai, Laser-induced breakdown spectroscopy analysis of the limestone Nuragic statues from Monte Prama site (Sardinia, Italy), Spectrochim. Acta Part B At. Spectrosc. 149 (2018) 62–70.

[127] S. Pagnotta, M. Lezzerini, L. Ripoll-Seguer, M. Hidalgo, E. Grifoni, S. Legnaioli, G. Lorenzetti, F. Poggialini, V. Palleschi, Micro-laser-induced breakdown spectroscopy (micro-LIBS) study on ancient Roman mortars, Appl. Spectrosc. 71 (2017) 721–727.

[128] S. Pagnotta, M. Lezzerini, B. Campanella, G. Gallello, E. Grifoni, S. Legnaioli, G. Lorenzetti, F. Poggialini, S. Raneri, A. Safi, V. Palleschi, Fast quantitative elemental mapping of highly inhomogeneous materials by micro-laser-induced breakdown spectroscopy, Spectrochim. Acta Part B At. Spectrosc. 146 (2018) 9–15.

[129] S. Raneri, S. Pagnotta, M. Lezzerini, S. Legnaioli, V. Palleschi, S. Columbu, N.F. Neri, P. Mazzoleni, Examining the reactivity of volcanic ash in ancient mortars by using a micro-chemical approach, Mediterr. Archaeol. Archaeom. 18 (2018) 147–157.

[130] B. Campanella, I. Degano, E. Grifoni, S. Legnaioli, G. Lorenzetti, S. Pagnotta, F. Poggialini, V. Palleschi, Identification of inorganic dyeing mordant in textiles by surface enhanced laser-induced breakdown spectroscopy, Microchem. J. 139 (2018) 230–235.

[131] M.A. Aguirre, S. Legnaioli, F. Almodóvar, M. Hidalgo, V. Palleschi, A. Canals, Elemental analysis by surface enhanced laser-induced breakdown spectroscopy combined with liquid-liquid microextraction, Spectrochim. Acta Part B At. Spectrosc. 79–80 (2013) 88–93.

[132] G. Agrosì, G. Tempesta, E. Scandale, S. Legnaioli, G. Lorenzetti, S. Pagnotta, V. Palleschi, A. Mangone, M. Lezzerini, Application of laser induced breakdown spectroscopy to the identification of emeralds from different synthetic processes, Spectrochim. Acta Part B At. Spectrosc. 102 (2014) 48–51.

[133] G. Tempesta, G. Agrosì, Standardless, minimally destructive chemical analysis of red beryls by means of laser induced breakdown spectroscopy, Eur. J. Mineral. 28 (2016) 571–580.

[134] S. Duchêne, V. Detalle, R. Bruder, J.B. Sirven, Chemometrics and laser induced breakdown spectroscopy (LIBS) analyses for identification of wall paintings pigments, Curr. Anal. Chem. 6 (2010) 60–65.

[135] D. Syvilay, B. Bousquet, R. Chapoulie, M. Orange, F.-X. Le Bourdonnec, Advanced statistical analysis of LIBS spectra for the sourcing of obsidian samples, J. Anal. At. Spectrom. 34 (2019) 867–873.

[136] D. Syvilay, V. Detalle, N. Wilkie-Chancellier, A. Texiera, L. Martinez, S. Serfaty, Novel approach of signal normalization for depth profile of cultural heritage materials, Spectrochim. Acta Part B At. Spectrosc. 127 (2017) 28–33.

[137] S. Pagnotta, E. Grifoni, S. Legnaioli, M. Lezzerini, G. Lorenzetti, V. Palleschi, Comparison of brass alloys composition by laser-induced breakdown spectroscopy and self-organizing maps, Spectrochim. Acta Part B At. Spectrosc. 103–104 (2015) 70–75.

[138] E. D'Andrea, B. Lazzerini, V. Palleschi, Combining multiple neural networks to predict bronze alloy elemental composition, in: S. Bassis, A. Esposito, F. Morabito, E. Pasero (Eds.), Advances in Neural Networks. WIRN 2015. Smart Innovation, Systems and Technologies, vol. 54, Springer, Cham, 2016, pp. 345–352, https://doi.org/10.1007/978-3-319-33747-0_34.

[139] S. Grégoire, M. Boudinet, F. Pelascini, F. Surma, V. Detalle, Y. Holl, Laser-induced breakdown spectroscopy for polymer identification, Anal. Bioanal. Chem. 400 (2011) 3331–3340.

[140] A. Arafat, M. Na'es, V. Kantarelou, N. Haddad, A. Giakoumaki, V. Argyropoulos, D. Anglos, A.-G. Karydas, Combined in situ micro-XRF, LIBS and SEM-EDS analysis of base metal and corrosion products for Islamic copper alloyed artefacts from umm Qais museum, Jordan, J. Cult. Herit. 14 (2013) 261–269.

[141] S. Hemeda, Laser induced breakdown spectroscopy and other analytical techniques applied on construction materials at Kom El-Dikka, Alexandria, Egypt, Mediterr. Archaeol. Archaeom. 13 (2013) 103–119.

[142] L. Pardini, A. El Hassan, M. Ferretti, A. Foresta, S. Legnaioli, G. Lorenzetti, E. Nebbia, F. Catalli, M.A. Harith, D. Diaz Pace, F. Anabitarte Garcia, M. Scuotto, V. Palleschi, X-ray fluorescence and laser-induced breakdown spectroscopy analysis of Roman silver denarii, Spectrochim. Acta Part B At. Spectrosc. 74-75 (2012) 156–161.

[143] S. Siano, J. Agresti, Archaeometallurgical characterisation of Donatello's Florentine copper alloy masterpieces using portable laser-induced plasma spectroscopy and traditional techniques, Stud. Conserv. 60 (1) (2015) 106–119.

[144] A. Brysbaert, P. Siozos, M. Vetters, A. Philippidis, D. Anglos, Materials analyses of pyrotechnological objects from LBA Tiryns, Greece, by means of laser-induced breakdown spectroscopy (LIBS): results and a critical assessment of the method, J. Archaeol. Sci. 83 (2017) 49–61.

[145] K. Hatzigiannakis, K. Melessanaki, A. Philippidis, O. Kokkinaki, E. Kalokairinou, P. Siozos, P. Pouli, E. Politaki, A.P. Aki, A. Dokoumetzidis, E. Katsaveli, E. Kavoulaki, V. Sithiakaki, Monitoring and mapping of deterioration products on cultural heritage monuments using imaging and laser spectroscopy, in: A. Moropoulou, M. Korres, A. Georgopoulos, C. Spyrakos, C. Mouzakis (Eds.), transdisciplinary multispectral modeling and cooperation for the preservation of cultural heritage TMM_CH 2018 Athens, Greece, October 10–13, 2018, Commun. Comput. Inf Sci. 962 (2018) 419–429.

[146] M. Veneranda, N. Prieto-Taboada, S. Fdez-Ortiz de Vallejuelo, M. Maguregui, H. Morillas, I. Marcaida, K. Castro, F.-J. Garcia-Diego, M. Osanna, J.M. Madariaga, In- situ multianalytical approach to analyze and compare the degradation pathways jeopardizing two murals exposed to different environments (Ariadne house, Pompeii, Italy), Spectrochim. Acta A Mol. Biomol. Spectrosc. 203 (2018) 201–209.

[147] I. Gaona, P. Lucena, J. Moros, F.J. Fortes, S. Guirado, J. Serrano, J.J. Laserna, Evaluating the use of standoff LIBS in architectural heritage: surveying the Cathedral of Málaga, J. Anal. At. Spectrom. 28 (6) (2013) 810–820.

[148] V. Lazic, LIBS analysis of liquids and of materials inside liquids, in: S. Musazzi, U. Perrini (Eds.), Laser-Induced Breakdown Spectroscopy. Theory and Applications, Springer Series in Optical Sciences, vol. 182, 2014, pp. 195–225.

[149] A. De Giacomo, M. Dell'Aglio, O. De Pascale, M. Capitelli, From single pulse to double pulse ns-laser induced breakdown spectroscopy under water: elemental analysis of aqueous solutions and submerged solid samples, Spectrochim. Acta Part B At. Spectrosc. 62 (2007) 721–738.

[150] M. López-Claros, M. Dell'Aglio, R. Gaudiuso, A. Santagata, A. De Giacomo, F.J. Fortes, J.J. Laserna, Double pulse laser induced breakdown spectroscopy of a solid in water: effect of hydrostatic pressure on laser induced plasma, cavitation bubble and emission spectra, Spectrochim. Acta Part B At. Spectrosc. 133 (2017) 63–71.

[151] A. De Giacomo, A. De Bonis, M. Dell'Aglio, O. De Pascale, R. Gaudiuso, S. Orlando, A. Santagata, G.S. Senesi, F. Taccogna, R. Teghil, Laser ablation of graphite in water in a range of pressure from 1 to 146 atm using single and double pulse techniques for the production of carbon nanostructures, J. Phys. Chem. C 115 (2011) 5123–5130.

[152] D.C.S. Beddows, O. Samek, M. Liška, H.H. Telle, Single-pulse laser-induced breakdown spectroscopy of samples submerged in water using a single-fibre light delivery system, Spectrochim. Acta Part B At. Spectrosc. 57 (2002) 1461–1471.

[153] T. Sakka, H. Oguchi, S. Masai, K. Hirata, Y.H. Ogata, M. Saeki, H. Ohba, Use of a long-duration ns pulse for efficient emission of spectral lines from the laser ablation plume in water, Appl. Phys. Lett. 88 (2006) 61120.

[154] S. Guirado, F.J. Fortes, V. Lazic, J.J. Laserna, Chemical analysis of archeological materials in submarine environments using laser-induced breakdown spectroscopy. On-site trials in the Mediterranean Sea, Spectrochim. Acta Part B At. Spectrosc. 74–75 (2012) 137–143.

[155] S. Guirado, F.J. Fortes, J.J. Laserna, Elemental analysis of materials in an underwater archeological shipwreck using a novel remote laser-induced breakdown spectroscopy system, Talanta 137 (2015) 182–188.

[156] M. López-Claros, F.J. Fortes, J.J. Laserna, Subsea spectral identification of shipwreck objects using laser-induced breakdown spectroscopy and linear discriminant analysis, J. Cult. Herit. 29 (2018) 75–81.

[157] F.J. Fortes, S. Guirado, A. Metzinger, J.J. Laserna, A study of underwater stand-off laser-induced breakdown spectroscopy for chemical analysis of objects in the deep ocean, J. Anal. At. Spectrom. 30 (2015) 1050–1056.

[158] B. Thornton, T. Sakka, T. Takahashi, A. Tamura, T. Masamura, A. Matsumoto, Spectroscopic measurements of solids immersed in water at high pressure using a long duration nano second laser pulse, Appl. Phys. Express 6 (2013), 082401.

[159] B. Thornton, T. Sakka, T. Masamura, A. Tamura, T. Takahashi, A. Matsumoto, Long-duration nanosecond single pulse lasers for observation of spectra from bulk liquids at high hydrostatic pressures, Spectrochim. Acta Part B At. Spectrosc. 97 (2014) 7–12.

[160] T. Sakka, A. Tamura, A. Matsumoto, K. Fukami, N. Nishi, B. Thornton, Effects of pulse width on nascent laser-induced bubbles for underwater laser-induced breakdown spectroscopy, Spectrochim. Acta Part B At. Spectrosc. 97 (2014) 94–98.

[161] T. Takahashi, B. Thornton, K. Ohki, T. Sakka, Calibration-free analysis of immersed brass alloys using long-ns-duration pulse laser-induced breakdown spectroscopy with and without correction for nonstoichiometric ablation, Spectrochim. Acta Part B At. Spectrosc. 111 (2015) 8–14.

[162] A. Matsumoto, A. Tamura, R. Koda, K. Fukami, Y.H. Ogata, N. Nishi, B. Thornton, T. Sakka, A calibration-free approach for on-site multi-element analysis of metal ions in aqueous solutions by electrodeposition-assisted underwater laser-induced breakdown spectroscopy, Spectrochim. Acta Part B At. Spectrosc. 118 (2016) 45–55.

[163] T. Takahashi, B. Thornton, T. Sato, T. Ohki, K. Ohki, T. Sakka, Temperature based segmentation for spectral data of laser-induced plasmas for quantitative compositional analysis of brass alloys submerged in water, Spectrochim. Acta Part B At. Spectrosc. 124 (2016) 87–93.

[164] S. Yoshino, B. Thornton, T. Takahashi, Y. Takaya, T. Nozaki, Signal preprocessing of deep-sea laser-induced plasma spectra for identification of pelletized hydrothermal deposits using artificial neural networks, Spectrochim. Acta Part B At. Spectrosc. 145 (2018) 1–7.

[165] B. Thornton, T. Takahashi, T. Sato, T. Sakka, A. Tamura, A. Matsumoto, T. Nozaki, T. Ohki, K. Ohki, Development of a deep-sea laser-induced breakdown spectrometer for in situ multi-element chemical analysis, Deep Sea Res. Part I Oceanogr. Res. Pap. 95 (2015) 20–36.

[166] R. Gaudiuso, M. Dell'Aglio, O. De Pascale, S. Loperfido, A. Mangone, A. De Giacomo, Laser-induced breakdown spectroscopy (LIBS) of archaeological findings with calibration-free inverse method: comparison with classical LIBS and conventional techniques, Anal. Chim. Acta 813 (2014) 15–24.

[167] M. Rossi, M. Dell'Aglio, A. De Giacomo, R. Gaudiuso, G.S. Senesi, O. De Pascale, F. Capitelli, F. Nestola, M.R. Ghiara, Multi-methodological investigation of kunzite, hiddenite, alexandrite, elbaite and topaz, based on laser induced breakdown spectroscopy and conventional analytical techniques for supporting mineralogical characterization, Phys. Chem. Miner. 41 (2014) 127–140.

[168] D. Syvilay, A. Texier, A. Arles, B. Gratuze, N. Wilkie-Chancellier, L. Martinez, S. Serfaty, V. Detalle, Trace element quantification of lead based roof sheets of historical monuments by laser induced breakdown spectroscopy, Spectrochim. Acta Part B At. Spectrosc. 103–104 (2015) 34–42.

[169] S.S. Rai, N.K. Rai, A.K. Rai, U.C. Chattopadhyaya, Rare earth elements analysis in archaeological pottery by laser induced breakdown spectroscopy, Spectrosc. Lett. 49 (2016) 57–62, https://doi.org/10.1080/00387010.2015.1072094.

[170] O. Kokkinaki, C. Mihesan, M. Velegrakis, D. Anglos, Comparative study of laser induced breakdown spectroscopy and mass spectrometry for the analysis of cultural heritage materials, J. Mol. Struct. 1044 (2013) 160–166.

[171] O. Syta, B. Wagner, E. Bulska, D. Zielińska, G.Z. Żukowskac, J. Gonzalez, R. Russo, Elemental imaging of heterogeneous inorganic archaeological samples by means of simultaneous laser induced breakdown spectroscopy and laser ablation inductively coupled plasma mass spectrometry measurements, Talanta 179 (2018) 784–791.

[172] T. Trejos, A. Flores, J.R. Almirall, Micro-spectrochemical analysis of document paper and gel inks by laser ablation inductively coupled plasma mass spectrometry and laser induced breakdown spectroscopy, Spectrochim. Acta Part B At. Spectrosc. 65 (2010) 884–895.

[173] M. Hoehse, A. Paul, I. Gornushkin, U. Panne, Multivariate classification of pigments and inks using combined Raman spectroscopy and LIBS, Anal. Bioanal. Chem. 402 (2012) 1443–1450.

[174] A. Metzinger, R. Rajkó, G. Galbács, Discrimination of paper and print types based on their laser induced breakdown spectra, Spectrochim. Acta Part B At. Spectrosc. 94–95 (2014) 48–57.

[175] H.E. Ahmed, O.A. Nassef, M.A. Harith, The eye of Horus viewed by the spectrochemical analytical eye of LIBS, Archaeol. Anthropol. Sci. 11 (2019) 5053–5063, https://doi.org/10.1007/s12520-019-00861-0.

[176] J. Qi, T. Zhang, H. Tang, H. Li, Rapid classification of archaeological ceramics via laser-induced breakdown spectroscopy coupled with random forest, Spectrochim. Acta Part B At. Spectrosc. 149 (2018) 288–293.

[177] R.-J. Lasheras, J. Anzano, C. Bello-Gálvez, M. Escudero, J. Cáceres, Quantitative analysis of roman archeological ceramics by laser-induced breakdown spectroscopy, Anal. Lett. 50 (2017) 1325–1334.

[178] P. Singh, E. Mal, A. Khare, S. Sharma, A study of archaeological pottery of Northeast India using laser induced breakdown spectroscopy (LIBS), J. Cult. Herit. 33 (2018) 71–82.

[179] S. Mahmood, S.A. Abbasi, S. Jabeen, M.A. Baig, Laser-induced breakdown spectroscopic studies of marbles, J. Quant. Spectrosc. Radiat. Transf. 111 (2010) 689–695.

[180] M. Fahad, M. Abrar, Laser-induced breakdown spectroscopic studies of calcite (CaCO3) marble using the fundamental (1064 nm) and second (532 nm) harmonic of a Nd:YAG laser, Laser Phys. 28 (2018), 085701.

[181] F. Colao, R. Fantoni, P. Ortiz, M.A. Vazquez, J.M. Martin, R. Ortiz, N. Idris, Quarry identification of historical building materials by means of laser induced breakdown spectroscopy, X-ray fluorescence and chemometric analysis, Spectrochim. Acta Part B At. Spectrosc. 65 (2010) 688–694.

[182] S. Awasthi, R. Kumar, G.K. Rai, A.K. Rai, Study of archaeological coins of different dynasties using libs coupled with multivariate analysis, Opt. Lasers Eng. 79 (2016) 29–38.

[183] Y. Li, D. Tian, Y. Ding, G. Yang, K. Liu, C. Wang, X. Han, A review of laser-induced breakdown spectroscopy signal enhancement, Appl. Spectrosc. Rev. 53 (2018) 1–35.

[184] Z. Tang, R. Zhou, Z. Hao, S. Ma, W. Zhang, K. Liu, X. Li, X. Zeng, Y. Lu, Microdestructive analysis with high sensitivity using double-pulse resonant laser-induced breakdown spectroscopy, J. Anal. At. Spectrom. 34 (2019) 1198.

[185] C. Goueguel, S. Laville, F. Vidal, M. Chaker, M. Sabsabi, Resonant laser-induced breakdown spectroscopy for analysis of lead traces in copper alloys, J. Anal. At. Spectrom. 26 (2011) 2452–2460.

[186] A. De Giacomo, R. Gaudiuso, C. Koral, M. Dell'Aglio, O. De Pascale, Nanoparticle-enhanced laser induced breakdown spectroscopy of metallic samples, Anal. Chem. 85 (2013) 10180–10187.

[187] A. De Giacomo, R. Gaudiuso, C. Koral, M. Dell'Aglio, O. De Pascale, Nanoparticle enhanced laser induced breakdown spectroscopy: effect of nanoparticles deposited on sample surface on laser ablation and plasma emission, Spectrochim. Acta Part B At. Spectrosc. 98 (2014) 19–27.

[188] C. Koral, M. Dell'Aglio, R. Gaudiuso, R. Alrifai, M. Torelli, A. De Giacomo, Nanoparticle-enhanced laser induced breakdown spectroscopy for the noninvasive analysis of transparent samples and gemstones, Talanta 182 (2018) 253–258.

Neutron diffraction

Winfried Kockelmann[a] and Evelyne Godfrey[b]

[a]*STFC Rutherford Appleton Laboratory, ISIS Neutron Facility, Oxfordshire, United Kingdom*
[b]*Uffington Heritage Watch Ltd, Faringdon, Oxfordshire, United Kingdom*

1 Introduction

Neutrons are present in all atoms except ordinary hydrogen. As constituents of the atomic nuclei, neutrons are stable. They have a mass, a magnetic moment, but no charge. When released by nuclear reactions, free neutrons become unstable and disintegrate on average in 15 minutes into electrons and antineutrinos. During their lifetime, neutrons can be used as a probe to study compositions, structures, magnetic properties, and collective motions of materials. Since neutrons are uncharged and as they interact via the short-range strong force with atomic nuclei, neutrons pass through substantial thicknesses of many materials without interaction. The probability for neutrons to be absorbed or scattered in a material is much lower than for many other particle probes such as X-rays and electrons.

Neutron diffraction is one of the methods for analysis of condensed materials in general, including cultural heritage materials. The first neutron diffraction experiment was reported by Mitchell and Powers in 1936 [1]. Systematic crystal and magnetic structure studies of materials started in the 1950s with the installation of intense neutron reactor sources. In comparison, the analysis of archaeological objects using neutron diffraction is rather recent [2–4]. Nowadays, neutron diffraction studies are performed at large-scale facilities such as nuclear reactors and accelerator-based neutron spallation sources.

Neutron diffraction is used for determining the bulk structures and microstructures of crystalline materials. The analysis is performed by noninvasive means, and is therefore consistent with one of the main ethical considerations of conservation science [5]. Assemblages of intact artifacts can be analyzed with no need to remove samples, and no special preparation or environment is required. A far greater number of objects can be analyzed than would ever be possible by metallographic techniques, as curators rarely allow destructive sampling for optical microscopy. The aim of crystallographic characterization studies is to establish the type of materials present and to look at particularities of the structures in order to reconstruct

253

Spectroscopy, Diffraction and Tomography in Art and Heritage Science. https://doi.org/10.1016/B978-0-12-818860-6.00006-4

technical details of fabrication methods, such as alloying, casting, hammering, and heat-treatment in early metalworking. Fig. 1 illustrates the underlying basic process of neutron diffraction and of the structural characterization options, based on Bragg's law and the concept of reflection of neutrons by crystal planes.

A main advantage of neutron diffraction is the high-penetration property of a neutron beam and the option for nondestructive analysis of materials that may be opaque to X-rays. The structure of an artifact or object is unaltered by the neutron beam. In any case, the X-ray penetration depths of materials are usually measured in microns whereas the penetration depths of neutrons are in the range of millimeters and centimeters. The method measures right through the thickness of an object, providing averaged, quantitative structure data from internal parts. Emerging neutron diffraction imaging techniques based on Bragg edge transmission provide new opportunities for structure studies of archaeological materials in a radiography- or tomography-type setup.

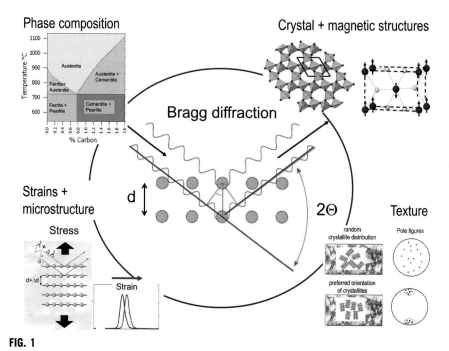

FIG. 1

Bragg diffraction: the basis of structural characterization methods using neutrons for phase separation, crystal and magnetic structure analysis, texture and strain analysis.

2 Basics of neutron diffraction

2.1 Bragg's law

Diffraction is a general phenomenon that occurs if waves impinge onto an obstacle; they can be sound waves, light waves, X-rays, or particle waves. Diffraction is based on the superposition of waves that re-enforce each other (constructive interference) or cancel each other out (destructive interference). Diffraction occurs for scattering of waves on periodic structures such as atoms in a crystal. Neutrons are particles, but according to the wave-particle dualism of quantum mechanics, they are also waves with an amplitude, wavelength and phase, and as such display interference phenomena. The prerequisite for the diffraction process in a material is that the radiation has a wavelength comparable in magnitude with the atomic spacing. Slow neutrons with wavelengths of a few angstroms (10^{-10} m) match interatomic distances and are therefore diffracted by crystals. Diffraction is synonymous with elastic coherent scattering or Bragg scattering, one of several interaction mechanisms that may occur when neutrons impinge on a material. Other scattering processes termed inelastic coherent, elastic incoherent, and inelastic incoherent will not be further considered.

Cultural heritage materials are usually polycrystalline and multiphase opposed to single-crystal and single-phase. Accordingly, experimental and data analysis approaches are adopted from established neutron powder diffraction techniques. Diffraction by a polycrystal is described by the fundamental relation of crystallography, Bragg's law:

$$2d_{hkl}\sin\theta = \lambda \tag{1}$$

where d_{hkl} are the distances between crystal planes with Miller indices h, k, l. θ is half of the scattering angle and λ is the neutron wavelength. The central part of Fig. 1 illustrates Bragg diffraction of neutrons as reflection by crystal planes.

There are two experimental setups, angle-dispersive and energy-dispersive diffraction as illustrated in Fig. 2. The angle-dispersive mode uses a monochromatic beam, i.e., neutrons of a single wavelength, λ, to obtain diffraction peaks at varying scattering angles 2θ, depending on the distribution of the lattice plane spacings d_{hkl}. The wavelength-dispersive mode requires a "white," polychromatic beam with a broad range of wavelengths, so that neutrons diffracted into a fixed angle 2θ yield peaks at varying wavelengths. For the latter, neutron wavelengths are usually measured using the time-of-flight (TOF) method. For both modes, the measured patterns of Bragg reflections exhibit peak positions, peak shapes, and peak intensities that are characteristic for one or more crystallographic phases in the material and that are analyzed in similar ways to X-ray diffraction patterns.

2.2 Analysis of a neutron diffraction pattern

Bragg peaks (or Bragg reflections) occurring at characteristic positions in a diffraction pattern are denoted by Miller indices (hkl), where h, k, l are integers. For example, (100), (110), and (111) are the cube, face diagonal, and space diagonal planes of

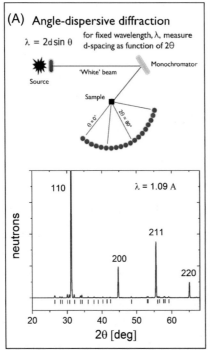
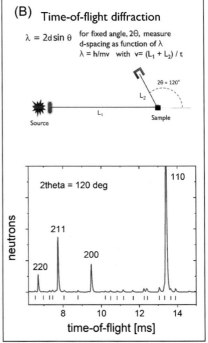

FIG. 2

Angle-dependent (A) and time-of-flight (B) neutron diffraction. Simulated neutron diffraction patterns for a fixed 2θ and fixed λ, respectively, for a ferritic steel with 0.8 wt% carbon. Ferrite peaks are labeled with (hkl) indices; positions of Fe_3C cementite peaks are indicated by tick marks.

a cubic lattice, but also denote corresponding peaks in a diffraction pattern. In fact, a peak in a powder diffraction pattern corresponds to a family of planes in the crystal, for example, {100} stands for a set of planes with sixfold multiplicity: (100), (010), (001), (-100), (0-10), (00-1). Diffraction analysis involves comparing a measured diffraction pattern with a calculated pattern, based on the most probable multiphase structure model of the material under investigation.

2.2.1 Neutron intensity from a multiphase material

For a material with NP crystalline phases, each phase generates a sequence of Bragg peaks. The intensity y_i for each point i in the diffraction pattern is calculated using [6]:

$$y_i(t) = yb_i + K \sum_{p=1}^{NP} s_p A_p \sum_{hkl} M_{hkl} |F_{hkl}|^2 P_{hkl} H(t - t_{hkl}) \tag{2}$$

where t is either 2θ or the time-of-flight; t_{hkl} is the Bragg peak position; yb_i is the background count; K is an overall scale factor related to, for example, the illuminated volume; s_p is the phase scale factor; A_p is an absorption factor; M_{hkl}, F_{hkl}, and P_{hkl} are the peak multiplicity, structure factor, and texture factor, respectively; H is the profile function describing the shape of the diffraction peak. Note that the intensity is obtained at every point of the pattern by summing diffraction contributions from every phase and from every crystal plane $\{hkl\}$ in case of peak overlap.

The phase scale factor, s_p, is converted to a weight fraction, w_p, of the pth phase by the normalized product [7]

$$w_p = \frac{s_p z_p M_p V_p}{\sum_{j=1}^{NP} s_j z_j M_j V_j} \tag{3}$$

where M_p, z_p, and V_p are the mass of the formula unit, the number of formula units per unit cell, and the unit cell volume, respectively. The summation in the denominator accounts for all crystalline phases, and accordingly the weight fractions add up to $100\,\mathrm{wt}\%$.

2.2.2 Bragg peak positions: Lattice parameters and indexing

The first task is to identify the phases that are accountable for the peaks in a diffraction pattern. Identification of metal or mineral phases is based on peak positions and intensities of the strongest Bragg peaks, leading to an estimation of the size of the unit cell, i.e., the lattice parameters, and an indexing scheme of all peaks with Miller indices. For a cubic lattice, for instance, spacing d_{hkl} and lattice parameter, a, are related by:

$$d_{hkl} = \frac{a}{\sqrt{h^2 + k^2 + l^2}} \tag{4}$$

On X-ray diffractometers, indexing is usually fully automated. For neutron diffraction, indexing is still often performed via a trial-and-error approach and by comparison with X-ray database patterns. In practice, it is useful to create a list of peak parameters including positions, widths, and intensities of the most intense peaks, and compare those with the simulated neutron diffraction patterns of guess-structures.

2.2.3 Bragg peak intensities: Structure factor

The neutron intensity formula (Eq. 2) contains one of the fundamental quantities of crystallography, the structure factor:

$$F_{hkl} = \sum_{j=1}^{N} b_j t_j \exp\left[2\pi i \left(hx_j + ky_j + lz_j\right)\right] \tag{5}$$

where the neutron scattering length b_j takes care of the interaction of the neutron with a nucleus in the crystal. The sum is over N atoms and $3N$ locations x, y, z of atoms in the unit cell. The Debye Waller factor (or temperature factor), t_j, accounts for the displacements of atoms from their idealized positions due to vibrations. With rising temperature, Debye Waller factors suppress Bragg intensities more. Temperature parameters are taken from the structure databases and CIF files (discussed in Section 2.2.4).

The neutron scattering length, b_j, given in Fermi (fm: 10^{-15} m) corresponds to the form factor in X-ray diffraction and represents the neutron-nucleus interaction. This interaction is short-range compared to the neutron wavelength, which is in the range of 10,000 fm. The diffraction process as such, and the form of intensity formulas and structure factors are very similar for neutrons and X-rays. The peak positions are the same, and X-ray and neutron diffraction patterns of the same substance may look similar, especially for single-atom structures. This is the justification for X-ray databases to be used for indexing neutron data. However, there are also distinct differences. Neutron scattering is sensitive to different isotopes of the same element or can be drastically different for neighboring elements in the periodic table. Interactions can be large for light elements and weak for heavy elements. For example, hydrogen atoms and water molecules are very visible in the vicinity of heavy elements, whereas lead is almost transparent for neutrons. In contrast, X-ray scattering power increases monotonically with atomic number as it is the electrons which scatter X-rays. This is one property which makes the two techniques complementary. Neutron scattering lengths were compiled by Sears [8] and are listed online at the NIST Centre for Neutron Research [9]. Many data analysis packages include tables of scattering lengths of isotopes, as well as weighted averages for natural abundances of isotopes in an element.

It is worth mentioning a second interaction mechanism of neutrons with matter: the magnetic interactions with the magnetic moments of partially occupied electron shells. Since diffraction requires periodicity, magnetic Bragg diffraction only occurs for compounds with ordered magnetic structures below Curie or Neel temperatures. The magnetic structure factor can be constructed in a similar manner than the nuclear structure factor mentioned previously. The neutron scattering length b_j is replaced by a magnetic form factor; a magnetic interaction vector is introduced to take account of the distinct directionality of magnetic diffraction. Magnetic diffraction needs to be considered when analyzing iron and/or iron oxides. Ferritic iron exhibits a long-range ordered ferromagnetic structure, adding magnetic intensities to the "nuclear" Bragg peaks. Magnetic Bragg peaks in the diffraction patterns from hematite and magnetite, due to antiferromagnetic and ferrimagnetic structures, respectively, can be used to increase the sensitivity for detecting those compounds.

2.2.4 Relevance of crystal symmetry

Symmetry is an important concept to describe a crystal structure that is constructed from repeating the smallest building block, the unit cell. The unit cell belongs to one of 14 Bravais lattice types: cubic, hexagonal, tetragonal, etc.; with centering or not: e.g., primitive, face centered cubic (fcc), body centered cubic (bcc). The size of the unit cell is given by the lattice parameters: $a, b, c, \alpha, \beta, \gamma$. For a cubic cell, there is only one parameter to be considered (i.e., a). The crystal structure is then fully described by one of 230 space groups listed in the International Tables of Crystallography [10] and by the types and positions of atoms in the unit cell.

The description of symmetry and atom positions in a space group massively simplifies the calculations of diffraction effects and facilitates the analysis of diffraction data. In practice, the diffraction analysis requires the input of: space group, lattice parameters, and positions of representative atoms on crystallographic sites from which all atoms are generated by symmetry elements (e.g., rotations, mirrors, translations) of the space group. Crystallographic analysis software packages take care of the symmetry calculations, and many allow importing the complete symmetry and structure information via the Crystallographic Interchange Format (CIF). CIF files for crystal structures can be obtained from, for example, ICSD [11] or COD [12], and are also used to simulate neutron diffraction patterns.

Diffraction peaks may be "allowed" or "extinct" due to crystal symmetry. Therefore a glance at a neutron or X-ray diffraction pattern often hints at the structure type of the material. Structures with the same space group appear similar in the diffraction image, for example, for fcc metals such as austenitic iron, copper, and aluminum which all crystallize in space group Fm-3m. A crystal with a small unit cell and high symmetry, for example, a cubic metal structure, produces a rather clear diffraction pattern with few peaks. In contrast, a structure with a large unit cell and low symmetry, for example, a monoclinic feldspar, has a complicated diffraction pattern with a large number of peaks.

2.2.5 Texture analysis

The analysis of crystallographic texture can provide information about the history of deformation and heat treatment of metals, alloys, and ceramics. If the microcrystallites in a polycrystalline material are oriented at random, i.e., all grain orientations are equally realized, then the material is said to have no texture. Distinct textures are produced by specific conditions, for example, during solidification, annealing, cold or hot working [13] leaving the crystallites with some preferred orientation with reference to a sample coordinate system. Preferential orientations in all or part of the sample are described by the orientation distribution function (ODF), which has a distinct statistical interpretation in that it specifies for the interrogated volume the number of crystallites with certain orientations.

Neutron texture analysis is an established method to determine these orientation distributions. Owing to the high penetration power of neutrons and to the large beam sizes used for neutron diffraction, textures for course-grained materials with grain sizes up to millimeters can be studied. For engineering or archaeological materials, the ODF is usually determined by collecting a set of pole figures, by recording intensities for several Bragg peaks and for many sample (or detector) angles as illustrated in Fig. 3A. The presence of texture manifests itself as intensity variations (valleys and troughs) in a pole figure for a certain crystal plane; pole densities are a measure of the number of crystal planes in a certain direction, given in multiples of a random distribution (m.r.d.). Pole figures of a texture-free material has m.r.d. $= 1$ pole densities everywhere. Maximum m.r.d. values and a texture index quantify the strength of a texture [13, 14]. Texture is presented using direct pole figures or inverse pole figures. A pole density in a direct pole figure represents the number of crystallites in the sample coordinate system. A pole density point in the equivalent inverse pole figures indicates how strongly a certain crystal direction (e.g., a diagonal ⟨111⟩) is aligned along a unique sample direction (e.g., a wire axis). A pole figure is a two-dimensional projection of the three-dimensional ODF. The ODF is reconstructed from a number of pole figures and vice versa; once the ODF is known, every pole figure can be recalculated. While pole figures are rather abstract maps, they are nonetheless convenient illustrations of crystallographic texture which, in turn, is characteristic for the material treatment. Fig. 3B shows three pole figures for four bronze reference samples, for which increasing working-hardening is reflected by more pronounced (110) fiber textures and higher texture strengths.

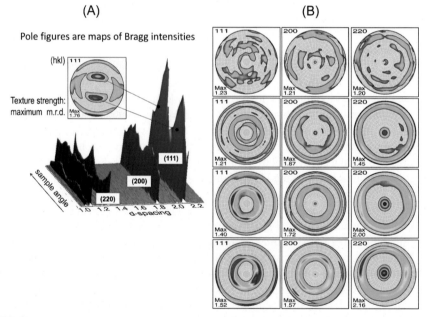

FIG. 3

Neutron texture analysis. (A) Pole figures are obtained by mapping Bragg peak intensities as a function of sample angle. (B) (111), (200), and (220) experimental pole figures of as-received (top) and work-hardened bronze specimens for thickness reductions of 8%, 20%, and 45%, respectively.

Data from W. Kockelmann, S. Siano, L. Bartoli, D. Visser, H. Hallebeek, R. Traum, R. Linke, M. Schreiner, A. Kirfel, Applications of TOF neutron diffraction in archaeometry, Appl. Phys. A 83 (2006) 175–182.

2.2.6 Strain and microstructure analysis

Diffraction allows for measurements of both macroscopic lattice deformation (residual strains) and of microscopic crystallite deformation resulting in peak shifts and peak broadenings, respectively. Residual strains are locked-in strains that are induced by macroscopic compressive or tensile stresses, or plastically or thermally induced misfits between different parts of a component. Atomic lattice planes are used as atomic strain gauge for residual stress analysis. The shifts of lattice spacings are compared to the spacing d_{hkl}^0 of the same planes measured in an unstressed material. The relative elastic strain ε_{hkl} of one strain component is obtained from the deformed lattice spacing d_{hkl} by:

$$\varepsilon_{hkl} = \frac{d_{hkl} - d_{hkl}^0}{d_{hkl}^0} \tag{6}$$

To calculate the full strain (tensor) properties at a selected analysis point in the sample, six independent strain directions are to be measured. This number can be reduced to three principal components if assumption on the sample symmetry is made. Residual stresses can then be calculated from the strains using the stiffness tensor.

Peak broadening generally indicates deviations from ideal structures in terms of finite sizes of crystallites and defects [14]. Slow heating and annealing reduce crystal defects and relieve lattice strains whereas a rough treatment like hammering and quenching in water generates microstrains. The evaluation of peak broadening may provide a handle to analyze microstrains, lattice deformations, particle or domain sizes, and compositional variations. Microstrains and lattice deformations induced by mechanical and thermal working processes lead to a measurable spread of lattice plane distances. Particle size broadening is caused by the average size of the crystallites being small (typically <1 μm for neutron diffraction); large crystal domains do not affect the peak width and thus cannot be analyzed. Variation of the material composition in the diffracting volume may cause peak broadening as well, of a type similar to microstrain broadening. Furthermore, for a very inhomogeneous material the spread of lattice parameters, for a dendritic microstructure, for example, may lead to markedly structured peak profiles [15]. Several broadening contributions are often superposed which makes it extremely challenging to analyze and interpret them. The differentiation between size and microstrain broadening may be attempted by a Williams-Hall plot analysis [16]. To study peak broadening effects, the profile function H in Eq. (2) needs to be calibrated with good accuracy, in order to separate intrinsic sample broadening effects from instrument broadening.

2.3 Neutron diffraction imaging

A recently developed energy-resolving neutron radiography method permits analysis of diffraction signals in transmission. The technique requires a pulsed neutron beam and a time-resolving imaging camera behind the object and thus is based on the measurement of TOF. Fig. 4 illustrates the setup for diffraction imaging, in comparison to the normal diffraction setup. Structure parameters can be mapped using sharp

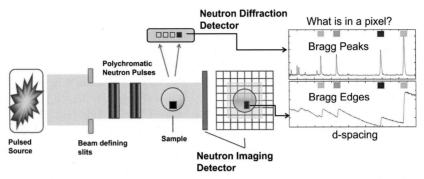

FIG. 4

Comparison of TOF Bragg diffraction and Bragg edge analysis with two detectors, at $2\theta \sim 90°$ and in transmission. There is a correspondence of positions of Bragg peaks and edges. *Colored squares* indicate that specific Bragg edges/peaks are recorded with different neutron wavelengths.

From W. Kockelmann, et al., Status of the neutron imaging and diffraction instrument IMAT, Phys. Procedia 69 (2015) 71–78.

features, the Bragg edges, which occur if neutrons are removed from the incident beam by Bragg scattering [4, 17, 18]. The positions, widths, and heights of Bragg edges are analyzed in a similar manner than Bragg peaks, thus leading to a 2D mapping (pixel-by-pixel) of the related structure properties (phase, strain, and texture) but with much higher spatial resolution than possible with diffraction. It is anticipated that this emerging method will extend the use of neutron diffraction techniques for archaeological science and conservation science.

The transmitted signal in a pixel (i, j) of the camera is given by Beer-Lambert's law:

$$p_{i,j}(\lambda) = p_{i,j}^0(\lambda) \exp(-\mu(\lambda)d) \tag{7}$$

$p_{i,j}$ and $p_{i,j}^0$ are the transmitted and incident neutron beam rays. μ is the averaged attenuation coefficient through the material along the path of the ray. d is the thickness of the material. The positions of the Bragg edges are defined by the condition $2\theta = 180°$, hence Bragg's law reduces to:

$$\lambda = 2d_{hkl} \tag{8}$$

The above considerations for diffraction with regard to structure factor, symmetry, texture, and signal broadening are also valid for the transmission case. The beam attenuation can be expressed using the sum of absorption, μ_{abs}, and diffraction, μ_{dif}, contributions, the latter being given by the Fermi relation for the coherent elastic scattering cross-section [19]. For a single phase:

$$\mu(\lambda) = \mu_{abs} + \mu_{dif} = \mu_{abs}(\lambda) + N\frac{\lambda^2}{4V}\sum_{d=0}^{2d<\lambda}|F_{hkl}|^2 d_{hkl}M_{hkl}P_{hkl} \tag{9}$$

where N is the particle density and V is the unit cell volume. The structure and texture properties are accounted for by the structure factor F_{hkl} and the pole density P_{hkl}.

2.4 Other diffraction modes

Neutron diffraction is, as discussed earlier, based on coherent scattering that requires long-range order of atoms or magnetic moments. As such, Bragg diffraction provides information about average structures and microstructures of crystalline components of a polycrystalline material. Diffraction at other levels, for example, single crystal diffraction, is beyond the scope of this text but two analysis methods shall be briefly mentioned.

Small angle neutron scattering (SANS) is a coherent scattering technique, where the interference effects are observed at very small scattering angles of fractions of a degree. This technique allows determining properties of "particles" in the size range from nanometers to micrometers, such as precipitates and voids in alloys, pores in ceramic, nanoparticles in suspension, and magnetic domains in ferromagnets. The primary interest is the determination of compositions, dimensions, shapes, and surface roughness properties of those "particles" in a statistical sense. Pore analysis of archaeological ceramics by SANS has been correlated to firing temperatures [20]. Materials need to be relatively thin, in the range of a few millimeters, thus normally the method cannot be applied on complex, composite objects.

Beyond the structure analysis using Bragg diffraction, a "total scattering" approach [21] analyses Bragg diffraction, diffuse scattering and small-angle scattering effects of a material together in one analysis step. In addition to the long-range average structure, local structure details are obtained from materials that are partially crystalline or have no long-range crystal structure, such as glasses. Although this technique has a promise to produce all-encompassing information, it is problematic to apply it to cultural heritage materials. Total scattering studies are mostly performed on single-phase materials, of basic shapes and with stringent requirements for normalization, background and absorption corrections, thereby largely precluding a nondestructive analysis of a composite object.

3 Instrumental and experimental considerations

3.1 Neutron sources

Neutron diffraction requires relatively slow ("thermal" or "cold") neutrons with velocities of a few thousand meters per second, corresponding to wavelengths in the range of a few angstroms. Intense neutron beams are produced by nuclear reactions, by fission or by spallation [22]. Nuclear fission of ^{235}U induced by slow neutrons is the underlying nuclear process in a research reactor. Diffractometers at reactors are usually operated in an angle-dispersive mode. Alternatively, neutrons can be produced by spallation using particle accelerators. At a spallation source, electrons or protons are accelerated to high energies and directed onto a metal target such as tungsten or lead. For a proton spallation source, the excited target nuclei release tens of neutrons per incident proton, mostly as a consequence of neutron "evaporation," creating in effect intense bursts of neutrons. Spallation sources produce continuous beams, but most of them are pulsed by nature of a pulsed proton beam. The ISIS facility in Oxfordshire, United Kingdom, is such a pulsed neutron source which uses an 800-MeV proton synchrotron and a composite tungsten/tantalum target to produce 50 sharp neutron pulses every second [23].

For both types of source, the generated neutrons are too fast to be used for diffraction. The energy of neutrons is reduced by collisions with nuclei in moderators consisting of hydrogenous materials, such as light or heavy water, liquid or solid methane, or liquid hydrogen. The slowed-down neutrons travel in vacuum tubes or supermirror guides to different stations downstream at distances between 10 and 100 m. Table 1 lists neutron diffractometers at large-scale facilities that have been used for cultural heritage materials studies. A reactor produces a continuous polychromatic beam from which a single wavelength is selected. In comparison, the neutron flux at a pulsed source is concentrated in short polychromatic bunches of fractions of a second, which are, for example, 20 ms and 100 ms long for target station 1 (TS1) and target station 2 (TS2) at the ISIS facility. Most of the instruments in Table 1 are neutron strain scanners, designed for studies on a variable range of small and large engineering components and well suited for analysis of museum objects.

3.2 TOF neutron diffraction

The overall majority of neutron diffraction studies on archaeological and historic objects have been performed at pulsed neutron spallation sources. At a pulsed source, a diffraction analysis has a particularly simple setup, and several structural aspects can be studied in one combined measurement. A pulsed source produces a sequence of polychromatic neutron bursts which contain shorter and longer wavelength neutrons that are almost produced at the same time. The initial sharp neutron pulse becomes broader, more spread out in time, when approaching the sample. The shorter wavelength neutrons in a pulse travel faster, therefore arrive earlier at the sample, whereas the longer wavelength neutrons travel with slower speed and arrive later at the sample. The pulsed neutron beam is an important feature of a spallation source, as it allows for the application of a TOF measurement.

In practice, the TOF is obtained from the difference between an electronic detector signal and a start signal provided by the particle accelerator at the time of neutron generation. Hence, neutrons with velocities v and kinetic energies E

$$E = \frac{1}{2}m_n v^2 = \frac{1}{2}m_n \frac{L^2}{t^2} \tag{10}$$

are separated by measuring their travel times, t, from source to detector given that the velocity is $v = L/t$, where $L = L_1 + L_2$ is the total flight path (see Fig. 2B) and m_n is the

Table 1 Neutron sources and neutron diffractometers for cultural heritage studies.

Center	Neutron source	Type	Diffracto-meter	Mode	Applications
Harwell Campus, United Kingdom	ISIS TS1 and TS2[a]	Spallation; pulsed	ENGIN-X, GEM, IMAT, INES	Polychromatic; TOF	Phase + structure; spatially-resolved strain/texture;diffraction imaging
Tokai Campus, Japan	J-PARC[b]	Spallation; pulsed	TAKUMI	Polychromatic; TOF	Spatially-resolved strain and texture
ANSTO, Australia	OPAL[c]	Reactor; continuous	KOWARI	Monochromatic; angle-dispersive	Spatially-resolved strain and texture
Oak Ridge National Lab, United States	SNS[d]	Spallation; pulsed	VULCAN	Polychromatic; TOF	Spatially-resolved strain and texture
Polygone Scientifique Grenoble, France	ILL[e]	Reactor; continuous	D20, SALSA	Monochromatic; angle-dispersive	Phase + structure; spatially-resolved strain and texture
Frank Laboratory Dubna, Russia	IBR-2[f]	Reactor; pulsed	DN-12	Polychromatic; TOF	Phase + structure; texture
TU Munich, Garching, Germany	FRM-II[g]	Reactor; continuous	STRESS-SPEC	Monochromatic; angle-dispersive	Spatially-resolved strain and texture

[a]www.isis.stfc.ac.uk.
[b]http://jparc.jp.
[c]www.ansto.gov.au.
[d]https://neutrons.ornl.gov/sns.
[e]www.ill.fr.
[f]http://flnph.jinr.ru/en/.
[g]www.frm-2.de. Accessed 15 January 2020.

mass of a neutron. The neutron velocity is related to the neutron wavelength, λ, via the de Broglie relation

$$\lambda = \frac{h}{m_n v} = \frac{ht}{m_n L} \tag{11}$$

where h is Planck's constant. Finally, diffraction involves reflection of neutrons by crystal planes. Substituting Bragg's equation into Eq. (11) and re-ordering yields:

$$d = \frac{ht}{2m_n L \sin\theta} = 1.978 \frac{t}{L \sin\theta} \tag{12}$$

which relates the measured quantity, the time-of-flight t, to interplanar spacings d for a neutron detector at 2θ, with t in s, L in m, and d in Å. This means d-spacings are directly proportional to t and consequently, a sequence of Bragg peaks is measured at a single scattering angle. The positions of Bragg peaks in d-spacing patterns identify lattice plane distances, d_{hkl}, and thereby symmetries and lattice parameters of phases and structures, which are associated with compositions and strains, and to some extent element concentrations via Vegard's rule. Fig. 2B shows an example of a simulated TOF diffraction pattern.

The resolution function of a TOF diffractometer defines the width δd of a diffraction peak with varying d-spacing:

$$\frac{\delta d}{d} = \sqrt{\left(\frac{\delta t}{t}\right)^2 + \left(\frac{\delta L}{L}\right)^2 + (\delta\theta \cot\theta)^2} \tag{13}$$

δt, δL, and $\delta\theta$ are the instrumental uncertainties related to pulse width, moderator depth, and size of detector, respectively. From Eq. (13) it is clear that diffraction peaks become sharp for long flight paths at tens of meters and for detectors with large 2θ, since $\cot(\theta) = 0$ for $2\theta = 180°$. The parameter δd is usually measured at the full width at half maximum (FWHM) of a Bragg peak.

3.3 Analysis setup on a neutron diffractometer

This section describes the experimental requirements for TOF neutron diffraction of cultural heritage objects.

3.3.1 Experimental requirements

General-purpose neutron diffractometers are designed for experiments in vacuum, for temperature and/or magnetic field–dependent studies and control of many other sample parameters such as pressure and gas load. The instrument environment does not necessarily suit analyses of archaeological and historic objects. Practical requirements for neutron diffraction analyses of cultural heritage materials are:

- The diffractometer needs to provide the infrastructure for safe handling and storage of materials and easy access to the sample area. Ideally, the environment should be temperature and humidity controlled.

- The instrument has sufficient spatial and spectral resolution, with variable beam sizes in the range of millimeters to centimeters (see Section 3.3.3).
- A wide *d*-spacing range is required for recording a good number of Bragg peaks for every phase. The measurement of long *d*-spacings up to 10 Å is needed to observe low-indexed peaks. A wide *d*-spacing range and analysis of peaks at short *d*-spacings is required for a robust Rietveld analysis (see Section 3.4.3).
- Objects are mounted and scans are set up in a way that minimizes risks of damage and collisions. The collection time is as short as possible to minimize activation of objects and materials (see Section 3.6).
- Focused analyses and detailed depth scans are best performed if guided by X-ray or neutron radiographies collected beforehand.
- The analysis should be planned and performed with a team that involves experienced object handlers and conservators.

A neutron diffraction analysis needs to be carefully considered, since beamtime is expensive and with risks involved for objects that are transferred to a facility. Usually the justifications for using neutrons are: the prospect of a nondestructive analysis, and to study the properties of materials deep within an object otherwise not accessible.

3.3.2 Instrument geometry and setup

A neutron strain scanner is chosen here as the preferred type of instrument for the investigation of archaeological and historical objects. The beneficial attributes of a strain scanner are:

- Ample operational space and easy access;
- Lattice parameters are determined with high accuracy and precision;
- Possibility of a spatially resolved diffraction analysis via selection of a defined analysis volume ("gauge volume") within an object, using beam shaping devices and radial collimators;
- Positioning table for setting up flexible scans for multipoint analyses;
- For TOF instruments: time-resolving neutron diffraction and transmission detectors.

Fig. 5A shows a schematic of the neutron diffractometer ENGIN-X [24] at ISIS, representing a generic TOF strain scanner. The instrument is on a 50-m flight path from a liquid methane moderator on TS1. Diffraction patterns are collected with two detectors at $2\theta = 90°$, to the left and right of the neutron beam. The incident beam is collimated using neutron-absorbing masks, usually made from boron carbide. Variable radial collimation between sample and detectors makes sure that only scattered neutrons from the gauge volume are recorded. The object is moved through the gauge volume, by automated step scans, with two diffraction patterns collected for every step. Diffraction scans are performed along quite arbitrary paths, for example, into the bulk of a material, or along the surface, to determine phase contents and strain distributions. Diffraction signals from the interior of the object, for instance from the

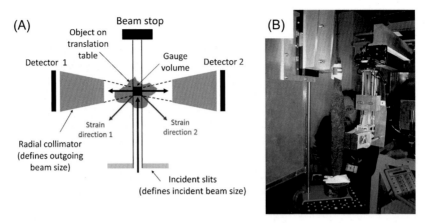

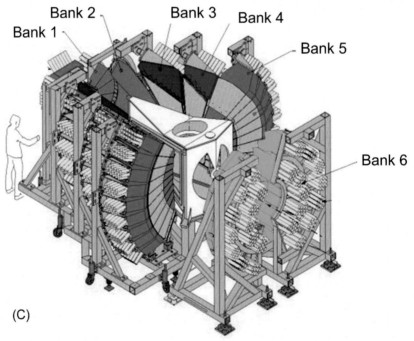

FIG. 5

TOF instrument setup. (A) Schematic of a strain scanner; (B) mounting of Merovingian single-edge sword (seax) on ENGIN-X; (C) GEM multidetector instrument.

original alloy, are not mixed up with scattered neutrons from surface corrosion layers. This is advantageous for the analysis of heavily restored or corroded objects covered with thick corrosion crusts.

An object is placed in air on the sample position stage (Fig. 5B) using safe suitable support and fixing materials, typically aluminum frames and foils as aluminum is almost transparent to neutrons. It can be emphasized that most of the neutrons exit a material without having interacted; they are stopped downstream in a beam catcher made of borated materials, cadmium and/or steel. For setting up automated scans, sample alignment is achieved with the help of theodolites and laser scanning tools. A 3D laser surface scan of an object can be used in conjunction with a virtual laboratory [25] to pinpoint analysis points in the interior of an object. Often wall-to-wall scans with a pencil-sized neutron beam are performed to find the edges and boundaries of an object's component in situ.

In practice, diffractometers not only use one or two, but hundreds and thousands of detectors to collect diffraction patterns simultaneously at many angles, thereby decreasing counting times drastically. ENGIN-X has 1200 elements on each of the two 90° detector arrays. The General Materials diffractometer GEM at ISIS has about 7000 detector elements arranged in a series of six banks around the sample chamber (see Fig. 5C). For such a multidetector TOF instrument, the diffraction elements for a subset ("bank") are "focused" into one histogram, usually postexperiment and during the data processing steps. In this way, ENGIN-X and GEM yield two and six diffraction patterns, respectively. For these two instruments, the detectors are ZnS/^{6}LiF scintillators light-coupled to photomultipliers, where the ^{6}Li acts as the neutron stopper material. The analysis time on multidetector instruments varies from a few minutes to several hours, where the counting time is proportional to the size of the interrogated volume. Two types of calibration measurements are performed: (i) Data from a solid vanadium rod is used to correct for detection efficiencies and to determine the incident neutron intensities as a function of wavelength. Vanadium is a suitable material for this purpose since it does not produce Bragg peaks. (ii) Calibration data on material standards with known structure and lattice parameters are used to calibrate the product $L \sin(\theta)$ (see Eq. 12) for each detector element. Suitable reference materials are silicon powder, yttrium oxide, and ceria. Furthermore, the profiles and widths of Bragg peaks of those standards are obtained as a function of d-spacing for each detector bank, thus providing the resolution function of the instrument.

3.3.3 Importance of instrument resolution

There are two important resolution properties of a neutron diffractometer to be considered: (i) the spatial resolution, i.e., the size of the diffracting volume; (ii) the spectral resolution, i.e., the sharpness of the diffraction peaks.

The spatial resolution is controlled by incident beam slits and collimators and determines the smallest spatial features that can be resolved. Defining a 3D gauge volume (Fig. 5A) is conceptually the simplest method for diffraction analysis. Spatial resolutions on strain scanners range from a few cubic millimeters to several cubic centimeters, i.e., for one analysis point a relatively large volume is probed with neutron diffraction. Normally, a large gauge volume is chosen as long it can be afforded on grounds of spatial resolution, in order to minimize collection time. Doubling the gauge volume allows to half the counting time for the same statistical precision. Many diffractometers do not have a sufficiently tight radial collimation. In such cases, the spatial resolution is controlled by the beam cross-section only whereas the spatial dimension along the beam is not resolved.

The spectral (or instrument) resolution is defined by the function $\delta d/d$ (Eq. 13) and measured with standard materials. The instrument resolution describes the potential of the instrument to resolve close-lying peaks which is crucial for analyzing multiphase materials and for studying broadening effects. The Bragg peaks shapes (profiles) are described by analytical functions such as Gaussian, Lorentzian, specific TOF profiles like Ikeda Carpenter, or convolutions of such functions [6].

For TOF instruments, the resolution function (Eq. 13) becomes minimal at backscattering angles, and has immense significance for TOF diffraction, and especially for cultural heritage material studies, because for $2\theta > 90°$ peak widths become small, almost independent of the beam size and of the shape of the diffracting part of the sample. This is a crucial feature which makes TOF diffraction superior to angle-dispersive diffraction. Objects with irregular shapes can be straightforwardly studied at these back-reflection angles.

3.3.4 Transmission setup

Bragg edge transmission analysis [4, 26] is performed in a radiography setup as shown in Fig. 4. There are few dedicated instruments to perform such measurements; one of them is IMAT at ISIS [17]. The technique requires an incident beam with a broad wavelength range, and a time-resolving transmission camera directly behind the object. First, an "open-beam" image is taken without sample to determine p^0 in Eq. (7) for every pixel. With the object inserted in the beam, neutrons are absorbed and scattered, according to Eq. (9), and the transmitted signals $p_{i,j}$ exhibit Bragg edges. By analyzing the positions, widths and heights of Bragg edges pixel-by-pixel, 2D maps of the associated lattice parameters, strains and phases are obtained with counting times of a few hours and spatial resolutions of a few hundred microns. In principle, the technique can be extended to produce 3D maps by recording a large number of radiographs for different orientations of the object in the beam.

3.4 Data processing and analysis

3.4.1 Data reduction

Individual TOF diffraction patterns are corrected for misalignment of detectors using data from powder standards, and then divided by the data collected from a solid vanadium rod to account for the wavelength dependence of the neutron flux and to correct for efficiency variations of detectors. A large number of TOF spectra of one detector bank are "focused" into a single histogram. This initial raw data reduction can be performed with a software package such as Mantid [27]. At this stage, diffraction patterns are inspected, and intense peaks are indexed by assigning *hkl* by comparison with peak position lists from pattern simulations or structure databases. Single-peak fitting of Bragg peaks that have not been identified is performed to obtain *d*-spacings as input into search-match ICSD or PDF [28] databases.

A quantitative analysis of the diffraction data is performed by various approaches, among them are single peak fitting, full pattern LeBail or Pawley fitting, or Rietveld analysis [6].

3.4.2 Strain analysis

For strain and stress analyses, the positions of peaks are determined by single-peak or full-pattern analysis. Different sample directions need to be analyzed to build up the strain and stress tensor for each analysis point of a scan. The TOF setup for strain scanning (Fig. 5A) with two detectors at 90 degrees allows for measuring two strain components simultaneously along the bisector of incident and diffracted beam; for the third or higher components, the sample needs to be reoriented on the diffractometer. A major problem for residual strain analysis of cultural heritage object is that compositional information required to produce adequate d_0-samples is often not available. In the absence of a d_0-sample, reference data for d_{hkl}^0 are taken from a part of the object that is expected to have the same composition and expected to be strain-free. With this d_{hkl}^0 the d_{hkl}-spacings are converted to strain (Eq. 6). Examples of residual strain analyses of archaeological objects are given in [15, 29].

Particle size and microstrain effects can be evaluated and separated using empirical approaches based on the Scherrer formula or the Williamson-Hall plot [6, 16]. Peak broadening is evaluated by analysis of individual Bragg peak widths or as part of the full-pattern Rietveld analysis (discussed in the following section) whereby one or more peak width parameters are fitted. In all cases the instrumental resolution is taken into account. An example based on a single-peak analysis is given in [30].

3.4.3 Rietveld analysis

The most important analysis approach is the Rietveld method [6, 31], a well-established method for the treatment of neutron and X-ray powder diffraction data. The Rietveld method is arguably the best way to achieve a combined analysis of phase, structure, texture, strain, and microstructure parameters from one or many diffraction patterns. In a Rietveld analysis, the parameters of a crystal model are varied until the theoretical pattern matches the measured pattern as well as possible. Given a suitable starting model for a multiphase mixture of a material, the Rietveld fit decreases the differences between the calculated and measured curves, for example, by minimizing a "reliability value" [6]

$$R_p = \sqrt{\frac{\sum_i \left(y_i^{obs} - y_i^{calc}\right)^2}{\sum_i \left(y_i^{obs}\right)^2}} \qquad (14)$$

via a nonlinear least squares fit, by iteratively adjusting the values of scale factors, lattice parameters, atomic positions, atomic displacement factors, phase fractions, texture, microstructure parameters, etc. The summation index i runs over all observed y_i^{obs} and calculated y_i^{calc} intensities using Eq. (2). Examples of software packages for Rietveld analysis are GSAS [32], FULLPROF, TOPAZ, and MAUD [33].

Benefit of a model-based Rietveld fit is that overlapping peaks are treated straightforwardly and that it does not require an internal standard for quantitative phase analysis. Table 2 lists the parameters that are typically refined. The detection limits for specific parameters cannot be given in general terms, as they depend on a number of factors: count rates; gauge volume size; data processing and data binning steps; adequacy of the structure model; skills of the software operator. The systematic and statistical error bars, and the detection limits have to be determined on a case-by-case basis.

For solid solutions of alloys, lattice parameters can be used to estimate the element concentration of the alloying component. Tin contents in α-bronze are estimated using experimental calibration curves [15, 34] based on the linear relationship between elemental fraction and lattice parameter known as Vegard's rule, which has its justification in the different atom radii of the elements. Vegard's rule can be applied to binary alloys with two elements in solid solution. The

Table 2 Parameters obtained from a combined Rietveld structure refinement.

Material parameter	Rietveld parameter	Symbol	Comments
Weight/phase fractions	Phase scale factors	s_p	Weight fractions add up to 100% Detection limits: sub%
Deviations from ideal phases; phase transformations	Symmetry; atom positions; occupancies	x, y, z f	Effects of alloying on symmetry
Lattice parameters	Lattice parameters	$a, b, c,$ α, β, γ	Identify phases Element contents via Vegard's rule Residual strains Thermal expansion
Structure and atom fractions; stoichiometry	Atom position in mixed alloys; occupancies	x, y, z f	Structure solutions for hydrogenous crystal structures
Magnetic structure	Magnetic moments	μ_x, μ_y, μ_z	Magnetic structures of iron oxides, dependent on impurities and alloying elements
Microstrain, domain/particle size; compositional variation	Peak shape + width parameters	H	Extracted from peak broadening; to be handled and interpreted cautiously
Texture	For example, texture coefficients of spherical harmonics expansion	P_{hkl}	ODF Pole figures Type and strength of texture

limitations for ternary bronzes with more than one element in solution was discussed by Gliozzo et al. [35]. It is worth noting that, once lattice parameters are determined to a first approximation, the percentage values of alloying elements from Vegard's rule need to be updated in the Rietveld model.

Fig. 6 displays examples of Rietveld refined TOF patterns collected from low and high tin leaded bronze tokens that were produced in the laboratory to support the analysis of Renaissance bronze statuettes [29]. The data display α- and δ-phases of Cu-Sn, and a Pb phase. In order to achieve good profile pattern fits of the broad peaks for 10% Sn, five α-bronze components (instead of one) with slightly different lattice parameters were included in the structure model following an approach proposed in [35, 36]. Average weight fractions, lattice parameter and average Sn contents were calculated. Moreover, the standard deviation of the spread of lattice parameters is taken as a measure of the degree of homogenization, as the broadening and splitting of the alpha peaks is clearly due to the dendritic tin segregation during solidification, increasing up to the separation into α and δ phase at higher Sn contents.

Microstructure parameters for microstrain and domain size analysis can be modeled with the analytical peak shape functions used in the Rietveld analysis. A peak broadening analysis can provide valuable clues to the working processes. Such an analysis is particularly useful if microstructural trends are revealed by comparison with suitable reference samples that are produced in a controlled way.

3.4.4 Texture analysis

Neutron texture analysis is used to determine the orientation distribution of crystallites created by plastic deformations from drawing, rolling, hammering, and recrystallization from heat treatment. Grain orientations are measured by recording diffraction patterns from many different sample angles by rotating the sample and/or by having detectors at many angles. For diffractometers with substantial detector coverage, data are collected for a small number of sample orientations, for example, 10–20 orientations on ENGIN-X or 1–3 orientations on GEM. For diffractometers with very large coverage (e.g., GEM in Fig. 5C) often one measurement in one orientation suffices. For the data analysis, each data set is subdivided into smaller groups of detector elements, for example, 20 groups for ENGIN-X and 150 groups for GEM detectors, with each such virtual detector covering about 10 degrees horizontally and vertically. Consequently, the number of patterns per texture data set is given by the number of virtual detectors times the number of sample orientations. The ODF can be evaluated by a combined Rietveld fit of all patterns. Examples of Rietveld texture analysis using the MAUD software [33] on data from archaeological copper axes [37] and bronze axes [38] as well as various studies on coins [39] have been reported. The advantage of the Rietveld approach is that data from multiphase materials can be analyzed even if there is a strong peak overlap; structure and texture parameters can be extracted in a combined refinement.

Recently, a new texture analysis strategy has been developed by F. Malamud [40] to perform spatially resolved texture analysis on neutron strain scanners. The

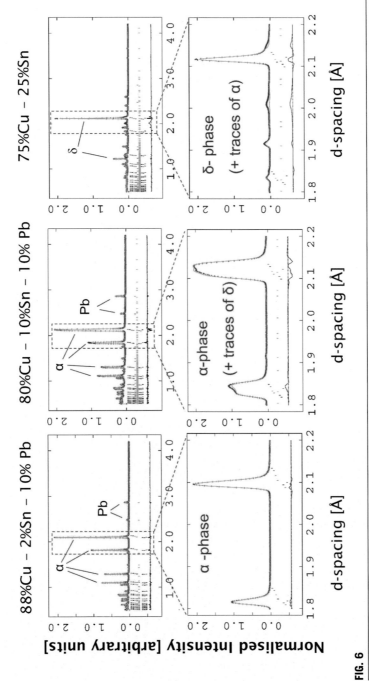

FIG. 6

Rietveld fitted diffraction patterns from POLARIS at ISIS, on three leaded bronze replica tokens. Data points (*red symbols*), calculated patterns (*green curve*), and differences between y^{obs} and y^{calc} (*black curve*) as obtained with GSAS. The patterns in the bottom row are enlarged regions of patterns above.

NyRTex [40] software tool facilitates the design of data collection scans on a diffractometer by calculating the sample angles required for ODF analysis for a given set of detector subgroups. Bragg intensities are extracted from the multitude of diffraction patterns to construct incomplete experimental pole figures. The ODF calculation is then accomplished by NyRTex by making use of texture analysis algorithms of MTEX [41]. Fig. 7 shows an example of this approach applied to a Napoleonic War era copper bolt on ENGIN-X [42]. Data on this 35 mm diameter object were collected for a total of 76 orientations and for a gauge volume of $4 \times 4 \times 4 \, mm^3$, for seven points across the cross-section of the bolt. Pole figure collected for points A–E across the bolt exhibit similar texture features. (001) and (122) pole density maxima are likely related to recrystallization from heat treatment whereas (111) maxima are related to mechanical working of the material along the bolt axis direction. Smaller copper-type (112) and Taylor-type (4 4 11) components are identified which are typical deformation textures in fcc metals [13].

Collection times for texture data range from a few minutes to some hours for one analysis point. Textures in ancient materials are often found to be very weak, with m. r.d. values close to one (e.g., [38]). In such cases, interpretation of the ODF is difficult and vague. It is worth noting that textures change even with time, at ambient temperatures [43], which poses additional challenges to the interpretation of texture data of archaeological materials.

3.5 Sources of errors in neutron diffraction

The analysis of neutron powder diffraction data relies on well characterized instrument and sample geometries, whereas the sizes and shapes of objects for cultural heritage studies are often far apart from the recommended geometry. Misalignments left uncorrected affect lattice parameters, and thus the compositions and strains derived from those. Sources of systematic errors for a Rietveld analysis are incomplete, oversimplified or inappropriate structure models. A further source of misinterpretation originates from gauge volumes if those are larger than the features to be studied in (typically) heterogeneous ancient artifacts. For example, a carbon content measured as indicator of intentional carburization may turn out incorrect if the interrogating volume is much larger than the range of the carburized region.

Even though it is possible, in principle, to extract the microstructure parameters from neutron diffraction data, the analysis is bedeviled by the number and complexity of line-broadening sources. For example, microstrain broadening and broadening due to compositional variations are difficult to distinguish. Different causes of peak broadening may be present (see Section 2.2.6) whilst there is usually no one-to-one correspondence between peak width and the mechanical or thermal treatment; in other words, different working cycles can lead to similar type and level of broadening. It is often not possible to separate the contributions, especially if diffraction patterns in a multiphase analysis exhibit substantial peak overlap. If too many levels of broadening are incorporated in the analytical peak

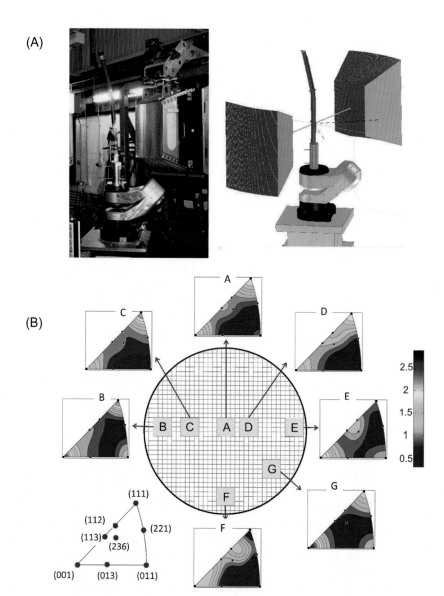

FIG. 7

Pole figure collection of a war-ship copper bolt from HMS Impregnable. (A) Setup of the bolt of 956 mm length and 35.5 mm diameter on ENGIN-X at ISIS, and laser surface-scan data of the object on a virtual instrument. (B) Analysis points A–G within a cross-section of the bolt as indicated in the center. Inverse pole figures are displayed for the bolt axis direction. A reference scheme for the pole figures is given on the lower left.

Courtesy of F. Malamud [42] and reproduced by permission of the International Union of Crystallography.

shape function, the extracted microstructure parameters may be wrong or ambiguous, leading to erroneous interpretations.

3.6 Neutron activation of objects

The energies of thermal and cold neutrons, and the beam fluxes available on diffractometers are too low to leave materials structurally damaged or to induce, for instance, decolorization effects. However, short-term activation of objects in the neutron beam have to be considered. When neutrons pass through matter, some of them interact with the atoms and may generate radio-isotopes. As a result, a material which is exposed to neutrons is activated. The unstable nuclides transform into stable nuclei by radioactive decay through the release of alpha, beta, and/or gamma radiation. After the exposure of an object on the diffractometer, the number of particles emitted is measured with radiation monitors in order to assess the activity of the material in units of counts per second (becquerel) or the equivalent dose in micro-Sievert per hour. The time for the activity of a radionuclide to fall to half of its original value is the half-life. In successive half-lives, the activity of a material is reduced by decay to 1/2, 1/4, 1/8, etc. This means that one can predict how long it takes for the activation to drop to an irrelevant value.

Overall, the short-time radio-activation of objects on neutron diffractometers is moderate. Before the neutron diffraction experiment, the likely activation of a material is calculated based on available compositional data, for example, from XRF. In most cases, objects can leave the neutron facility after a few days after which any residual activity has disappeared. Materials with elements such as Ag, Au, Co, and Mn can be activated for weeks rather than days if left in the neutron beam for hours.

4 Case study: Neutron diffraction on coining dies

Here we present a study on historical steel tools at the ISIS facility to illustrate the use and advantages of neutron diffraction. This project was brought to ISIS by S. Payne, University of Lancaster, aimed at analyzing about 30 coining dies and other steel tools dating from the late-18th and early-19th centuries. The objects originate from the Royal Mint and from the Soho Mint in Birmingham, and are now held in the Royal Mint Museum, Birmingham Museums & Art Gallery, and the British Museum. Neutron diffraction aimed at shedding light on the materials and processes involved in manufacturing the dies and punches, and to establish whether or not the dies are composite tools consisting of more than a single piece of metal. While previous studies had addressed the questions with metallographic analyses, extracting samples was not an option for most of the objects. Here we present neutron diffraction and transmission results on three selected objects. Fig. 8A shows two of the objects mounted on the ENGIN-X diffractometer at ISIS. The objects were not

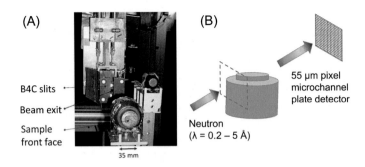

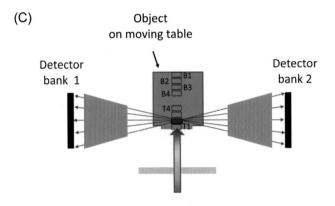

FIG. 8

Setup of steel tool for neutron diffraction analysis. (A) Mounting of objects on ENGIN-X at ISIS; (B) schematic of the diffraction imaging geometry; (C) overview of diffraction scans on ENGIN-X.

specially prepared for the analysis; they were merely wrapped in aluminum foil and placed onto aluminum supports.

TOF Bragg edge transmission analysis was performed on beamport N2 (ROTAX) at the ISIS facility, to determine regions of interest for the diffraction scans. The beamport views the cold 110 K liquid methane moderator and operates at a source frequency of 40 pulses per second. A time-resolving imaging detector based on microchannel plates (MCP) [44] was used at $L = 15.93$ m from the moderator, with 2678 time channels for each of the 55 μm pixels. As the sensitive area of the MCP camera of 28×28 mm^2 was smaller than the object size, two or three data sets were collected for different parts of an object and afterwards merged during data processing. Fig. 8B depicts the experimental setup and Fig. 9 shows Bragg edge spectra from the interior and the surface region of one of the tools. The positions, widths, and heights of the (110) Bragg edge of the ferrite structure were analyzed pixel-by-pixel, using a suitable Bragg edge fitting function [26, 45]. Fig. 9 shows the lattice parameter maps for the three tools, where red/white colors indicate larger values, thus

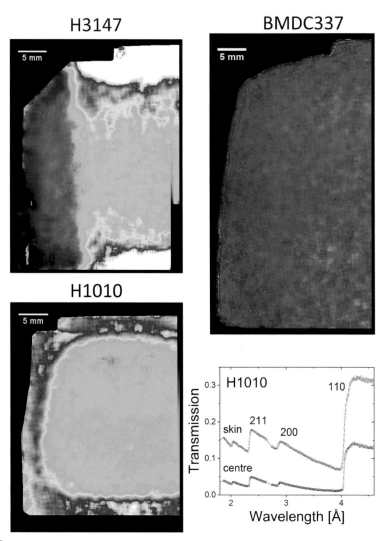

FIG. 9

Lattice parameter maps of coining tools obtained from Bragg edge analysis indicating transitions between ferritic and martensitic regions. Bottom right: Bragg edge spectra from two regions on H1010, from the outer skin and from the interior.

indicating expansion of the ferrite lattice near surface regions for H3147 and H1010, whereas the structure map for BMDC337 is quite homogeneous.

Neutron diffraction patterns were recorded on ENGIN-X with detectors at 90 degrees (Fig. 8C) at a distance of 1.5 m from the object. ENGIN-X is also installed on the methane moderator at ISIS, on a flight path of 50 m. The wavelength range was set to 1–3 Å for the measurements. Data were collected with a gauge volume of $6 \times 10 \times 4\,\text{mm}^3$, with four analysis points (T1–T4) near the top working surface and four points (B1–B4) near bottom side of a tool; T1 and B1 were just beneath the respective surfaces (Fig. 8C). Bragg peaks for ferrite, austenite, and cementite phases were observed. A significant overall shift of ferrite peaks for the border points, as well as a massive peak broadening were taken as indicators for the presence of a martensite structure. The peak positions of ferrite and martensite cannot be separated by neutron diffraction. Thus one ferrite structure (cubic, Im-3m) with variable lattice parameter was used in the Rietveld analysis, in addition to austenite (cubic, Fm-3m) and cementite (orthorhombic, Pnma). The magnetic diffraction component of ferrite was ignored in the analysis. Fig. 10 shows examples of GSAS-fitted diffraction patterns and Table 3 reports the resulting parameters.

The diffraction scans for H1010 and H3147 showed a variation of the structure parameters from outside to inside of an object whereas the object BMDC337 is fairly homogeneous, in agreement with the Bragg edge maps in Fig. 9. For H1010 and H3147, significant peak shifts and peak broadening of the ferrite (α-Fe) peaks at the surface were found to be correlated with the presence of retained austenite (γ-Fe) peaks. For BMDC337, no such behavior was observed, and the amount of

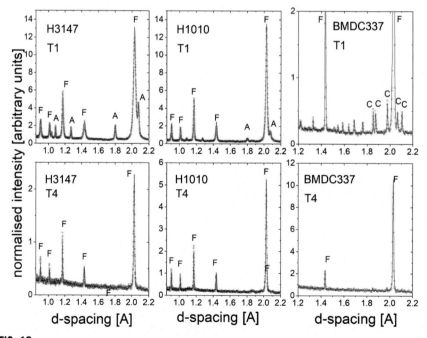

FIG. 10

Rietveld-fitted diffraction patterns of selected analysis points. The phases are marked as: ferrite (F), austenite (A), and cementite (Fe_3C) (C).

Table 3 Rietveld analysis results of steel dies: lattice parameters (a) of ferrite and austenite; "*" indicates: not refined; weight (w) fractions and calculated carbon contents; Lorentzian peak (γ_1) width as indicator of the martensitic phase, corrected for the instrument resolution.

		a α-Fe (Å)	a γ-Fe (Å)	w α-Fe (wt%)	w γ-Fe (wt%)	w Fe$_3$C (wt%)	γ_1 (μs/Å)	C in Fe$_3$C (wt%)	C in martensite (wt%)	C total (wt%)
H3147	T1	2.8706	3.5984	83.2	16.7	0	225.9	0.00	0.23	0.23
	T2	2.8670	3.5921	92.2	5.2	2.5	98.3	0.17	0.05	0.22
	T3	2.8664	3.5921*	95.7	0.3	3.9	73.8	0.26	0.02	0.28
	T4	2.8661	3.5921*	95.8	0.02	4.2	77.9	0.28	0.01	0.29
	B1	2.8704	3.5921*	84.1	15.8	0.05	219.2	0.00	0.22	0.23
	B2	2.8669	3.5921*	92.2	4.2	3.6	87.1	0.24	0.05	0.29
	B3	2.8663	3.5921*	95.1	0.8	4.1	73.8	0.27	0.02	0.29
	B4	2.8666	3.5921*	95.2	0.5	4.3	74.0	0.29	0.03	0.32
H1010	T1	2.8682	3.6009	91.4	7.1	1.1	133.1	0.07	0.11	0.19
	T2	2.8667	3.5921*	95.5	0.23	4.1	67.7	0.27	0.04	0.31
	T3	2.8673	3.5921*	95.4	0.02	4.6	64.7	0.31	0.06	0.37
	T4	2.8680	3.5921*	95.1	0.02	4.9	64.6	0.33	0.10	0.43
	B1	2.8677	3.6017	93.1	5.6	1.3	127.0	0.09	0.09	0.17
	B2	2.8673	3.5919	94.7	0.16	5.1	65.6	0.34	0.06	0.40
	B3	2.8685	3.5921*	95.1	0.02	4.9	59.4	0.33	0.12	0.45
	B4	2.8686	3.5921*	95.1	0.02	4.9	65.9	0.33	0.13	0.46
BMDC337	T1	2.8667	3.5921*	90.9	0.37	8.7	12.3	0.58	0.04	0.62
	T2	2.8661	3.5921*	93.2	0.34	6.5	19.4	0.43	0.01	0.44
	T3	2.8664	3.5921*	92.6	0.19	7.3	27.4	0.49	0.02	0.50
	T4	2.8664	3.5921*	95.1	0.19	4.9	33.7	0.33	0.02	0.35
	B1	2.8664	3.5921*	93	0.32	6.8	9.3	0.45	0.02	0.47
	B2	2.8663	3.5921*	93.6	0.36	6.0	24.9	0.40	0.01	0.42
	B3	2.8666	3.5921*	93.1	0.3	6.6	30.9	0.44	0.03	0.47
	B4	2.8666	3.5921*	93.6	0.02	6.4	33.6	0.43	0.03	0.46

The error bars of the weight fractions are in the order of 0.2%. Estimated standard deviations (e.s.d.) of lattice parameters: 0.0001 Å.

retained austenite anywhere in the object is small. Inside the tools, ferrite and cementite are observed consistent with the expected pearlite structure for carbon steel. The carbon fraction in cementite (Fe_3C) is 6.67 wt% carbon. The carbon content in martensite was estimated from the lattice parameter shift relative to 2.8664 Å using a Vegard's type relation by assuming expansion by 0.02 Å per wt% carbon [46].

The three tools are made of low-carbon steel with average carbon contents of 0.25 0.35, and 0.5 wt% for H1347, H1010, and BMDC337, respectively. The observed martensite and retained austenite phases at the surfaces of H13147 and H1010 are indicative of quenched carbon steel, extending up to several millimeters into the tools, while pearlite was found in the center. The formation of martensite, associated with supersaturation of carbon in the ferrite lattice, increases the hardness of the material. The distributions of the martensite zones are easily visible in the transmission data, revealing a quenched zone all around H1010, whereas for H3147 top and bottom of the tool was affected. For BMDC337, insignificant amounts of martensite and austenite were found, i.e., the tool is a rather soft iron throughout. The deviating behavior of the lattice parameters for H1010 and residual strain profiles for all tools will be revisited in the context of the total number of objects studied.

The transmission data are less sensitive to minor phases whereas Bragg diffraction is much more sensitive to structure details. Austenite and cementite were not observed by Bragg edge imaging, which, however, is useful to provide structure variations of the main ferrite/martensite component over larger areas and with much higher spatial resolution than possible with neutron diffraction. It can be mentioned that X-ray and white-beam neutron radiography failed to provide any appreciable contrast for the studied steel punches.

5 Summary

Neutron diffraction is used to determine the bulk structures and microstructures of cultural heritage materials without removing samples. Neutrons penetrate easily through thick coatings and corrosion layers and are able to access the inner parts of objects which favors, in particular, the analysis of metal objects. The aim of the crystallographic characterization is to reveal technical details in early metalworking, such as alloying, casting, hammering, and heat treatment.

The technique provides data complementary to that obtained by the conventional analytical methods, such as optical and scanning electron microscopy, and X-ray diffraction of mineral and metal phases. Large neutron beams access a considerable volume portion of the object. As a result, averaged and statistically representative structural information is obtained, and problems associated with single-spot analyses used by other metallographic techniques are avoided. A far greater number of points on objects can be analyzed than would ever be possible with metallographic techniques, as curators rarely allow destructive sampling. The method overcomes the interpretive problem of relying on just one or two very small metallography sections in order to characterize a whole—typically heterogeneous—ancient artifact. Neutron diffraction is often used in conjunction with other nondestructive methods such as radiography and tomography. Spatiallyresolved elemental data can be acquired from objects using complementary neutron-based techniques such as prompt gamma activation analysis or neutron resonance capture analysis.

Neutron diffraction at a pulsed spallation source has many advantages, such as ease of sample positioning and short data collection times. For TOF diffraction with detector arrays comprised of hundreds and thousands of individual elements, a convenient analysis angle can be chosen, a particular advantage for archaeological studies. The patterns from different detector elements are normally summed up to increase counting statistics, or they can be analyzed separately, as is the case of texture studies. It is likely that TOF Bragg edge imaging will extend the use of neutron diffraction for archaeological science and conservation science.

The downside of the high penetration is a relatively coarse millimeter-sized spatial resolution for diffraction, and hundreds of microns for diffraction imaging. If a spatial resolution of microns is required, then the method is unsuitable. Neutron diffraction is not normally applied to glassy and amorphous materials. The presence of wet and hydrogenous materials such as plastic or wood may preclude a neutron analysis.

Data interpretation relies to a large extent on well-characterized standards that have been produced according to ancient technological practices, which should be analyzed alongside the archaeological material. A conservator, conservation scientist or archaeological scientist specializing in the material should be closely involved from the proposal stage through to the practical application of neutron analyses of cultural heritage objects.

Acknowledgments

We wish to thank J. Payne (Royal Mint Museum) and R. van Langh (Rijksmuseum Amsterdam) for providing materials, historic background and for help with the data analysis. We thank S. Kabra (ISIS) for help with the ENGIN-X experiments. We are very grateful to F. Malamud (Centro Atomico Bariloche, Argentina) for the provision of figures and for her important development of novel texture analysis methods.

References

[1] D.P. Mitchell, P.N. Powers, Bragg reflection of slow neutrons, Phys. Rev. 50 (1936) 486–487.

[2] W. Kockelmann, A. Kirfel, E. Haehnel, Non-destructive phase analysis of archaeological ceramics using TOF neutron diffraction, J. Archaeol. Sci. 28 (2001) 213–222.

[3] G. Artioli, M. Dugnani, T. Hansen, L. Lutterotti, A. Pedrotti, G. Sperl, Crystallographic texture analysis of the iceman and coeval copper axes by non-invasive neutron powder

diffraction, in: A. Fleckinger (Ed.), Die Gletschermumie aus der Kupferzeit, Folio Verlag, Vienna, 2003, pp. 9–22.

[4] S. Siano, L. Bartoli, J.R. Santisteban, W. Kockelmann, M.R. Daymond, M. Miccio, G. DeMarinis, Non-destructive investigation of bronze artefacts from the Marches National Museum of Archaeology using neutron diffraction, Archaeometry 48 (1) (2006) 77–96.

[5] W. Kockelmann, E. Godfrey, Neutron diffraction, in: S.L. López Varela (Ed.), The Encyclopedia of Archaeological Sciences, John Wiley & Sons Inc., Hoboken, 2018, https://doi.org/10.1002/9781119188230.saseas0405.

[6] R.A. Young, Introduction to the Rietveld Methods, in: R.A. Young (Ed.), The Rietveld Method, International Union of Crystallography, Oxford, University Press, Oxford, 1993, pp. 1–38.

[7] R.J. Hill, C.J. Howard, Quantitative phase analysis from neutron powder diffraction data using the Rietveld method, J. Appl. Crystallogr. 20 (1987) 467–474.

[8] V.F. Sears, Neutron scattering lengths and cross sections, Neutron News 3 (1992) 26–37.

[9] A. Munter. https://www.ncnr.nist.gov/resources/n-lengths/. (Accessed 15 January 2020).

[10] M.I. Aroyo (Ed.), International Tables for Crystallography Volume A – Space Group Symmetry, sixth ed., International Union of Crystallography, Wiley, New Jersey, United States, 2016.

[11] ICSD, Inorganic Crystal Structure Database, FIZ Karlsruhe, Germany and National Institute of Standards and Technology, USA, 2014.

[12] P. Murray-Rust, Crystallography Open Database, 2020. http://www.crystallography.net/cod/. (Accessed 15 January 2020).

[13] U.F. Kocks, C.N. Tomé, H.-R. Wenk, Texture and Anisotropy: Preferred Orientations in Polycrystals and Their Effect on Materials Properties, Cambridge University Press, Cambridge, United Kingdom, 1998.

[14] D. Chateigner, Combined Analysis, Wiley, New Jersey, United States, 2010.

[15] S. Siano, L. Bartoli, M. Zoppi, W. Kockelmann, M.R. Daymond, J. Dann, M. Garagnani, M. Miccio, Microstrucural bronze characterisation by time of flight neutron diffraction, in: Proc. Archaeometallurgy in Europe, vol. 2, 2003, pp. 319–330.

[16] G.K. Williamson, W.H. Hall, X-ray line broadening from filed aluminium and wolfram, Acta Metall. 1 (1953) 22–31.

[17] W. Kockelmann, et al., Status of the neutron imaging and diffraction instrument IMAT, Phys. Procedia 69 (2015) 71–78.

[18] R. Woracek, J. Santisteban, A. Fedrigo, M. Strobl, Diffraction in neutron imaging – a review, Nucl. Instrum. Methods Phys. Res. A 878 (2018) 141–158.

[19] E. Fermi, W.J. Sturm, R.G. Sachs, Transmission of slow neutrons through microcrystalline materials, Phys. Rev. 71 (1947) 589–594.

[20] A. Botti, M.A. Ricci, G. DeRossi, W. Kockelmann, A. Sodo, Methodological aspects of SANS and TOF neutron diffraction measurements on pottery: the case of Miseno and Cuma, J. Archaeol. Sci. 33 (2006) 307–319.

[21] M.T. Dove, M.G. Tucker, D.A. Keen, Neutron total scattering method: simultaneous determination of long-range and short-range order in disordered materials, Eur. J. Mineral. 14 (2002) 331–348.

[22] B.T.M. Willis, C.J. Carlile, Experimental Neutron Scattering, Oxford University Press, Oxford, UK, 2009.

[23] J.W.G. Thomason, The ISIS Spallation Neutron and Muon Source – the first thirty-three years, Nucl. Instrum. Methods Phys. Res. A 917 (2019) 61–67.

[24] J.R. Santisteban, M.R. Daymond, J.A. James, L. Edwards, ENGIN-X: a third-generation neutron strain scanner, J. Appl. Crystallogr. 39 (2006) 812–825.

[25] R. van Langh, J. James, G. Burca, W. Kockelmann, S.Y. Zhang, E. Lehmann, M. Estermann, A. Pappot, New insights into alloy compositions: studying renaissance bronze statuettes by combined neutron imaging and neutron diffraction techniques, J. Anal. At. Spectrom. 26 (2011) 949–958.

[26] J.R. Santisteban, L. Edwards, A. Steuwer, P.J. Withers, Time-of-flight neutron transmission diffraction, J. Appl. Crystallogr. 34 (2001) 289–297.

[27] MANTID: Manipulation and Analysis Toolkit for Instrument Data, Mantid Project (2015), doi:https://doi.org/10.5286/SOFTWARE/MANTID3.5 (Accessed 15 January 2020).

[28] International Centre for Diffraction data ICDD®, Powder Diffraction File™ (PDF®) Search, 2020. http://www.icdd.com/pdfsearch/. (Accessed 15 January 2020).

[29] R. van Langh, L. Bartoli, J. Santisteban, D. Visser, Casting technology of Renaissance bronze statuettes: the use of TOF-neutron diffraction for studying afterwork of Renaissance casting techniques, J. Anal. At. Spectrom. 26 (2011) 892–898.

[30] E. Pantos, W. Kockelmann, L.C. Chapon, L. Lutterotti, S.L. Bennet, M.J. Tobin, J.F.W. Mosselmans, T. Pradell, N. Salvadó, S. Butí, R. Garner, A.J.N.W. Prag, Neutron and X-ray characterisation of the metallurgical properties of a 7th century BC Corinthian-type bronze helmet, Nucl. Instrum. Methods. Phys. Res. B 239 (2005) 16–26.

[31] H.M. Rietveld, A profile refinement method for nuclear and magnetic structures, J. Appl. Crystallogr. 2 (1969) 65–71.

[32] A.C. Larson, R.B. Von Dreele, General Structure Analysis System (GSAS), Los Alamos National Laboratory Report LAUR 86-748, New Mexico, USA, 2004.

[33] H.R. Wenk, L. Lutterotti, S.C. Vogel, Rietveld texture analysis from TOF neutron diffraction data, Powder Diffract. 25 (2010) 283–296. Material Analysis Using Diffraction (MAUD) http://maud.radiographema.eu/. (Accessed 15 January 2020).

[34] F. Grazzi, L. Bartoli, S. Siano, M. Zoppi, Characterization of copper alloys of archaeometallurgical interest using neutron diffraction: a systematic calibration study, Anal. Bioanal. Chem. 397 (2010) 2501–2511.

[35] E. Gliozzo, G. Artioli, W. Kockelmann, Neutron diffraction of Cu–Zn–Sn ternary alloys, J. Appl. Crystallogr. 50 (2017) 49–60.

[36] M. Mödlinger, E. Godfrey, W. Kockelmann, Neutron diffraction analyses of Bronze Age swords from the Alpine region: benchmarking neutron diffraction against laboratory methods, J. Archaeol. Sci. Rep. 20 (2018) 423–433.

[37] G. Artioli, Crystallographic texture analysis of archaeological metals: interpretation of manufacturing techniques, Appl. Phys. A Mater. Sci. Process. 89 (2007) 899–908.

[38] R. Arletti, L. Cartechini, R. Rinaldi, S. Giovannini, W. Kockelmann, A. Cardarelli, Texture analysis of bronze age axes by neutron diffraction, Appl. Phys. A Mater. Sci. Process. 90 (2008) 9–14.

[39] M. Griesser, W. Kockelmann, K. Hradil, R. Traum, New insights into the manufacturing technique and corrosion of high leaded antique bronze coins, Microchem. J. 126 (2016) 181–193.

[40] F. Malamud, J.R. Santisteban, M.A. Vicente Alvarez, R. Bolmaro, J. Kelleher, S. Kabra, W. Kockelmann, Texture analysis with a time-of-flight neutron strain scanner, J. Appl. Crystallogr. 47 (2014) 1337–1354.

[41] R. Hilcher, H. Schaeben, A novel pole figure inversion method: specification of the MTEX algorithm, J. Appl. Crystallogr. 41 (2008) 1024–1037.

[42] F. Malamud, S. Northover, J. James, P. Northover, S. Nneji, J. Kelleher, Spatially resolved texture analysis of Napoleonic War era copper bolts, J. Appl. Crystallogr. 50 (2017) 1359–1375.

[43] E. Jansen, W. Schaefer, A. Kirfel, J. Palacios, On the long term stability of copper rolling texture analysed by neutron diffraction pole figures, Mat. Sci. Forum 278–281 (1998) 502–507.

[44] A.S. Tremsin, J.B. McPhate, A. Steuwer, W. Kockelmann, A.M. Paradowska, J.F. Kelleher, J.V. Vallerga, O.H.W. Siegmund, W.B. Feller, High-resolution strain mapping through time-of-flight neutron transmission diffraction with a microchannel plate neutron counting detector, Strain 48 (2012) 296–305.

[45] A.S. Tremsin, T.Y. Yau, W. Kockelmann, Non-destructive examination of loads in regular and self-locking Spiralock® threads through energy-resolved neutron imaging, Strain 52 (2016) 548–558.

[46] L. Xiao, Z. Fan, Z. Jinxiu, Z. Mingxing, K. Mokuang, G. Zhenqi, Lattice parameter variation with carbon content of martensite, Phys. Rev. 52 (1995) 9970–9978.

Laboratory and synchrotron X-ray spectroscopy

Laszlo Vincze[a], Pieter Tack[a], Brecht Laforce[a], Ella De Pauw[a], Stephen Bauters[b], Geert Silversmit[c], and Bart Vekemans[a]

[a]*Department of Chemistry, Ghent University, Ghent, Belgium* [b]*European Synchrotron Radiation Facility (ESRF), Helmholtz-Zentrum Dresden-Rossendorf CRG-Beamline, Grenoble Cedex 9, France* [c]*Research Department, Belgian Cancer Registry, Koningsstraat, Brussels, Belgium*

1 Introduction

X-ray fluorescence (XRF) and X-ray absorption spectroscopic (XAS) techniques represent important analytical tools to determine the elemental composition and speciation within the investigated materials down to trace-level concentrations for a majority of chemical elements, in both qualitative and quantitative sense. The detectable atomic number range typically starts from $Z > 10$, but the range can be extended depending on the excitation/detection conditions, associated with the use of vacuum and/or He atmosphere as sample environment, thin-window detectors and the accessibility of low-energy excitation [1].

Conceptually the XRF methodology is very simple: a collimated or focused X-ray beam is employed to irradiate the sample and the resulting element specific XRF radiation is detected by either energy-dispersive or wavelength-dispersive detection systems, as shown in Fig. 1. While XRF can be performed using either poly- or monochromatic excitation, most implementation of XAS requires a highly monochromatic, finely tunable incident beam to scan the excitation energy across selected absorption edges (e.g., K or L_3), corresponding to the element of interest, either in transmission or emission mode. Because of the energy tunability requirement for the incident beam, apart from a few experimental laboratory setups, XAS is performed almost exclusively at synchrotron facilities [2–10].

What makes these X-ray spectroscopic techniques extremely valuable with respect to analyzing precious cultural heritage materials and works of art is their nondestructive/noninvasive nature, enabling to perform elemental analysis on objects in situ, combined with the ability to obtain bulk or spatially resolved information about the elemental distributions in 2D or 3D, with lateral resolution reaching the (sub)microscopic scale [11–20].

Spectroscopy, Diffraction and Tomography in Art and Heritage Science. https://doi.org/10.1016/B978-0-12-818860-6.00007-6

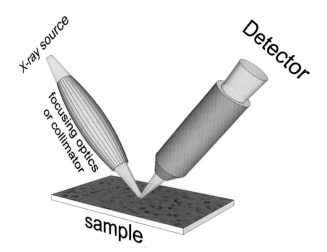

FIG. 1

Schematic illustration of XRF measurements: a collimated or focused (e.g., by polycapillary optics as shown) X-ray beam is employed to irradiate the sample and the resulting element-specific XRF radiation is detected by energy or wavelength dispersive detection systems.

Regarding microanalysis, until the 1980s, XRF and XAS suffered from an important constraint with respect to their usability to obtain spatially resolved information with high spatial resolution: the apparent difficulty to focus the X-ray beam efficiently to the microscopic scale for the analysis of small and/or heterogeneous samples. Early attempts of micro-XRF analysis for profiling heterogeneous samples were based on the use of pinholes used in conjunction with X-ray tubes but the resulting flux was often too low for practical applications [21]. Direct focusing of X-rays was made quasiimpossible by the high absorption coefficient for many of the materials applicable for focusing, the near unity refractive index, and the *mrad* range of critical angle of total reflection. It was over the last 30–35 years that methods of practical use for X-ray beam confinement/focusing gradually appeared on the basis of diffraction, total-reflection or refraction phenomena [22–34].

The rapid evolution in X-ray focusing optics and X-ray sources during the last few decades led to the new scanning/imaging methods in XRF and XAS that are based on the confinement of the interaction volume of the primary X-ray beam with the material being analyzed, paving the way to inventive 3D approaches of analysis, including confocal XRF/XAS [15, 31–34] and XRF/XAS tomography [20, 35–43]. In general, micro-XRF and micro-XAS analysis is based on the localized excitation of a microscopic volume within a larger sample. It provides information on the distribution of major, minor, and trace elements in heterogeneous materials or can be used for the analysis of samples of microscopic dimensions. Micro-XRF is typically exploited with laboratory X-ray sources but is considerably more sensitive and can be readily combined with micro-XAS when applied with X-rays emitted from a synchrotron radiation (SR) source [21].

Due to their high intensity and directionality, SR sources are ideal for the generation of micro- and nanoscopic X-ray beams. This is especially true for the emerging 4th generation SR facilities, such as the European Synchrotron Radiation Facility (ESRF), Grenoble, France, which started its user operation in September 2020. This facility now offers X-ray sources with a brightness which is nearly two orders of magnitude higher than that of 3rd generation devices [44].

Combined with advanced X-ray optics, such as compound refractive lenses, Kirkpatrick-Baez mirrors, or Fresnel zone plates, the extremely high brightness of SR sources allows the straightforward realization of monochromatic (sub)micron size X-ray beams from the emitted radiation from the storage ring. In addition, the polarization can be used to reduce the relative contribution of scattered radiation reaching the detector and thus to enhance considerably the signal-to-background ratio of fluorescence spectra, improving significantly the detection limits.

SR-based micro-XRF and micro-XAS offer a number of advantages compared to other microprobe techniques (EPMA, SIMS, PIXE): it combines high spatial resolution with excellent sensitivity, they can be used under atmospheric conditions and is relatively insensitive to beam damage to the sample. Their relative simplicity and the good understanding of the physics of the processes involved in X-ray-matter interactions make these techniques also more adaptable for quantitative analysis compared to a number of other beam methods of analysis [21].

This chapter will describe the basic principles of XRF in the context of cultural heritage applications, presenting basic figures of merit and application examples based on the use of advanced laboratory and synchrotron sources. Related methods of analysis making use of absorption edge phenomena by X-ray absorption spectroscopy (XAS), including X-ray absorption near-edge structure (XANES) and extended X-ray absorption fine structure (EXAFS) spectroscopy will also be discussed.

2 Principles of XRF

XRF is based on the excitation of elemental constituents, atoms, of the investigated materials by a mono-, or more commonly, a polychromatic X-ray beam. The most important X-ray-matter interaction type resulting in the excitation process is the photoelectric effect, which in turn results in the emission of XRF radiation with energies characteristic to the chemical elements present in the irradiated sample or Auger electron emission (see Fig. 2).

The energy-dispersive (ED) or wavelength-dispersive (WD) detection of these element-specific XRF lines constitutes the basis of elemental analysis by XRF spectroscopy. WD-XRF spectrometers can be applied to analyze elements down to Be. The more compact and simpler ED-XRF spectrometers, except for a few specialized setups developed specifically for the analysis of light elements [45], are able to (quantitatively) determine elements with atomic numbers above 10.

The discovery of characteristic X-ray emission by the chemical elements, studied first by Moseley in his early 20s shortly after the discovery of X-rays, laid down the

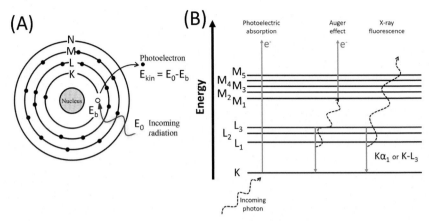

FIG. 2

Schematic illustration of (A) the photoelectric effect and the relevant processes playing role in the generation of XRF radiation and (B) the photoelectric effect, followed by electron transitions resulting in the emission of either XRF photons or Auger electrons.

fundamental principles of chemical composition determination of materials by XRF [46, 47]. Moseley measured the characteristic emission line frequencies of about 40 elements, and he found that the square root of the measured frequencies (energies) was linearly proportional with the atomic number of the emitting element. The linear relationship between the square root of the emission energies and the corresponding atomic numbers is illustrated in Fig. 3 using values obtained by the software

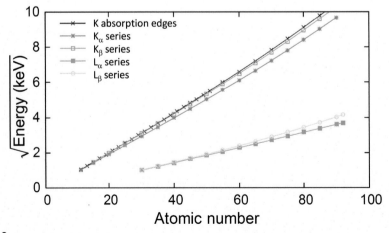

FIG. 3

Square root of XRF line (K- and L-series) and K-absorption edge energies as a function of atomic number, calculated by XRAYLIB [48, 49].

XRAYLIB [48, 49]. Fig. 2 (right) also shows a simplified scheme indicating the atomic transitions corresponding to the main detected fluorescence lines.

As mentioned earlier, the main X-ray-matter interaction type relevant for XRF is the photoelectric effect (see Fig. 2, left), during which an X-ray photon is absorbed by the sample atom, coupled with the ejection of an atomic electron referred to as photoelectron. In the process, the X-ray photon transfers all of its energy (E_0) to the photoelectron (with binding energy E_b), which is ejected from the atom with a kinetic energy of $E_{kin} = E_0 - E_b$, while creating a vacancy in the electronic shell it originates from. A necessary condition to excite a given shell with binding energy E_b is that the photon energy E_0 must exceed the binding energy of the shell. The original vacancy is subsequently filled up by an electronic transition from one of the higher electronic shells, in its turn creating a subsequent vacancy and ultimately resulting in a cascade of electronic transitions associated with the emission of a series of XRF lines or Auger electrons while the excited atom returns to its ground state. The signal of interest arises from the XRF emission, with energies corresponding to the difference of the shell binding energies involved in the electron transitions. XRF emission competes with Auger electron emission, the latter having increasing probability with decreasing atomic numbers. This represents one of the fundamental reasons for the generally poor detection limits for the low atomic number elements, in addition to the increased absorption effects of their low energy emission lines and the reduced excitation efficiency as a result of their lower photoelectric cross-section compared to high-Z atoms.

In general, for typical XRF experiments with excitation/detection energies in the range of 1–100 keV, the relevant interaction types contributing to the spectral features are (i) photoelectric absorption followed by XRF or Auger electron emission, (ii) Rayleigh, and (iii) Compton scattering. Fig.4 illustrates the relative magnitude of these interactions for Pb, showing calculated cross-sections as a function of X-ray excitation energy for each of these interaction types. As illustrated, in the X-ray energy range of 1–100 keV, the photoelectric cross-section is orders of magnitude larger than that of the scattering type of interactions. This high probability of the photoelectric effect, which is proportional with $\sim Z^4/E^3$ for atomic number Z and photon energy E, compared to competing interactions, explains the high sensitivity of both XRF and XAS. Also note the presence of the M_{1-5}, L_{1-3} and K absorption edges, corresponding to the respective binding energies of the electrons. The analysis of these absorption edges is performed using XAS, as discussed in Section 7.

3 Quantitative considerations in XRF analysis

The above interactions are quantitatively characterized by the so-called atomic interaction cross-sections, which are available through a number of databases and software [48–54] and can be used for applications in e.g. fundamental parameter or Monte Carlo (MC) simulation-based quantitative approaches in XRF spectroscopy [55–60]. For a given sample atom at a given excitation energy, the relative

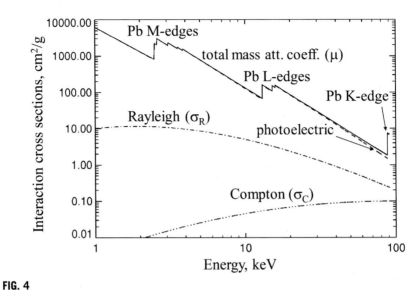

FIG. 4

Calculated interaction cross-sections for Pb, showing the total, phototelectric, Rayleigh scattering and Compton scattering cross-sections obtained by XRAYLIB [49].

magnitudes of the photoelectric, Compton and Rayleigh scattering cross-sections determine the probability of each of these interactions to occur, and ultimately their contribution to the various spectral features observed in the detected XRF spectra. These spectral features include the fluorescence lines and the elastic (Rayleigh) and inelastic (Compton) scattering peaks or scattering background, determining to a large extent the detected XRF spectra in addition to detector artifacts (e.g., escape-peaks, sum-peaks, incomplete charge-collection effects in ED-XRF) [61], as well as (minor) contributions associated with in-sample electron-matter interactions by photo and to a lesser extent Auger electrons generated via photoelectric interactions.

3.1 Fundamental parameter equation

The foundations of the fundamental parameter method-based quantification of XRF spectra was laid down by Sherman in the 1950s, a method which exploits the theoretical relationship between the net XRF line intensities and the elemental concentrations [62].

The example below handles the simplest case of calculating the intensities of K_α emission lines generated by monochromatic excitation of a homogeneous sample with a given thickness T, density ρ, composed of n elements of atomic numbers $\{Z_i\}_{i=1}^n$ with weight fractions $\{w_i\}_{i=1}^n$. Fig. 5 shows the beam-sample-detector arrangement for the modeled system with a single layer. The model below can be easily generalized for multilayer samples and polychromatic excitation.

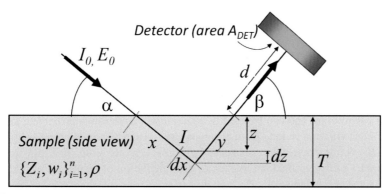

FIG. 5

The beam-sample-detector arrangement for the derivation of the fundamental parameter equation, showing the modeled system with a single layer.

Enhancement effects, which refer to the excitation of lower atomic number elements by fluorescent lines originating from elements of higher atomic numbers, are not considered and the reader is referred to the literature detailing the effects of secondary and tertiary excitation by sample constituents [63].

Detected by an ideal detector with an active area of A at a sample-detector distance of d, the theoretical elemental intensity of the K_α line of a given element i having concentration of w_i, originating from a sample of thickness T and density ρ is given by Eq. (1):

$$I_{i,K\alpha} = I_0 G w_i Q_{i,K\alpha} \rho T A_{i,K\alpha} \tag{1}$$

where I_0 is the intensity of the incident monochromatic beam and $G = \Omega_{Det}/(4\pi \sin(\alpha)) \approx A_{Det}/(4\pi d^2 \sin(\alpha))$ is the so-called geometry factor determined by the detector active area (A_{Det}), sample-detector distance d and angle of incidence α. Furthermore, w_i is the weight fraction and $Q_{i,K\alpha}$ is the production cross-section of element i, $A_{i,K\alpha}$ is the absorption correction factor for the K_α line of element i. The production cross-section $Q_{i,K\alpha}$ can be calculated as:

$$Q_{i,K\alpha} = \tau_{i,K} \omega_{i,K} p_{i,K\alpha} \tag{2}$$

where $\tau_{i,K}$ is the photoelectric cross-section of the K-shell of element i, $\omega_{i,K}$ is the K-shell fluorescence yield and $p_{i,K\alpha}$ is the total K_α transition probability.

The element- and matrix-dependent absorption correction factor is dependent on the sample thickness, density and matrix composition, given by:

$$A_{i,K\alpha} = \frac{1 - e^{-\chi \rho T}}{\chi \rho T} \quad \text{with} \chi = \frac{\mu(E_0)}{\sin(\alpha)} + \frac{\mu(E_{i,K\alpha})}{\sin(\beta)} \tag{3}$$

The mass attenuation coefficients of the sample at the incident energy and at the energy of the K_α line if element i are given by:

$$\mu(E_0) = \sum_{i=1}^{n} w_i \mu_i(E_0) \quad \text{and} \mu(E_{i,K\alpha}) = \sum_{i=1}^{n} w_i \mu_i(E_{i,K\alpha}) \tag{4}$$

Eq. (1) is simplified to the so-called thin-sample approximation $I_{i,\ K\alpha}=I_0Gw_iQ_{i,K\alpha}\rho T$ in which case the fluorescence line intensity is linearly proportional with thickness and density for $\rho T \ll 1/\chi$. For infinitely thick samples, i.e. when $\rho T \gg 1/\chi$, Eq. (1) reduces to $I_{i,\ K\alpha}=I_0Gw_iQ_{i,K\alpha}/\chi$.

The above equations are extremely useful for estimating the expected intensities of XRF-lines for given sample types when using monochromatic excitation, encountered at SR sources, and can be used in an iterative manner to approximate the elemental concentration values from the measured intensities, after establishing the value I_0G via external calibration standards for a given setup. For thin samples and samples with intermediate thickness, the calculation of elemental weight fractions assumes knowledge of ρT.

Please note that the equations above only cover the primary excitation case, representing the leading term, without considering interelement enhancement effects (and interlayer enhancement in case of layered samples), i.e., when sufficiently energetic XRF-lines of heavier elements excite elements of lower atomic numbers. Enhancement correction terms take these effects into account and are discussed extensively in the literature, covering secondary and even higher-order enhancement effects [63].

3.2 MC simulation of ED-XRF spectra

MC simulations offer a more powerful solution for the calculation of the complete response of XRF spectrometers, simulating entire XRF-spectra, by considering all relevant X-ray-matter interaction types in the simulated sample. In this way the contribution of enhancement and absorption effects, as well as the influence of in-sample scattering is considered. MC models can also be easily generalized toward heterogeneous samples, such as layered structures.

With respect to quantification, even simple MC models, which only take the major fluorescent lines of each element into account and dispense with simulation of scattering phenomena altogether, are useful. For example, quantification can be achieved by establishing the X-ray intensities corresponding to a number of standard samples via MC simulation. The composition of the simulated standard can be chosen to be close to that of the unknown samples so that simple calibration relations can be employed. A more complete MC simulation of an energy-dispersive (ED)-XRF spectrometer also covers scattering of the primary radiation and includes second- and higher-order effects such as the enhancement of fluorescent lines by higher-energy fluorescent or scattered radiation. As phenomena that contribute to the background of ED-XRF spectra can be accounted for (e.g., low-energy multiple scattering tails of the scatter peaks in case of monochromatic excitation and the scattering of the primary spectrum in the case of polychromatic excitation), it is possible to 'predict' the complete spectral response of an ED-XRF spectrometer [58, 59]. Fig. 6 shows an example of a simulated XRF spectrum for a glass standard (Multicomponent glass NIST SRM 1412) together with its experimental equivalent, showing the excellent

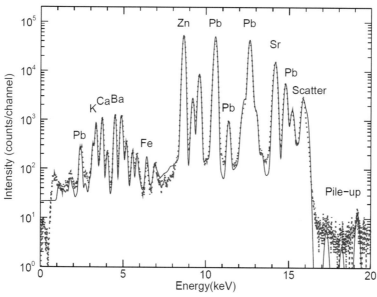

FIG. 6

Example of a simulated XRF spectrum (*blue*) for a glass standard (Multicomponent glass NIST SRM 1412) together with its experimental equivalent (*red*), showing the excellent agreement between the theory and experiment. Simulated by the XMI-MSIM code developed by Schoonjans et al. [58, 59].

Adapted from T. Schoonjans, V.A. Sole, L. Vincze, M.S. del Rio, K. Appel, C. Ferrero, A general Monte Carlo simulation of energy-dispersive X-ray fluorescence spectrometers – Part 6. Quantification through iterative simulations, Spectrochim. Acta Part B At. Spectrosc. 82 (2013) 36–41.

agreement between the theory and experiment simulated by the XMI-MSIM code developed by Schoonjans et al. [58, 59], making use of the XRAYLIB library [49].

A significant advantage of MC simulation-based quantification schemes compared to other quantification methods, such as those relying on the fundamental parameter (FP) algorithms, is that the simulated spectrum can be compared directly to the experimental data in its entirety, considering not only the fluorescence line intensities, but also the scattered background of the XRF spectra. This is coupled with the fact that MC simulations are not limited to first- or second-order approximations and to ideal geometries.

In the past, MC models have been applied in an iterative manner to quantification of XRF data corresponding to homogeneous or simple heterogeneous samples in much the same way as other quantification algorithms based on the fundamental parameter approach. Relative deviations in the range of 1%–15% have been achieved by the MC quantification scheme, depending on the analyzed element and sample type [55, 56, 58, 59] which illustrates the high potential of the method as an XRF quantification tool. Errors are mostly due to the uncertainties in the physical constants (cross-sections, fluorescence yields, transition probabilities, etc.) applied in the simulations.

A number of models exist for solving photon-transport problems for both unpolarized and polarized incident beams, their direct usability is often limited when dealing with the case of conventional (X-ray-tube based) or synchrotron radiation XRF on multielement samples due to the lack of a sufficiently optimized photoelectric/fluorescence and/or scattering models for the relevant X-ray energies (1–100 keV).

4 Laboratory scale instrumentation

4.1 Portable and handheld XRF

The in situ analysis of precious works of art and cultural heritage objects in general was in many ways revolutionized by the significant number of handheld and/or portable ED-XRF systems developed and commercialized in recent years. This emerging class of portable and handheld XRF devices rely, on the one hand, on the appearance of miniature X-ray generators with power levels of the order of 5–10 W and, on the other hand, on the availability of small Si pin-diode or SDD detectors with crystal thicknesses in the range of 300–500 μm, usable up to ~30 keV detection energies with energy resolutions of approximately 140–170 eV at 5.9 keV [64].

The analytical characteristics of portable systems in terms of achievable detection limits are comparable with a low-power desktop micro-XRF spectrometer operated in air. Fig. 7 (bottom-right) shows the limit of detection (LOD) values for a glass matrix (NIST SRM 1412) for a measuring time of 300 s for a handheld XRF spectrometer (Olympus Innov-X Delta, top-left), a portable XRF/XRD system (Assing Surface Monitor, top-right) and a laboratory micro-XRF spectrometer (EDAX Eagle III, bottom-left) in the atomic number range of 12–38 using their optimal operational settings [65]. Photos of these instruments are also shown in Fig. 7. Due to the short excitation/detection air-path achievable when using the handheld instrument, the obtained LOD values for Mg and Al, the lowest atomic numbers this instrument could detect, are only slightly worse than that achievable by a laboratory micro-XRF spectrometer using a Si(Li) detector operated at an X-ray tube power of 40 W under vacuum conditions.

In addition to their portability, their noninvasive character and high sensitivity, multielement detection capability and their simplicity in use led to a large number of innovative investigations in recent years, covering topics from the analysis of historical wall paintings [66], cave art [67], ceramics and glass fragments [68–70], pottery [71], sculptures [72], medieval metallic artifacts [73–75], historical musical instruments [76], historical marble artifacts [77], etc.

Fig. 7 (top-left) shows an in situ hand-held XRF analysis of a unique, 16th century majolica tile floor in the chapel of the Rameyenhof castle (Gestel, Belgium), which is the only known example of its kind in Belgium [78]. The Rameyenhof castle is located on a private domain, with its chapel possessing a majolica tile floor that

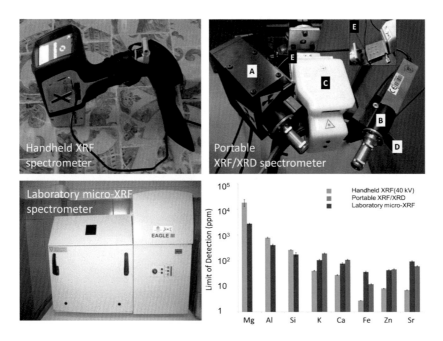

FIG. 7

Examples of laboratory (bottom left), handheld (top left) and portable (top right) XRF spectrometers together with measured detection limits from a NIST SRM 1412 glass standard, corresponding to a measuring time of 300s.

Adapted from L. Van de Voorde, Chapter 2, in: Optimization and Applications of Mobile, Laboratory and Synchrotron X-Ray Based Non-Destructive Microanalysis Techniques for the Study of Cultural Heritage Objects (PhD thesis), Ghent University, Belgium, 2015, p. 7–62.

is a magnificent representation of the majolica production which was once so popular in the 16th–17th century Antwerp. The handheld XRF measurements were performed in situ using the above-mentioned Olympus InnovX Delta instrument, producing an X-ray beam of approximately 5 mm diameter on the sample. The spectra shown in Fig. 8 correspond to a measuring time of 300s, X-ray tube voltage of 40kV, tube current 79 μA. The portable XRF measurements allowed to identify the pigments used for the production of these majolica tiles, earthenware covered with a white opaque glaze (tin glaze) and decorated with various metal oxides-based pigments to obtain a multicolored painted surface.

The XRF measurements revealed the characteristic elemental signatures of the various pigments used for decorating the floor tiles, and confirmed that the tiles show essentially identical elemental fingerprint typical for majolica tiles manufactured in the 16th–17th century Antwerp [79]. Furthermore, comparative trace-level XRF analysis indicated that a medallion, currently stored in the Rubens House Museum in Antwerp, may have been the original medallion of the Rameyenhof's majolica tile floor [79].

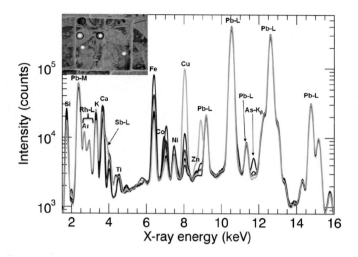

FIG. 8

XRF spectra corresponding to measurements on areas of different colors on a 16th-century majolica tile with locations of point measurements indicated on the inset. The tile is part of the floor in the chapel of the Rameyenhof castle (Gestel, Belgium). The in situ measurement is shown in Fig. 7 (top left).

Adapted from L. Van de Voorde, M. Vandevijvere, B. Vekemans, J. Van Pevenage, J. Caen, P. Vandenabeele, P. Van Espen, L. Vincze, Study of a unique 16th century Antwerp majolica floor in the Rameyenhof castle's chapel by means of X-ray fluorescence and portable Raman analytical instrumentation, Spectrochim. Acta Part B At. Spectrosc. 102 (2014) 28.

4.2 Macro-XRF

Noninvasive elemental imaging type of studies on historical paintings have been greatly benefitted by the recently appearing transportable XRF scanning systems, which are able to scan large areas of or even entire paintings. Such systems are commonly referred to as macro-XRF (MA-XRF) setups, allowing to record the distribution of a broad range of elements from the surface of a painting in a fast, in situ manner nondestructively.

During the last decade, a number of self-built devices and even commercial systems have been presented, able to analyze areas of thousands of cm^2 with spot sizes down to 100 µm [80]. Although these scanning systems can be installed at SR sources, and in fact one of the first demonstrations of the principle has been performed at HASYLAB [81], a more practical implementation is based on the use of relatively low-power micro-focus X-ray sources, as shown by the examples in Fig. 9 [80]. Beam confinement/focusing in such systems is either achieved by simple pinholes or by the use of polycapillary-based focusing with working distances in the mm range. Making use of high-throughput silicon drift detectors (SDD) in conjunction with high-speed digital data acquisition systems, scanning speeds of up to 100 mm/s can be achieved, as demonstrated by the commercial version of the MA-XRF spectrometer, the Bruker M6 Jetstream [82].

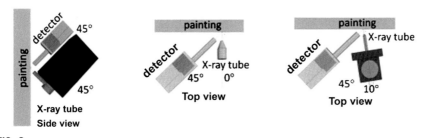

FIG. 9

Various implementations of macro-XRF (MA-XRF) systems as proposed by Alfeld et al.

Reproduced from M. Alfeld, K. Janssens, J. Dik, W. de Nolf, G. van der Snickt, Optimization of mobile scanning macro-XRF systems for the in situ investigation of historical paintings, J. Anal. At. Spectrom. 26(5) (2011) 899–909, with permission from The Royal Society of Chemistry.

The possibility to use MA-XRF scanning to visualize elemental distributions from a painting's surface and from layers below makes it especially useful to reveal hidden subsurface paint layers, potential alterations and to study painting techniques.

A particularly beautiful and famous example of discovering a hidden, and therefore before the detailed macro-scanning XRF experiments effectively lost, portrait of a woman by MA-XRF scanning has been published by Dik et al. [81], which clearly demonstrates the unique capabilities of the method. The first-time use of synchrotron-based MA-XRF enabled to visualize a woman's head with unprecedented details hidden under the surface of the *Patch of Grass* by Van Gogh. MA-XRF imaging opened a unique spectroscopic window through the surface of the painting to see the invisible, which was especially important to understand the evolution of Van Gogh's works throughout his initial years as an artist. About one-third of his early paintings reused canvas from his earlier abandoned, overpainted works. In comparison with commonly applied X-ray transmission radiographic images, the achievable contrast for hidden layers is increased tremendously under favorable conditions in which the subsurface paint layers contain (noninterfering) high atomic number elemental constituents. The high-energy fluorescent lines of high-Z elements can escape through the cover layers toward the detector without significant absorption effects. In this case two elemental signals, namely Sb-K_α and Hg-L_α associated respectively with Naples yellow and vermillion, could be used to visualize the hidden portrait with excellent quality. Fig. 10 shows the pseudo-color reconstruction based on the distribution of the above elements obtained by MA-XRF scanning on the relevant area ($17.5 \times 17.5\,\text{cm}^2$) of the *Patch of Grass*.

Using a dwell-time of 2 s per pixel ($0.5 \times 0.5\,\text{mm}^2$), the scan of the corresponding area took approximately 2 days. The reconstructed image is obtained by assigning realistic skin-like color to Sb and red to Hg, corresponding to the expected colors of the hidden face. The discovered image is believed to represent the missing link between a series of similar paintings from the artist's period in the village of Nuenen

FIG. 10

Pseudo-color reconstruction of a woman's face hidden under the surface of the *Patch of Grass* by Van Gogh [81], based on a MA-XRF measurement. Choosing realistic colors (skin-like and *red*) for the detected Sb-Kα and Hg-Lα intensities, the portrait could be reconstructed with excellent quality.

Reproduced from J. Dik, K. Janssens, G. Van der Snickt, L. van der Loeff, K. Rickers, M. Cotte, Visualization of a lost painting by Vincent van Gogh using synchrotron radiation based X-ray fluorescence elemental mapping, Anal. Chem. 80(16) (2008) 6436–6442, with permission from the American Chemical Society.

(The Netherlands), where he painted portraits of peasant women in dark settings in the period of 1884–85 [83].

A more recent example shows the use of a laboratory source-based MA-XRF for large area scans of van Eyck's world renowned Ghent Altarpiece, illustrating how the technique could be used to visualize underlying paint layers and images below the surface, not only to discover the painting's history but to aid the applied conservation strategy as well. The world-famous piece of art by the van Eyck brothers is often considered to be the pinnacle of European mediaeval painting techniques, attracting up to 200,000 visitors per year [84]. Considering its vast size, elemental imaging of the entire altarpiece was particularly challenging and, up to 2017, compositional imaging type information was collected from eight wing panels measuring approximately 8 m^2 divided in 37 separate scan areas [85].

By taking advantage of the penetrating character of X-rays and the noninvasive elemental imaging capability of XRF, the authors could visualize the original paint layers by van Eyck, hidden beneath the overpainted surface and, in addition, to assess their condition. As shown in Fig. 11, the distribution of the relatively high-energy Pb-L and Hg-L emission lines revealed the location of hidden areas with paint losses,

FIG. 11

(A) Portrait of Joos Vyd. from van Eyck's renowned Ghent Altarpiece. (B) Corresponding composite elemental image showing the distribution of Hg *(red)*, Fe *(green)*, and Pb *(white)*.

Reproduced from G. Van der Snickt, H. Dubois, J. Sanyova, S. Legrand, A. Coudray, C. Glaude, M. Postec, P. Van Espen, K. Janssens, Large-area elemental imaging reveals Van Eyck's original paint layers on the Ghent Altarpiece (1432), rescoping its conservation treatment, Angew. Chem. Int. Ed. 56(17) (2017) 4797–4801, with permission from Wiley-VCH GmbH, Weinheim.

and the Fe-K maps (indicated by green) demonstrated how these gaps were filled in the past using an iron-containing material.

The success of MA-XRF for the analysis of invaluable historical paintings is clearly demonstrated by the significant number of publications appearing in the literature in recent years, investigating high-profile works of art of unmeasurable value and complementing more conventional analytical and/or imaging techniques, see e.g., [86–89]. Alfeld and de Viguerie have recently published a comprehensive overview on the status and use of MA-XRF and complementary spectroscopic imaging techniques for the analysis of historical paintings [90].

4.3 Confocal-XRF

The relatively large information depth encountered in XRF analysis, combined with the use of (sub)micron X-ray beams and appropriate detection methodologies, offers an excellent starting point to expand the technique toward a three-dimensional (3D), and potentially quantitative analytical method with a depth-resolution level of the order of 10 µm. Due to the penetrating character of X-rays in case of a regular scanning micro-XRF instrument, the detected analytical information represents a line-integral along the beam path up to the information depth within the sample. A typical scanning XRF experiment therefore results in a two-dimensional (2D) projection of the elemental distributions within the examined sample, without any, or very limited, depth information.

The most frequently used and very popular approach to achieve depth selectivity by micro-XRF in cultural heritage studies is based on the excitation with an X-ray microbeam combined with a confocal detection scheme using a polycapillary half-lens (see Fig. 12). Based on this type of confocal XRF, 3D distributions of (trace) elements can be obtained, but also 1D depth profiles across a layered sample or 2D cross-sections across arbitrary scanning planes can be acquired, with information depths defined by the element- and matrix-dependent absorption length of characteristic X-rays. Confocal-XRF was initially developed for depth-resolved, 3D elemental mapping, however, it also allows the collection of depth-resolved XANES information [15, 35–39].

This combination of polycapillary half-lens coupled with an energy-dispersive detector for 3D-XRF was first proposed in the early 1990s by Gibson and Kumakhov, however, no experimental demonstration was given in this work for the proposed principle [91]. As the result of the rapid development of monolithic polycapillary optics at the end of the 1990s, the principle was applied in practice by several authors starting from 2002 [15, 35, 36, 92]. Havrilla and Gao demonstrated the use of polycapillary detection optics for scatter reduction in scanning micro-XRF experiments

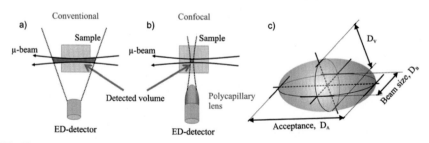

FIG. 12

Illustration of confocal 3D-XRF measuring techniques. (A) Conventional fluorescence detection, (B) confocal fluorescence detection, and (C) ellipsoidal detection volume for the confocal detection mode.

Adapted from G. Silversmit, B. Vekemans, S. Nikitenko, S. Schmitz, T. Schoonjans, F.E. Brenker, L. Vincze, Spatially resolved 3D micro-XANES by a confocal detection scheme, Phys. Chem. Chem. Phys. 12(21) (2010) 5653–5659.

with laboratory sources [92]. Kanngiesser et al. demonstrated the principle of localized detection and sample depth-scanning for synchrotron radiation-induced XRF [36]. Full characterization of polycapillary-based confocal XRF and its applications for 3D trace-element analysis is given in [15].

The polycapillary detector lens represents a significant modification to the regular scanning micro-XRF spectrometer, i.e., defining a specific micro-volume from which the XRF-signal is detected in the coinciding focii of the detection-side polycapillary and the incoming microbeam. The detection-side polycapillary, which acts as an efficient detector collimator, accepts an energy-dependent section from the incoming beam path as it intersects the sample: this defines a microscopic volume element of detection within the investigated object.

The laboratory-based realization of confocal-XRF is fairly rare, and such instruments are mainly developed in-house. A recent example of such development is a novel instrument for 3D elemental and morphological analysis presented by Laforce et al., combining X-ray computed tomography (μCT), XRF tomography, and confocal XRF analysis in a single laboratory instrument named Herakles [20]. Each module of Herakles (μCT, XRF-CT, and confocal XRF) represents the state-of-the-art of currently available laboratory X-ray techniques. The integration of these techniques enables linking the (quantitative) spatial distribution of chemical elements within the investigated materials to their 3D internal morphology/structure down to 10 μm resolution level, which has not been achieved sofar using the laboratory XRF techniques. The integration of the three different imaging end-stations is realized through a very high-precision motor stage in order to actively couple the data from the different techniques, allowing for optimal use of the measured data.

Previous efforts at integrating multiple X-ray-based analysis techniques into a single instrumental setup had a fixed sample stage with all components (e.g., sources/detectors for CT and XRF) arranged around it. This allows for a compact setup but unavoidably forces some compromises to the design. Therefore the Herakles 3D X-ray scanner uses a fundamentally different approach, where the sample stage is no longer a fixed point in the setup. On the contrary, long-travel range high-precision *xyz* motor stages allow movement of the sample stage from one analysis station to the other. The sample is aligned with the rotational axis using piezoelectric motors on top of these main motor stages, which causes the sample coordinate system to be unchanged when translating it between the end-stations of Herakles (Fig. 13). Since the three modules (μCT, XRF-CT, and cXRF) are independent entities in this instrument, their components and settings are optimized for the envisaged methodology without compromising the performances of the other techniques. The modular setup of the scanner permits a relatively easy change of components. This is an important asset, which allows upgrading the scanner with new components or testing of novel sources and/or detectors.

A typical experiment on the 3D X-ray scanner starts at the centrally positioned μCT station, where a high-resolution 3D image of the scanned objected is captured. On the basis of this image and the specific research question, in-house developed software allows the user to select certain points, lines, areas, slices, or even the complete sample to be investigated for their/its elemental composition using one

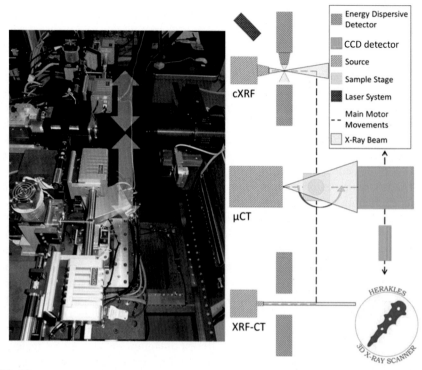

FIG. 13

The Herakles 3D-XRF scanner instrument, combining absorption micro-CT (μCT), X-ray fluorescence CT (XRF-CT), and confocal XRF (cXRF).

Adapted from B. Laforce, B. Masschaele, M.N. Boone, D. Schaubroeck, M. Dierick, B. Vekemans, C. Walgraeve, C. Janssen, V. Cnudde, L. Van Hoorebeke, L. Vincze, Integrated three-dimensional microanalysis combining X-ray microtomography and X-ray fluorescence methodologies, Anal. Chem. 89(19) (2017) 10617–10624.

(or both) of the XRF end-stations. The control software then autonomously moves the sample to the right position, considering the optimal angle-of-attack for the confocal imaging station. Since the coordinate system is retained between the two scans, coupling the data sets becomes relatively easy and straightforward.

5 Synchrotron-based X-ray micro-spectroscopy

Compared to the emission characteristics of conventional X-ray sources, such as X-ray tubes, synchrotron radiation facilities offer a number of advantages concerning X-ray micro-spectroscopy. This is the result of the unique properties of synchrotron radiation summarized below:

1. High brightness/brilliance (emission intensity normalized by source area and emission solid-angle), exceeding the output of conventional X-ray tubes by about 15 orders of magnitude: 3rd generation sources typically have a brilliance higher than 10^{18} photons/s/mm^2/mrad^2/0.1%BW. This property is associated with the low emittance of these sources, i.e., the product of source size and angle of emission is small. The new ESRF EBS 4th generation source is expected to have a brilliance between 10^{21} and 10^{22} photons/s/mm^2/mrad^2/0.1%BW, depending on the X-ray energy [44].

2. High natural collimation, or small angular divergence of the beam.
3. Widely tunable in energy/wavelength by monochromatization (from infrared to hard X-ray regime), although most of the large storage rings are tuned for the hard X-ray region.
4. High degree of polarization (linear or elliptical).
5. Pulsed X-ray emission with pulse durations down to about 100 ps.

A detailed review by Cotte et al. [6] gives an overview of the current status of X-ray nanoprobes applicable for cultural heritage studies together with application examples on the use of X-ray micro/nano-spectroscopic techniques.

In a micro/nanoprobe shown in Fig. 14, the SR beam is focused on the sample by a suitable X-ray optical element (e.g., Kirkpatrick-Baez mirrors as indicated on the figure, Fresnel zone plates, compound refractive lenses). A typical setup furthermore consists of a mechanical sample stage with computer-controlled positioning stages with (sub)micron precision for X, Y, Z and (optionally) rotational movement of the sample in the beam path, a semiconductor type detector for the measurement of the generated fluorescence radiation, different visualization tools for observation and positioning of the sample and, finally, a range of diagnostic and control tools.

A particular advantage of synchrotron radiation in comparison with the conventional X-ray tube sources for micro-XRF is the extremely high brilliance available and excellent focusing potential. In addition, in the plane of the storage ring the radiation is linearly polarized with the electric field vector parallel and the magnetic field vector normal to the ring plane. The polarization of the incident radiation can be used to reduce the relative contribution of scattered radiation reaching the XRF detector, as the angular distribution of scattering intensities is dependent on the polarization

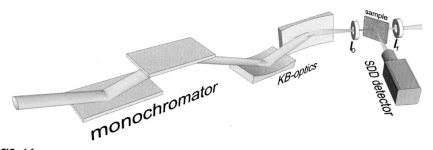

FIG. 14

Schematic representation of a synchrotron X-ray micro/nanoprobe facility, showing the main components of a beamline.

whereas the XRF emission is not. When performing the measurements in the plane of the SR source, under 90 degrees detection relative to the incoming beam, this increases the signal-to-background ratios by as much as two orders of magnitude compared to the unpolarized case, depending on the source characteristics and the particularities of the XRF setup (e.g., detector solid angle, degree of polarization).

The high intensity and directionality imply that SR is ideally suited for the generation of X-ray micro/nanobeams with very high intensity, exceeding now considerably 10^{10} photons/s/μm^2. Thanks partially also to the high directionality of the beam, quasimonochromatic X-ray microbeams can be generated from the polychromatic SR spectrum through the use of X-ray monochromators. By tuning the energy with a monochromator over a given energy range, the strong energy dependence of the inner shell photoelectric cross-sections can be exploited to either increase specificity of measurements or, else, to obtain speciation information in the XAS application mode (see Section 6).

A fixed-exit double crystal monochromator comprising a pair of crystals is a standard item in most monochromatic micro-XRF setups with exit direction and beam position kept constant during energy scanning. The energy resolution of the double crystal Si(111) monochromator is of the order of $\Delta E/E \sim 10^{-4}$, and this is sufficient for XRF coupled with absorption spectroscopic applications.

In its primary utilization mode as a tool for elemental analysis by micro/nano-XRF spectroscopy, often a high photon flux rather than high-energy resolution is required and an energy resolution of the order of $\Delta E/E \sim 10^{-2}$ is sufficient for the purpose. Synthetic multilayers, made by vacuum deposition of alternate thin layers of two materials with a different electron density, provide this ("pink beam") resolution while, through a wide energy band-pass, yielding a photon flux one to two orders of magnitude higher than available with a high-resolution double crystal monochromator [21].

As mentioned earlier, the extreme brightness, small source size and high directionality of SR is ideally suited for generating intense X-ray micro/nanobeams, based on the use of various types of X-ray optics. Grazing incidence bent mirrors in various configurations and geometries using crystals and multilayers, several types of glass mono-capillaries, complex polycapillary lens systems, diffractive lenses (Fresnel zone plates), Bragg-Fresnel lenses, 1D or 2D waveguides and refractive lenses have been developed and tested for use in micron size to submicron focusing at several synchrotron beamlines. At present, it is possible to obtain a sufficient beam intensity on microscale samples to allow reliable sub-ppm level determinations of a large number of elements. In particular circumstances spot sizes of less than 10 nm can be achieved [25]. A very popular design is based on Kirkpatrick-Baez focusing mirrors. Such systems are achromatic (the focal position is independent of the incident energy) and provide a long working distance. They are quite popular for micro-XAS applications, as the beam position can be maintained with a high-degree of positional accuracy while scanning over a specific energy range of an absorption edge using a fixed-exit monochromator. At the most advanced 3rd/4th generation SR facilities, such optical elements can provide beam sizes down to the 10–100-nm level, opening up possibilities for XRF nano-analysis.

When micron-sized focus is sufficient for the applications, typically in conjunction with bending magnet sources, micro-XRF/XAS installations in some cases make use of capillary optics focusing devices [39] because of their inherent constructional simplicity. This type of polycapillary optics-based focusing devices provide moderate lateral resolution (*c.* 10 μm) and can be used for both polychromatic and monochromatic X-rays, but suffer from the short working distance (typically <5 mm) between capillary tip and the sample.

The detection part of a micro-XRF setup is mostly based on a semiconductor detector, either high throughput (multielement) silicon drift detectors (SDD) or (multi-element) high purity germanium (HPGe) detectors, the latter mainly for XAS applications. Wavelength-dispersive spectrometers are seldom used because of the relatively low detection solid-angle and the resulting loss in sensitivity and speed.

The limited energy resolution of the energy dispersive detection systems gives rise to complex XRF spectra with multiple spectral interferences. Also, the high count rates must be adequately handled e.g. by using high-speed digital pulse processing. To take full advantage of the high degree of linear polarization of the SR source, to increase signal-to-noise ratios and to avoid overload of the detector by scattering, the X-ray detectors are typically positioned in the plane of the storage ring under an angle of 90 degrees to the incident beam direction, however, some micro and nanoprobes deviate from this arrangement. This is to optimize scanning geometry (e.g., sample surface irradiated under normal incidence) and to optimize detection solid angles in e.g. back-scattering geometry. Techniques were developed for fast and reliable nonlinear least squares deconvolution of XRF spectra that efficiently handles spectral interferences [93–95]. Multivariate statistical techniques for data reduction of image scans are available [96, 97].

For the measurement of elemental distribution maps, spectra are taken as the sample is moved over the exciting beam path. Contrary to the vacuum requirements of most other microprobe methods, samples can be observed in air, allowing the measurement of samples under their natural ambient conditions. 2D mapping of the elemental distributions in larger objects is possible, including the analysis of e.g. buried interfaces/heterogeneities, with relative detection limits in the ppb region and absolute detection limits for many elements well below the attogram.

Currently, a large number of existing storage rings are involved in performing micro and nano-XRF experiments. They combine the advantages of XRF as an elemental analytical tool with the unique possibilities of SR. Of special significance are the new 3rd and 4th generation storage rings that are specifically designed to obtain unprecedented intensity and brilliance, flexible choice and tunability of X-ray energy and nearly 100% linear polarization of the micro/nanobeams. A number of these are now routinely operational in Europe, the ESRF EBS, a recently upgraded 4th generation SR source (Grenoble, France) with unmatched brilliance at this moment among storage ring-based facilities, PETRA III (Hamburg, Germany), MAX IV in Lund, Sweden, Soleil (Saint-Aubin, France), Swiss Light Source (SLS, Villigen, Switzerland) and the Diamond Light Source (Oxfordshire, United Kingdom).

All of these SR facilities have at least one but mostly more micro and nanoprobes for multidisciplinary applications, some of which with very important contributions to research on cultural heritage materials. Highlighted examples are ESRF ID21 beamline with an exceptional engagement and success in studies on artistic materials [98], ESRF ID16B for nanoscale X-ray spectroscopy and imaging [99], PETRA III P06 Hard X-ray Micro/Nanoprobe [100] and the PUMA beamline [101] at Soleil, dedicated to cultural heritage applications.

Initiated at Soleil, IPANEMA ("European Institute for the nondestructive photon-based analysis of ancient materials") is a research platform that is devoted to the study of ancient samples and artifacts, facilitating the access of users to synchrotron beamlines at SOLEIL and at other European large-scale facilities [102].

6 X-ray absorption spectroscopy

XAS is a spectroscopic technique which is based on the absorption of X-rays in matter. The photon energy range used in XAS is of the order of 0.1–100 keV, able to access either the K, L, or M absorption edges of all elements in the periodic table. In this energy range the primary interaction type is the absorption of the photons by the photoelectric effect. The photon energy is too low to create an electron-positron pair and the scattering processes are typically 100–1000 times less probable than the photoelectric process around their absorption edges. As explained earlier, during the photoelectric effect, the incoming photon is fully absorbed by an atom and its energy is transferred to a bound electron that will leave the absorbing atom as a photoelectron.

In case a mono-energetic X-ray beam with intensity I_0 that travels through a given material, the X-ray beam will be attenuated exponentially along its path, therefore after passing through a sample of thickness d the intensity can be given as:

$$I = I_0 e^{-\mu_L d} = I_0 e^{-\mu_\rho \rho d} \tag{5}$$

with μ_L the linear attenuation coefficient. In practice the mass attenuation coefficient $\mu_\rho = \mu_L/\rho$ (with ρ the density of the material) is being used. The linear attenuation coefficient of a sample composed of different atom types at a certain photon energy E, can easily be calculated from

$$\mu_L(E) = \rho \sum_i w_i \cdot \mu_{\rho,i}(E) \tag{6}$$

with w_i the weight fraction of element i in the sample. The measurement of the linear attenuation coefficient $\mu_L(E)$ as a function of energy around the absorption edge (c. 100 eV below to 1 keV above the edge) of an element of interest directly (transmission mode) or indirectly (emission modes) is the basis of XAS.

As an example, the measured attenuation coefficient of Fe_2O_3 is shown in Fig. 15, indicating the relevant energy regions involved in XAS measurements, namely the XANES and the EXAFS regions.

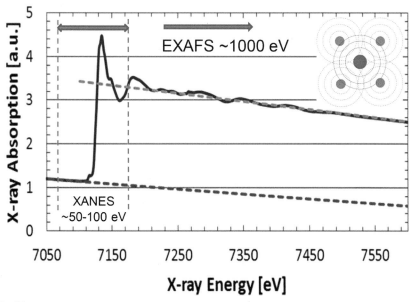

FIG. 15

Measured attenuation coefficient of $\alpha\text{-}Fe_2O_3$ as a function of energy, indicating the relevant energy regions involved in XAS measurements, namely the XANES and the EXAFS regions. The inset shows the origin of EXAFS oscillations due to the interference effects between incident and outgoing photoelectron waves.

The linear attenuation coefficient μ_L decreases monotonically before the edge, apart from preedge features, as matter becomes more and more transparent for X-rays with increasing X-ray energy. At the edge, a sudden increase in absorption can be observed as the linear attenuation coefficient increases suddenly. This jump is a so-called "absorption edge." The energy position of the absorption edges for a specific element corresponds to the binding energies of the different electron levels in that element. Imagine that we change the energy of the incoming photon beam from an energy value below an absorption edge for a certain element to an energy value above the edge. As soon as the energy of the photon becomes equal or larger than the binding energy of that particular electron energy level in that atom type, the electrons of that level can contribute to the absorption via the photoelectric effect. A new channel for absorption opens up and the total absorption for that element increases suddenly.

The XAS spectra contain valuable information on the local atomic environment of the central atom (selected by the absorption edge) upto a distance of about 5–15 Å, providing data on the chemical/oxidation state, coordination chemistry, bond-length values and disorder factors, and allow chemical fingerprinting of molecular species.

Typically accessible elements for XAS are $Z \geq 15$, from major to trace concentration levels, applicable to both crystalline and noncrystalline materials.

6.1 X-ray absorption near edge structure

XANES involves the excitation of photoelectrons to the unoccupied or partially filled levels from the Fermi level on, up to the EXAFS limit in the continuum. These levels thus contain the states that bind the system under study together and compose its electronic structure. For the rare earth metals for instance the XANES states include the transition toward the narrow unoccupied f bands (for the transition metals to the narrow d bands) and the less tightly bound s and p states. For molecules the empty bound levels (Rydberg series) and low-lying continuum states can be reached.

XANES requires a multiple scattering theory to describe the interaction of the photoelectron in the absorption process. The XANES spectrum is not only sensitive to the radial distribution of local structure around the absorber, but also to the relative orientation of the ligands, to the symmetry of the ligand coordination, to bond angles and so on. The position of the absorption edge shifts toward higher energies with increasing oxidation state, with a typical value of 1–2 eV per oxidation state. Furthermore, XANES spectra can be used efficiently for chemical fingerprinting of different molecular species of a given element down to trace concentration levels.

6.2 Extended X-ray absorption fine structure

To understand the EXAFS oscillations, one needs to consider the quantum mechanical nature of the electrons, namely the wave character of the electron. In matter, atoms are surrounded by neighboring atoms. The outgoing spherical electron wave can be scattered by these neighbors. Each scattering atom creates a new outgoing spherical wave centered on it, see inset in Fig. 15. The total final state is formed by the interference of all waves: the original outgoing spherical photoelectron wave and all scattered waves. In the case of constructive interference, photoelectron wave amplitude will be higher compared to the value for an isolated atom, so the absorption probability will be higher and thus also the absorption coefficient. For the destructive interference case, the absorption will be smaller compared to the isolated atom.

During the recording of an EXAFS spectrum, the absorption coefficient is measured as a function of photon energy. With increasing photon energy, the energy and impulse of the photoelectron increase and its De Broglie wavelength decreases. As a function of the wavelength, the interference pattern changes and with varying incident energy constructive or destructive interference can follow up. This explains the oscillating fine structure of EXAFS as a function of photon energy.

The EXAFS oscillations are taken relative to the spectrum for an isolated atom. The EXAFS signal $\chi(E)$ is therefore defined as:

$$\mu(E) = \mu_0(E)(1 + \chi(E)) \tag{7}$$

with $\mu(E)$ the measured absorption coefficient for the probed atom and $\mu_0(E)$ the (hypothetical) attenuation coefficient for the isolated atom.

The oscillating part of the EXAFS signal $\chi(E)$ can be seen as the sum of sine functions over the coordination shells around the absorbing atom, leading to the so-called EXAFS Eq. (3):

$$\chi(E) = \sum_{j=1}^{\text{shells}} A_j(k) \sin\left[2kR_j + \varphi_j(k)\right] \tag{8}$$

The summation is taken over the coordination shells, consisting of all atoms situated approximately at the same distance R_j around the absorbing atom. The phase shift function $\varphi(k)$ depends on the type of absorber-scatterer pair. The amplitude $A_j(k)$ is given by [3]:

$$A_j(k) = \frac{S_0^2(k)e^{-2R_j/\lambda}}{kR_j^2} \times N_j F_j(k) \times e^{-2\sigma_j^2 k^2} \tag{9}$$
$$= \text{damping factor} \times \text{scattering power} \times \text{disorder}$$

The first factor in the amplitude $A_j(k)$ describes the attenuation of the electron wave. The exponential term $e^{-2R_j/\lambda}$ takes the finite mean free path (λ) of the photoelectrons into account. $S_0^2(k)$ represents the amplitude damping corresponding to the energy loss of the photoelectron due to the interaction of many particles as shake-up/shake-off processes in the absorbing atom. The second factor $N_j F_j(k)$ is the total scattering power of the photoelectron at the jth shell consisting of N_j atoms of the same element type at a distance R_j and with a backscattering amplitude $F_j(k)$ per atom. N_j is the coordination number for shell j. The third factor represents a Debye Waller factor which considers the structural disorder due to thermal vibrations and/or any other cause that introduces a Gaussian disorder.

The EXAFS formulas (8) and (9) contain structural parameters such as bond distances (R_j), coordination numbers (N_j) and structural disorder (σ_j^2). These useful parameters can thus be extracted from an EXAFS spectrum, if the parameters $F_j(k)$, φ_j, and λ_j are known. These parameters can be calculated or be obtained from the spectra of reference materials (with known structural parameters R_j and N_j). The EXAFS signal is proportional with $\frac{1}{R_j^2}$ and thus decays quickly with the distance from the absorber. Therefore only the local structural arrangement up to the order of $10\,\text{Å}$ around the absorber can be determined with EXAFS. The advantage is that Eqs. (8) and (9) are valid for a wide variety of samples: gases, solids, liquids, crystalline and amorphous materials, bulk layers, surfaces, adsorbed atoms, interfaces, etc. Eqs. (8) and (9) represent the basis for the general EXAFS data analysis. The goal is to determine from the experimental spectrum the bond distance R_j, coordination number N_j, and structural disorder σ_j^2 for the first shells around the absorbing atom type.

7 Applications

7.1 Determination of ink composition of Herculaneum papyrus scrolls

Combined micro-XRF and micro-XANES spectroscopy was used by Brun et al. [17] and Tack et al. [103] to investigate the written text on carbonized Herculaneum scrolls, specifically on fragments partially destroyed by Mount Vesuvius' eruption in 79 AD, found in the Villa dei Papiri at Herculaneum between 1752 and 1754. Some of the fragments from Villa dei Papiri were analyzed on several occasions, bearing unreadable text by optical means. The text was first revealed by X-ray phase-contrast tomography [104], potentially paving the way to read carbonized papyri by virtual unrolling the sensitive scrolls. Unfortunately, some of the text was difficult to read based on phase-contrast tomography alone due to the interference of the papyrus fibers with the written text in terms of absorption patterns. Surprisingly, during these investigations, lead was found as an important elemental constituent in the writing itself, indicating the use of a metal in Greco-Roman inks several centuries earlier than previously thought [17]. The finding of lead at the detected high-concentration levels was so unexpected, that several hypotheses were postulated regarding its origin in the papyrus writing. Multiscale XRF micro-imaging, MC-based quantification, and X-ray absorption microspectroscopy experiments were used to provide additional information on the ink composition, in an attempt to determine the origin of the lead in the Herculaneum scrolls and validate the postulated hypotheses.

The presence of lead was not only interesting from a cultural heritage-historical point-of-view, but it also enabled the clear visualization of the lost text on the carbonized fragments by recording the spatial variation of Pb-M and L lines by scanning micro-XRF spectroscopy at the ESRF ID21 and ESRF DUBBLE beamlines, respectively.

Fig. 16 shows an optical image of a large fragment together with the Pb distribution detected by scanning XRF, showing the clearly readable characters from this carbonized fragment based on the Pb image. Also shown is a measured XRF spectrum corresponding to a Pb-rich area, together with its quantified equivalent obtained by MC simulation using the XMI-MSIM software [58, 59].

These simulations can be regarded as a "no-compromise" solution for quantification, as they model all relevant photon-matter interactions, simulating the trajectories of a large number of photons originating from an X-ray source and undergoing interactions in the sample, to the point of detection by the detector. Considering the total area covered by the writing on this fragment and the average sample density, the Pb concentration in the ink was found approximately $84 \pm 5\,\mu g/cm^2$. Based on this value and also considering the works by Kim et al. [105] and Delile et al. [106], it was clear that the Pb content was too high to be caused by e.g. the contamination of lead in the water ($<1.5\,mg/L$), used as a solvent for the ink, or as a contaminant from a bronze container. Alternatively, Pb could have been intentionally introduced

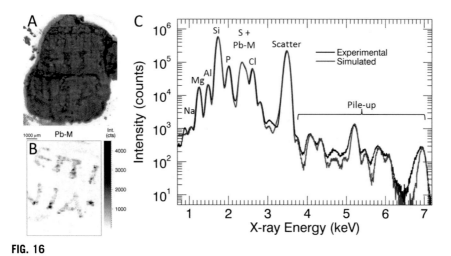

FIG. 16

A large fragment of a carbonized Herculaneum scroll (A) and written text visualized by scanning micro-XRF based on the Pb distribution (B). Also shown is a quantified measured XRF spectrum, based on the use of Monte Carlo simulation (C).

Adapted from P. Tack, M. Cotte, S. Bauters, E. Brun, D. Banerjee, W. Bras, C. Ferrero, D. Delattre, V. Mocella, L. Vincze, Tracking ink composition on Herculaneum papyrus scrolls: quantification and speciation of lead by X-ray based techniques and Monte Carlo simulations, Sci. Rep. 6 (2016) 20763.

to the ink in a controlled fashion. Lead-based pigments, being a black mineral galena (PbS) or lead white (different mixtures of cerusite ($PbCO_3$) and hydrocerusite ($2PbCO_3,Pb(OH)_2$)), were frequently used in ancient times as a pigment for cosmetic products [107–109]. Galena has been proposed as a pigment in black inks in Egyptian papyrus before [110], whereas minium ($Pb_2^{2+}Pb^{4+}O_4$) has been reported as a red pigment in Roman writing [111, 112]. Pb could also originate from a binding medium in the ink: Pb compounds have been used extensively as dryers in paintings as they speed up the process of oil drying [113, 114]. The use of litharge (PbO) as oil drier is already mentioned by Galen at the 2nd C. A.D. and by Marcellus at the 4th C. A.D [115, 116]. Furthermore, the use of a lipid-based ink to draft the papyrus writing can be hypothesized [117].

Fig. 17 shows the Pb distribution on the large fragment on both macroscopic and microscopic levels, as well as the distribution of Al, P, and S at different zoom levels down to the microscopic levels. The results show that Pb, P, S, and Al exhibit a partial colocalization on a scale of hundreds of microns. However, from the high-resolution micron-scale images one can conclude that Pb is not associated chemically with Al due to their lack of co-distribution at the submicron level. This indicates that the used ink has a complex composition, with different ingredients of which the distribution appears homogeneous at the millimeter scale, but not at the microscopic level. The S signal shows a fairly homogeneous distribution over the papyrus, with occasional hot spots which are only partially co-distributed with Pb. It is thus

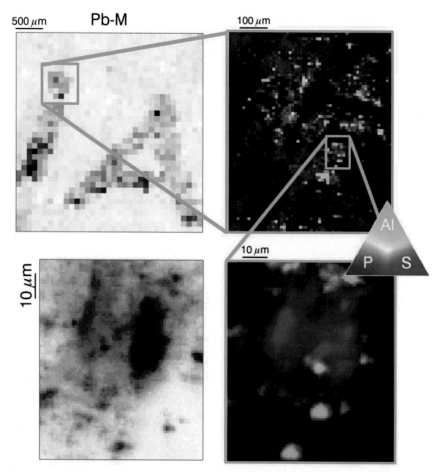

FIG. 17

Pb distribution on the large fragment on both macroscopic- (mm-scale) and microscopic-resolution levels (top and bottom left), as well as the distribution of Al, P, and S at different zoom levels, reaching (sub)microscopic spatial resolution.

Adapted from P. Tack, M. Cotte, S. Bauters, E. Brun, D. Banerjee, W. Bras, C. Ferrero, D. Delattre, V. Mocella, L. Vincze, Tracking ink composition on Herculaneum papyrus scrolls: quantification and speciation of lead by X-ray based techniques and Monte Carlo simulations, Sci. Rep. 6 (2016) 20763.

unlikely that PbS is the main constituent of the papyrus ink. This is also confirmed by the S K-edge XANES scan shown in Fig. 18, indicating complex features, characteristic of a mixture of different species. The primary information is obtained by comparing the detected signal from areas in and next to the ink. The spectra are quite similar, demonstrating that most of the signal in the area covered by the ink is most probably due to the papyrus itself, behind the ink. The spectra exhibit features similar to those recorded on plants or wood, with both reduced and oxidized sulfur species.

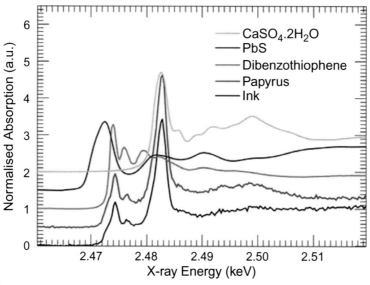

FIG. 18

S K-edge XANES spectra from various locations of the ink and papyrus, compared to reference compounds. The XANES spectra from the sample indicate complex features, characteristic of a mixture of different species.

Reproduced from P. Tack, M. Cotte, S. Bauters, E. Brun, D. Banerjee, W. Bras, C. Ferrero, D. Delattre, V. Mocella, L. Vincze, Tracking ink composition on Herculaneum papyrus scrolls: quantification and speciation of lead by X-ray based techniques and Monte Carlo simulations, Sci. Rep. 6 (2016) 20763, licensed under a Creative Commons Attribution 4.0 International License.

The presence of PbS in the papyrus writing is not indicated due to the absence of a sulfide peak at 2.4725 keV.

Similar studies have been performed by Christiansen et al. [118], investigating the composition of ancient Egyptian red and black inks on papyri by synchrotron-based microanalyses. They found that lead is regularly present in both the red and black inks, associated with phosphate, sulfate, chloride, and carboxylate ions. Their studies indicate that lead was probably used as a drier agent rather than as a pigment, similarly to its application in 15th century Europe during the development of oil paintings [118].

7.2 Studies on Iron Gall ink

A large number of X-ray spectroscopy-based works have been devoted to the investigation of Iron Gall inks in historical documents [119–128]. The origin of Iron Gall ink can be traced back as far as Antiquity with Gaius Plinius Secundus, describing the reaction of an iron salt soaked in a tannin solution with papyrus [129]. Studies either focus on issues related to degradation of historical manuscripts as a result of catalytic reactions induced by iron-gall inks, which is a major conservation issue, or on the investigation of the chemical composition of the ink for the better understanding of manufacturing processes, or for e.g. provenancing purposes.

In the example below, a letter written by the court of King Philip II of Spain to the archbishop of Mexico around 1580 was examined using scanning micro-XRF and EXAFS, in an attempt to link the current chemical composition of the ink with the geographical location of the original iron-nickel ores used [39]. The XRF maps of two letters (see inset in Fig. 19A) show Fe and K distributions associated with the ink. The lack of a strong preedge peak in the XANES spectrum (see inset in Fig. 19B) suggests a symmetrical octahedral direct coordination of the Fe atoms by O. The EXAFS first shell fit confirms a sixfold coordination with O atoms, as is expected for Iron Gall ink [123]. The XRF analysis revealed a lack of Ni in the used Iron Gall ink, a fact strengthened by the lack of any Fe-Ni interactions detectable in the EXAFS fit (until the third fitted shell). This could indicate that the original ores used to produce the Iron Gall ink did not contain Ni.

All XRF scans were performed at 8.5 keV, high enough to excite both Fe and Ni. A micro-focused beam of $13 \times 13 \, \mu m^2$ was used to allow for the investigation of several letters within a reasonable time frame with 1 s dwell time and a 20 µm step size. The Fe-K edge XANES was measured up to 7.26 keV with 0.5 eV energy resolution at the edge and 1 eV resolution postedge, while the EXAFS was measured until a k value of 12 (7.66 keV) and analyzed with the VIPER [130] and FEFF9 [131] software packages.

Fig. 20 shows the results of two consecutive XANES scans at the same point in the ink. The repeated XANES measurements expose the ink to high doses of X-rays, causing the formation of Fe(II) (and thus increasing the potential degradation damage, a major issue for historical documents). This form of radiation damage is becoming more prevalent with the use of micro-focus and high-intensity beamlines [39]. To combat these effects, the possibility to e.g. perform quick-EXAFS micro-focus experiments with reduced absorbed dose needs to be considered.

7.3 The investigation of Chinese blue-and-white porcelain dating from the Ming dynasty

A considerable attention in the literature is devoted to the analysis of porcelain and earthenware, performed for provenance determination [132–134], for conservation studies [135], and to determine the elemental composition in the different porcelain layers [136, 137], i.e., body, pigment, and glaze, which can also be determined for instance by SEM- [137] or PIXE-based [138, 139] analysis. To obtain additional information on the used pigment, next to the elemental composition by XRF [140], XANES measurements are often performed [141].

In the case study below by De Pauw et al. [142], two groups of Chinese blue and white porcelain samples, believed to originate from the Ming dynasty, were investigated using XRF and XAS: the first group (A, Fig. 21, left) consists of 10 shards

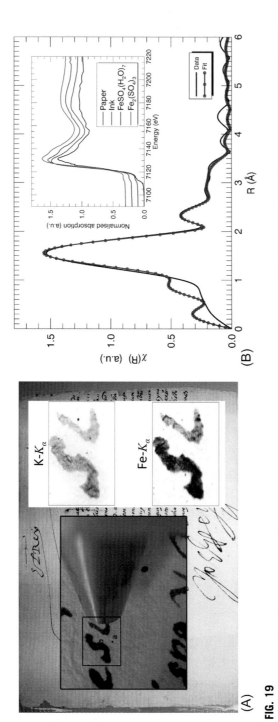

FIG. 19

XRF investigation of a letter written by the court of King Philip II of Spain to the archbishop of Mexico around 1580, showing the polycapillary optics focusing the beam in the investigated area (A). XRF maps of two letters (inset in A), showing Fe and K distributions associated with the ink. EXAFS result (B) and XANES scan on the Fe-K edge (inset in B).

Adapted from S. Bauters, P. Tack, J.H. Rudloff-Grund, D. Banerjee, A. Longo, B. Vekemans, W. Bras, F.E. Brenker, R. van Silfhout, L. Vincze, Polycapillary optics based confocal micro X-ray fluorescence and X-ray absorption spectroscopy setup at The European Synchrotron Radiation Facility Collaborative Research Group Dutch-Belgian Beamline, BM26A, Anal. Chem. 90(3) (2018) 2389–2394.

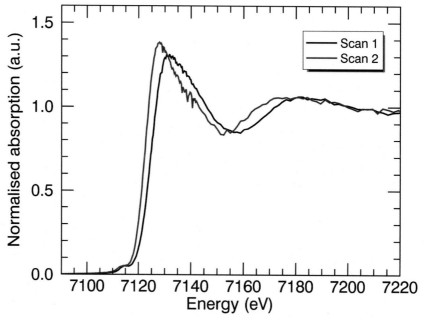

FIG. 20

Two consecutive XANES scans on the same Iron Gall ink spot. A clear edge energy shift indicates a higher presence of Fe^{2+} as a result of reduction from Fe^{3+}.

Adapted from S. Bauters, P. Tack, J.H. Rudloff-Grund, D. Banerjee, A. Longo, B. Vekemans, W. Bras, F.E. Brenker, R. van Silfhout, L. Vincze, Polycapillary optics based confocal micro X-ray fluorescence and X-ray absorption spectroscopy setup at The European Synchrotron Radiation Facility Collaborative Research Group Dutch-Belgian Beamline, BM26A, Anal. Chem. 90(3) (2018) 2389–2394.

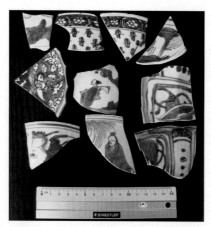
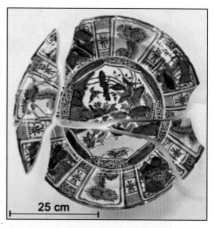

FIG. 21

Two groups of *Chinese blue and white* porcelain samples, believed to originate from the Ming dynasty, investigated by micro-XRF and XANES by De Pauw et al. [142].

Adapted from E. De Pauw, P. Tack, E. Verhaeven, S. Bauters, L. Acke, B. Vekemans, L. Vincze, Microbeam X-ray fluorescence and X-ray absorption spectroscopic analysis of Chinese blue-and-white kraak porcelain dating from the Ming dynasty, Spectrochim. Acta Part B 149 (2018) 190–196.

discovered close to the Malaysian coast in 1997 and retrieved by a team of under-water archaeologists from a Portuguese shipwreck in 2003 [142]. The second group (B, Fig. 21, right) consists of pieces from a plate belonging to a private collection. Via art-historical analyses, both groups were determined as so-called kraak porcelain (i.e., porcelain destined for trade to Europe) originating from the Ming dynasty (1368–1644). The word "kraak" itself is likely to originate from the Dutch word for the Portuguese merchant ships or "carracks" [143].

Fig. 22A shows elemental maps as an example acquired by micro-XRF scans on one of the samples belonging to group A, obtained using a laboratory micro-XRF spectrometer [142]. The elemental distributions over the different regions of interest

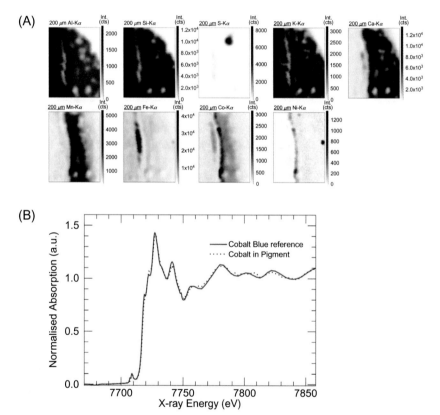

FIG. 22

Elemental maps acquired by micro-XRF scans on sample 10 belonging to group A, showing the layered structure of the porcelain's cross-section (A). XANES spectrum from the *blue region* compared to cobalt-blue reference (B).

Adapted from E. De Pauw, P. Tack, E. Verhaeven, S. Bauters, L. Acke, B. Vekemans, L. Vincze, Microbeam X-ray fluorescence and X-ray absorption spectroscopic analysis of Chinese blue-and-white kraak porcelain dating from the Ming dynasty, Spectrochim. Acta Part B 149 (2018) 190–196.

display a high similarity between the kraak porcelain shards that were recovered from the sunken merchant ship and the shard known to have originated from the Ming dynasty, preserved in a private collection. Analysis of all samples indicate that both groups have similar Si and K distributions over the entire porcelain cross-section, in the body and glaze high concentration of Al and Ca can be observed, respectively, and the underglaze pigment is characterized by the presence of Co and Ni.

The blue pigment was further studied by Co K-edge XANES measurements, taken in the pigment layer for each sample. In Fig. 22B the result for the Co-rich pigment layer is plotted together with the response of a measured reference sample for the Cobalt Blue pigment. The blue pigment in the investigated porcelain fragments shows essentially identical XANES spectra with that of a Cobalt Blue standard, confirming the identity of the blue pigment used in these materials.

8 Novel detection methods

8.1 Full-field XRF/XAS

As opposed to scanning type of micro/nano-XRF techniques, based on the use of conventional ED-detectors and X-ray microbeams for scanning the sample under investigation, new developments in imaging type of detection technology paves the way toward the so-called full-field XRF (FF-XRF) and XAS (FF-XAS) methodologies for elemental/chemical microanalysis. Such approaches require no beam focusing and scanning of the sample, simplifying equipment and subjecting the investigated objects to considerably lower flux-densities compared to nano/microbeam irradiation. Imaging is achieved on the detection side by combining X-ray fiber optics as an objective lens with energy-dispersive CCD detection.

A full-field detection-based 3D-XRF and emission-mode XANES approach have been demonstrated by Garrevoet et al. [144] and Tack et al. [40] using the SLcam (Strüder-Langhoff camera, or Color X-ray Camera) which is a Si-based pnCCD-type energy dispersive X-ray detector, consisting of a 450-μm thick Si layer with 264 by 264 $48 \times 48 \, \mu m^2$ pixels. The camera was developed by PNSensor GmbH (Munich, Germany) in collaboration with the Institute for Scientific Instruments GmbH (IFG, Berlin, Germany), the BAM Federal Institute for Materials Research and Testing (Berlin, Germany), and the Institut für Angewandte Photonik e.V. (IAP, Berlin, Germany) [145]. The camera can be equipped by polycapillary optics: a straight 1:1 magnification optic resulting in $48 \times 48 \, \mu m^2$ spatial resolution and a conical 6:1 magnification optic for $8 \times 8 \, \mu m^2$ spatial resolution.

Fig. 23A shows a photograph of FF-XRF/XANES investigations at the synchrotron facility PETRA III P06, corresponding to the analysis of a deep-earth diamond containing Fe-rich inclusions. The experiment shown here used a vertically oriented sheet-beam, indicated by green, to excite a given cross-section (and the inclusion cloud within) of the diamond sample. By recording a sequence of XRF images by

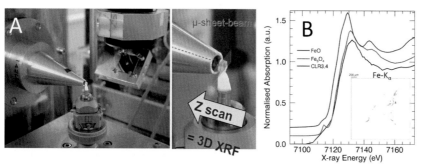

FIG. 23

Photograph of FF-XRF/XANES investigations at the synchrotron facility PETRA III P06 (A), and (average) XANES spectra measured by FF-XANES (B) corresponding to the analysis of a deep-earth diamond containing Fe-rich inclusions.

Adapted from P. Tack, B. Vekemans, B. Laforce, J. Rudloff-Grund, W.Y. Hernandez, Y. Wanton, J. Garrevoet, G. Falkenberg, F. Brenker, P. Van der Voort, L. Vincze, Application toward confocal full-field microscopic X-ray absorption near edge structure spectroscopy, Anal. Chem. 89 (3) (2017) 2123–2130.

a linear scan in the indicated "Z-scan" direction, one can achieve a full 3D-XRF measurement of the sample. Another possibility is to select a given sample cross-section, and record XRF images as a function of excitation energy across a given absorption edge. The results of such FF-XRF/XANES experiment are shown in Fig. 23B, showing a recorded Fe distribution in the inset corresponding to a given cross-section at an excitation energy just above the Fe-K absorption edge, and the (mean) Fe-K fluorescence mode XANES spectrum originating from the Fe-rich inclusions (black curve). The latter was obtained by cluster analysis-based averaging of the recorded pixel-spectra, corresponding to the image stack recorded as a function of excitation energy. Also shown are the recorded reference XANES spectra corresponding to FeO (blue) and Fe_3O_4 (red). The Fe-rich phase in the inclusions has the same or very similar chemical signature, which is according to the expectation for this particular specimen. The obtained XANES curve (black) is very similar to that of a magnetite (Fe_3O_4) reference compound (red), allowing the authors to hypothesize that the Fe in the investigated diamond inclusion cloud consists mainly of a mixture of Fe^{2+} and Fe^{3+} ions ordered in a magnetite-like structure.

8.2 X-ray excited optical luminescence

X-ray excited optical luminescence (XEOL) originates from X-ray excitation of material surfaces which, next to the emission of fluorescence and scattered X-rays, also results in the emission of low-energy electromagnetic radiation in the visible and near-visible bands [146]. The intensity of this optical light emission is dependent on the probability of absorption of the X-rays by core levels, hence it can provide information which is typically obtained by XANES and EXAFS spectroscopies with high surface sensitivity and the ability to be detected by adapted CCD-based microscopes in a full-field detection scheme using broad X-ray beam illumination.

A novel X-ray-excited optical emission microscope has been reported by Dowsett et al., optimized specifically for chemical imaging of heritage metal surfaces [146]. The instrument was demonstrated using copper test surfaces with a spatially varying patination, illustrating the imaging capability for copper, cuprite, nantokite and atacamite/paratacamite. The imaging software developed together with the microscope can process XEOM image stacks to obtain reduced data sets characterizing the surface chemical maps, including edge-shift (oxidation state) images and edge-height (concentration) images and XANES spectra from user defined regions of interest. Fig. 24 shows an example of a detected image stack as the energy of the broad incoming X-ray beam is scanned across an absorption edge of interest, showing the type of information that can be extracted from ROIs (summed XANES

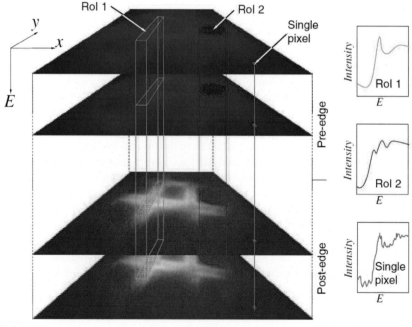

FIG. 24

Example of a detected image stack by XEOL as the energy of the broad incoming X-ray beam is scanned across an absorption edge of interest together with information that can be extracted from ROIs (summed XANES spectra) or from a series of individual pixels (pixel-level XANES spectrum) using the developed software [146].

spectra) or from a series of individual pixels (pixel level XANES spectrum) using the developed software [146].

As opposed to micro/nanobeam type of imaging by scanning methods, full-field approaches make use of orders of magnitude lower flux densities by unfocused, broad-beam excitation, which may reduce the risk of radiation damage or subtle changes of chemistry of the examined materials.

9 Radiation-induced changes and impact of X-ray irradiation

Although most variants of X-ray spectroscopic analysis are assumed to be nondestructive under reasonable conditions of use, their application to cultural heritage materials, especially at high-intensity SR sources, needs to be coupled with effective mitigation and monitoring strategies to limit radiation-induced alteration and/or damage to the studied artifacts. There are a surprisingly low number of studies concerning X-ray radiation effects in the field of cultural heritage compared to the considerable number of X-ray spectroscopic applications in the field appearing in the literature. The ongoing developments of upgrading existing 3rd generation SR sources and improving beam focusing leading to extreme flux densities for micro and nano-imaging applications make the topic of X-ray radiation-induced short- and long-term changes more important than ever.

Moini et al. assessed the impact of synchrotron X-ray irradiation on organic specimens at macro and molecular levels [147] and Bertrand et al. [148] discussed in detail the mitigation strategies concerning radiation damage in the analysis of ancient materials for various beam methods of analysis, and reviewed the types of direct and indirect radiation-induced effects on the main classes of cultural heritage materials [148]. Beam damage can manifest itself in various forms, including changes in visual appearance and optical properties, color, changes in morphology and mechanical properties, becoming brittle and even mechanical decomposition, etc. Even in cases when radiation-induced damage may not be detected by visual inspection, irradiation may still result in immediate or delayed atomic and molecular changes which can cause damage on the long term and/or bias subsequent analyses.

In case of X-ray irradiation, the primary source of radiation-induced change/damage originates on the one hand from the photoelectric effect itself, as high energy photons are absorbed by the atoms, generating high kinetic energy photoelectrons. The other mechanism of generating in-sample electrons is through Compton scattering, during which the X-ray photon transfers part of its energy to an electron of the target atom, generating a high kinetic energy free electron. The probability of Compton scattering is especially high for high X-ray energies and low Z matrices, such as organic materials. The above processes yield high-energy electrons within the irradiated volume which lose their kinetic energy through a large number of inelastic scattering events in the sample, causing secondary electron cascades. These processes are ultimately responsible for e.g. potential photoreduction of investigated species under prolonged X-ray beam irradiation, especially during XAS investigations. For example, in their studies of Iron Gall ink, Kanngiesser et al. [120], Wilke et al. [123], and Bauters et al. [39] observed a clear a shift of the Fe K absorption edge toward lower energies as a function of time, as revealed by a time series of XANES spectra collected at the same position. This is a clear indication of a reduction of Fe^{3+} ions to Fe^{2+} due to the (lengthy) exposure to radiation.

Very recently Monico et al. performed systematic studies on radiation damage induced by SR microanalysis in chrome yellow paints and related Cr compounds [149]. The combined use of Cr K-edge XANES, Cr-K_β X-ray emission spectroscopy, and μ-XRD, allowed to monitor and quantify the X-ray-induced damage in these compounds as a function of sample types for various chrome yellows ($PbCrO_4$ and $PbCr_{0.2}S_{0.8}O_4$), and found that modifications, including reduction of Cr(VI) species to Cr(III), are more pronounced for the $PbCr_{0.2}S_{0.8}O_4$ type especially when the pigment is mixed with a binder. Their results illustrate the necessity to evaluate X-ray sensitivity on systems as similar as possible to the real paints in advance, and to set upper limits of radiation fluence for the X-ray spectroscopic analysis, in this particular case recommending fluences not exceeding 10^{10}–10^{11} photons/μm^2.

It is clear that X-ray spectroscopy using highly focused SR beams at micro and especially at nanoprobes requires the implementation of appropriate mitigation strategies, such as avoidance of unnecessary beam exposure, improvement of detection efficiency and data collection (e.g., fast continuous scans) reducing exposure dwell time, reduction of the flux(density) if necessary, optimization of the sample environment e.g. by using He atmosphere or vacuum conditions for certain sample types [148]. In the fields of biology and protein crystallography, maintaining the sample in cryogenic conditions is a standard method to reduce radiation damage. This strategy is rarely applied to cultural heritage samples as it poses risks of damage itself to the precious artifacts. However, for certain types of materials e.g. micro-samples, this strategy could be considered in the future, in particular at X-ray nanoprobes [148, 149]. In addition to these mitigation efforts, standardized monitoring and recording strategies are recommended to evaluate the impact of the X-ray beam on the sample [148]. In spite of these efforts, the prediction of (subtle) long-term effects remains very challenging over extended multigeneration time-scales, considering that the studied invaluable pieces of art and cultural heritage materials are to be preserved for future generations.

References

[1] B. Beckhoff, B. Kanngießer, N. Langhoff, R. Wedell, H. Wolff (Eds.), Handbook of Practical X-Ray Fluorescence Analysis, Springer-Verlag, Berlin, Heidelberg, 2006.

[2] J.A. van Bokhoven, C. Lamberti (Eds.), X-Ray Absorption and X-Ray Emission Spectroscopy: Theory and Applications, John Wiley & Sons, Chichester, UK, 2016.

[3] D.E. Sayers, E.A. Stern, F. Lytle, New technique for investigating noncrystalline structures: Fourier analysis of the extended X-ray-absorption fine structure, Phys. Rev. Lett. 27 (1971) 1204–1207.

[4] D.C. Koningsberger, R. Prins (Eds.), X-Ray Absorption Principles, Application, Techniques of EXAFS, SEXAFS and XANES, John Wiley & Sons, Inc., 1988.

[5] M. Cotte, et al., Synchrotron-based X-ray absorption spectroscopy for art conservation: looking back and looking forward, Acc. Chem. Res. 43 (6) (2010) 705–714.

[6] M. Cotte, et al., Applications of synchrotron X-ray nano-probes in the field of cultural heritage, C. R. Physique 19 (2018) 575–588.

[7] K. Janssens, et al., Non-invasive and non-destructive examination of artistic pigments, paints, and paintings by means of X-ray methods, Top. Curr. Chem. 374 (6) (2016) 81.

[8] L. Bertrand, et al., Development and trends in synchrotron studies of ancient and historical materials, Phys. Rep. 519 (2) (2012) 51–96.

[9] F. Farges, M. Cotte, in: J.A. van Bokhoven, C. Lamberti (Eds.), X-Ray Absorption and X-Ray Emission Spectroscopy: Theory and Applications, John Wiley & Sons, Chichester, UK, 2016, pp. 609–636.

[10] K. Janssens, M. Cotte, Using synchrotron radiation for characterization of cultural heritage materials, 2457–2483, in: E.J. Jaeschke, S. Khan, J.R. Schneider, J.B. Hastings (Eds.), Synchrotron Light Sources and Free-Electron Lasers, Springer, Cham, 2020.

[11] K. Janssens, F. Adams, Applications in art and archaeology, in: K.H.A. Janssens, F.C.V. Adams, A. Rindby (Eds.), Microscopic X-Ray Fluorescence Analysis, John Wiley & Sons, Chichester, UK, 2000.

[12] K. Janssens, J. Dik, M. Cotte, J. Susini, Photon-based techniques for nondestructive subsurface analysis of painted cultural heritage artifacts, Acc. Chem. Res. 43 (6) (2010) 814–825.

[13] R. Terzano, M.A. Denecke, G. Falkenberg, B. Miller, D. Paterson, K. Janssens, Recent advances in analysis of trace elements in environmental samples by X-ray based techniques (IUPAC technical report), Pure Appl. Chem. 91 (6) (2019) 1029–1063.

[14] K. Janssens, G. Vittiglio, I. Deraedt, A. Aerts, B. Vekemans, L. Vincze, F. Wei, I. Deryck, O. Schalm, F. Adams, A. Rindby, A. Knochel, A. Simionovici, A. Snigirev, Use of microscopic XRF for non-destructive analysis in art and archaeometry, X-Ray Spectrom. 29 (1) (2000) 73–91.

[15] L. Vincze, B. Vekemans, F.E. Brenker, G. Falkenberg, K. Rickers, A. Somogyi, M. Kersten, F. Adams, Three-dimensional trace element analysis by confocal X-ray microfluorescence imaging, Anal. Chem. 76 (22) (2004) 6786–6791.

[16] B. De Samber, G. Silversmit, R. Evens, K. De Schamphelaere, C. Janssen, B. Masschaele, L. Van Hoorebeke, L. Balcaen, F. Vanhaecke, G. Falkenberg, L. Vincze, Three-dimensional elemental imaging by means of synchrotron radiation micro-XRF: developments and applications in environmental chemistry, Anal. Bioanal. Chem. 390 (1) (2008) 267–271.

[17] E. Brun, M. Cotte, J. Wright, M. Ruat, P. Tack, L. Vincze, C. Ferrero, D. Delattre, V. Mocella, Revealing metallic ink in Herculaneum papyri, PNAS 113 (14) (2016) 3751–3754.

[18] G. Silversmit, B. Vekemans, S. Nikitenko, W. Bras, V. Czhech, G. Zaray, I. Szaloki, L. Vincze, Polycapillary-optics based micro-XANES and micro-EXAFS at a third-generation bending-magnet beamline, J. Synchrotron Radiat. 16 (2) (2009) 237–246.

[19] G. Silversmit, B. Vekemans, S. Nikitenko, S. Schmitz, T. Schoonjans, F.E. Brenker, L. Vincze, Spatially resolved 3D micro-XANES by a confocal detection scheme, Phys. Chem. Chem. Phys. 12 (21) (2010) 5653–5659.

[20] B. Laforce, B. Masschaele, M.N. Boone, D. Schaubroeck, M. Dierick, B. Vekemans, C. Walgraeve, C. Janssen, V. Cnudde, L. Van Hoorebeke, L. Vincze, Integrated

three-dimensional microanalysis combining X-ray microtomography and X-ray fluorescence methodologies, Anal. Chem. 89 (19) (2017) 10617–10624.

[21] F. Adams, L. Vincze, B. Vekemans, Synchrotron radiation for microscopic X-ray fluorescence analysis, in: K. Tsuji, J. Injuk, R. Van Grieken (Eds.), X-Ray Spectrometry: Recent Technological Advances, John Wiley & Sons, 2004.

[22] A. Snigirev, et al., A compound refractive lens for focusing high-energy X-rays, Nature 384 (6604) (1996) 49.

[23] E. Di Fabrizio, et al., High-efficiency multilevel zone plates for keV X-rays, Nature 401 (6756) (1999) 895.

[24] G.E. Ice, J.D. Budai, J.W. Pang, The race to x-ray microbeam and nanobeam science, Science 334 (6060) (2011) 1234–1239.

[25] H. Mimura, S. Handa, T. Kimura, H. Yumoto, D. Yamakawa, H. Yokoyama, S. Matsuyama, K. Inagaki, K. Yamamura, Y. Sano, K. Tamasaku, Y. Nishino, M. Yabashi, T. Ishikawa, K. Yamauchi, Breaking the 10 nm barrier in hard-X-ray focusing, Nat. Phys. 6 (2) (2010) 122–125.

[26] C.G. Schroer, et al., Hard x-ray nanoprobe based on refractive x-ray lenses, Appl. Phys. Lett. 87 (2005) 124103.

[27] C.G. Schroer, B. Lengeler, Focusing hard x rays to nanometer dimensions by adiabatically focusing lenses, Phys. Rev. Lett. 94 (2005), 054802.

[28] V. Nazmov, R. Simon, E. Reznikova, J. Mohr, V. Saile, Polymer refractive crossed long lens: a new optical component for nanoimaging and nanofocussing in the hard X-ray region, J. Instrum. 7 (2012), 07019.

[29] V. Nazmov, E. Reznikova, J. Mohr, V. Saile, L. Vincze, B. Vekemans, S. Bohic, A. Somogyi, Parabolic crossed planar polymeric X-ray lenses, J. Micromech. Microeng. 21 (1) (2011), 015020.

[30] S. Gorelick, M.D. De Jonge, C.M. Kewish, A. De Marco, Ultimate limitations in the performance of kinoform lenses for hard x-ray focusing, Optica 6 (6) (2019) 790–793.

[31] T. Goto, S. Matsuyama, H. Hayashi, H. Yamaguchi, J. Sonoyama, K. Akiyama, H. Nakamori, Y. Sano, Y. Kohmura, M. Yabashi, T. Ishikawa, K. Yamauchi, Nearly diffraction-limited hard X-ray line focusing with hybrid adaptive X-ray mirror based on mechanical and piezo-driven deformation, Opt. Express 26 (13) (2018) 17477–17486.

[32] N. Moldovan, R. Divan, H.J. Zeng, L.E. Ocola, V. De Andrade, M. Wojcik, Atomic layer deposition frequency-multiplied Fresnel zone plates for hard x-rays focusing, J. Vac. Sci. Technol. A 36 (1) (2018), 01A124.

[33] X.P. Sun, X.Y. Zhang, Y. Zhu, Y.B. Wang, H.Z. Shang, F.S. Zhang, Z.G. Liu, T.X. Sun, 13.1 micrometers hard X-ray focusing by a new type monocapillary X-ray optic designed for common laboratory X-ray source, Nucl. Instrum. Methods Phys. Res. A 888 (2018) 13–17.

[34] J. Patommel, S. Klare, R. Hoppe, S. Ritter, D. Samberg, F. Wittwer, A. Jahn, K. Richter, C. Wenzel, J.W. Bartha, M. Scholz, F. Seiboth, U. Boesenberg, G. Falkenberg, C.G. Schroer, Focusing hard x rays beyond the critical angle of total reflection by adiabatically focusing lenses, Appl. Phys. Lett. 110 (10) (2017) 101103.

[35] K. Janssens, K. Proost, G. Falkenberg, Confocal microscopic X-ray fluorescence at the HASYLAB microfocus beamline: characteristics and possibilities, Spectrochim. Acta Part B At. Spectrosc. 59 (10–11) (2004) 1637–1645.

[36] B. Kanngiesser, W. Malzer, I. Reiche, A new 3D micro X-ray fluorescence analysis setup – first archaeometric applications, Nucl. Instrum. Methods Phys. Res. B 211 (2) (2003) 259–264.

[37] B. Kanngiesser, W. Malzer, A.F. Rodriguez, I. Reiche, Three-dimensional micro-XRF investigations of paint layers with a tabletop setup, Spectrochim. Acta Part B At. Spectrosc. 60 (1) (2005) 41–47.

[38] B. Kanngiesser, W. Malzer, I. Mantouvalou, D. Sokaras, A.G. Karydas, A deep view in cultural heritage-confocal micro X-ray spectroscopy for depth resolved elemental analysis, Appl. Phys. A Mater. Sci. Process. 106 (2) (2012) 325–338.

[39] S. Bauters, P. Tack, J.H. Rudloff-Grund, D. Banerjee, A. Longo, B. Vekemans, W. Bras, F.E. Brenker, R. van Silfhout, L. Vincze, Polycapillary optics based confocal micro X-ray fluorescence and X-ray absorption spectroscopy setup at the European Synchrotron Radiation Facility Collaborative Research Group Dutch-Belgian Beamline, BM26A, Anal. Chem. 90 (3) (2018) 2389–2394.

[40] P. Tack, B. Vekemans, B. Laforce, J. Rudloff-Grund, W.Y. Hernandez, Y. Wanton, J. Garrevoet, G. Falkenberg, F. Brenker, P. Van der Voort, L. Vincze, Application toward confocal full-field microscopic X-ray absorption near edge structure spectroscopy, Anal. Chem. 89 (3) (2017) 2123–2130.

[41] B. Laforce, B. Vermeulen, J. Garrevoet, B. Vekemans, L. Van Hoorebeke, C. Janssen, L. Vincze, Laboratory scale X-ray fluorescence tomography: instrument characterization and application in earth and environmental science, Anal. Chem. 88 (6) (2016) 3386–3391.

[42] G. Silversmit, B. Vekemans, F.E. Brenker, S. Schmitz, M. Burghammer, C. Riekel, L. Vincze, X-ray fluorescence nanotomography on cometary matter from comet 81P/Wild2 returned by Stardust, Anal. Chem. 81 (15) (2009) 6107–6112.

[43] D.F. Sanchez, A.S. Simionovici, L. Lemelle, V. Cuartero, O. Mathon, S. Pascarelli, A. Bonnin, R. Shapiro, K. Konhauser, D. Grolimund, P. Bleuet, 2D/3D microanalysis by energy dispersive X-ray absorption spectroscopy tomography, Sci. Rep. 7 (2017) 16453.

[44] G. Admans, P. Berkvens, A. Kaprolat, J. Revol (Eds.), ESRF Upgrade Programme Phase II (2015–2022) Technical Design Study, ESRF, 2014. http://www.esrf.eu/Apache_files/Upgrade/ESRF-orange-book.pdf.

[45] S. Smolek, C. Streli, N. Zoeger, P. Wobrauschek, Improved micro x-ray fluorescence spectrometer for light element analysis, Rev. Sci. Instrum. 81 (5) (2010), 053707.

[46] H.G.J. Moseley, The high-frequency spectra of the elements, Lond. Edinb. Dubl. Phil. Mag. J. Sci. 6 (26) (1913) 1024–1034.

[47] H.G.J. Moseley, The high-frequency spectra of the elements. Part II, Lond. Edinb. Dubl. Phil. Mag. J. Sci. 6 (27) (1914) 703–713.

[48] A. Brunetti, M.S. del Rio, B. Golosio, A. Simionovici, A. Somogyi, A library for X-ray-matter interaction cross sections for X-ray fluorescence applications, Spectrochim. Acta Part B At. Spectrosc. 59 (10–11) (2004) 1725–1731.

[49] T. Schoonjans, A. Brunetti, B. Golosio, M.S. del Rio, V.A. Sole, C. Ferrero, L. Vincze, The xraylib library for X-ray-matter interactions. Recent developments, Spectrochim. Acta Part B At. Spectrosc. 66 (11–12) (2011) 776–784.

[50] D.E. Cullen, The Evaluated Photon Data Library '97 Version (EPDL97), Lawrence Livermore National Laboratory, 1997. UCRL-50400, 6.

[51] L. Gerward, N. Guilbert, K.B. Jensen, H. Levring, X-ray absorption in matter. Reengineering XCOM, Radiat. Phys. Chem. 60 (1–2) (2001) 23–24.

[52] M.J. Berger, J.H. Hubbell, S.M. Seltzer, J. Chang, J.S. Coursey, R. Sukumar, D.S. Zucker, K. Olsen, XCOM: Photon Cross Sections Database, NIST Standard Reference Database 8 (XGAM), 2010. NBSIR 87-3597.

[53] D.E. Cullen, EPICS2014: Electron Photon Interaction Cross Sections (Version 2014), Document IAEA-NDS-218, 2014.

[54] F.C. Hila, A.V. Amorsolo, A.M.V. Javier-Hila, N.R.D. Guillermo, A simple spreadsheet program for calculating mass attenuation coefficients and shielding parameters based on EPICS2017 and EPDL97 photoatomic libraries, Radiat. Phys. Chem. 177 (2020), 109122.

[55] L. Vincze, K. Janssens, F. Adams, A general Monte-Carlo simulation of energy-dispersive X-ray fluorescence spectrometers. 1. Unpolarized radiation, homogeneous samples, Spectrochim. Acta Part B At. Spectrosc. 48 (4) (1993) 553–573.

[56] L. Vincze, K. Janssens, F. Adams, M.L. Rivers, K.W. Jones, A general Monte-Carlo simulation of ED-XRF spectrometers 2. Polarized monochromatic radiation, homogeneous samples, Spectrochim. Acta Part B At. Spectrosc. 50 (2) (1995) 127–147.

[57] U. Bottigli, A. Brunetti, B. Golosio, P. Oliva, S. Stumbo, L. Vincze, P. Randaccio, P. Bleuet, A. Simionovici, A. Somogyi, Voxel-based Monte Carlo simulation of X-ray imaging and spectroscopy experiments, Spectrochim. Acta Part B At. Spectrosc. 59 (10–11) (2004) 1747–1754.

[58] T. Schoonjans, V.A. Sole, L. Vincze, M.S. del Rio, K. Appel, C. Ferrero, A general Monte Carlo simulation of energy-dispersive X-ray fluorescence spectrometers – part 6. Quantification through iterative simulations, Spectrochim. Acta Part B At. Spectros. 82 (2013) 36–41.

[59] T. Schoonjans, L. Vincze, V.A. Sole, M.S. del Rio, P. Brondeel, G. Silversmit, K. Appel, C. Ferrero, A general Monte Carlo simulation of energy dispersive X-ray fluorescence spectrometers – part 5 polarized radiation, stratified samples, cascade effects, M-lines, Spectrochim. Acta Part B At. Spectrosc. 70 (2012) 10–23.

[60] A. Brunetti, B. Golosio, T. Schoonjans, P. Oliva, Use of Monte Carlo simulations for cultural heritage X-ray fluorescence analysis, Spectrochim. Acta Part B At. Spectrosc. 108 (2015) 15–20.

[61] A. Longoni, C. Fiorini, X-ray detectors and signal processing, in: B. Beckhoff, B. Kanngießer, N. Langhoff, R. Wedell, H. Wolff (Eds.), Handbook of Practical X-Ray Fluorescence Analysis, Springer-Verlag, Berlin, Heidelberg, 2006, p. 221.

[62] J. Sherman, The theoretical derivation of fluorescent X-ray intensities from mixtures, Spectrochim. Acta 7 (5) (1955) 283–306.

[63] M. Mantler, Basic fundamental parameter equations, in: B. Beckhoff, B. Kanngießer, N. Langhoff, R. Wedell, H. Wolff (Eds.), Handbook of Practical X-Ray Fluorescence Analysis, Springer-Verlag, Berlin, Heidelberg, 2006, pp. 311–327.

[64] S. Ridolfi, Portable systems for energy-dispersive X-ray fluorescence analysis, in: Encyclopedia of Analytical Chemistry, JohnWiley & Sons, Ltd, 2017.

[65] L. Van de Voorde, X-ray Fluorescence Spectrometers, in: Optimization and Applications of Mobile, Laboratory and Synchrotron X-Ray Based Non-Destructive Microanalysis Techniques for the Study of Cultural Heritage Objects, Ghent University, Belgium, 2015, p. 51 (PhD Thesis).

[66] C. Germinario, I. Francesco, M. Mercurio, A. Langella, D. Sali, I. Kakoulli, A. De Bonis, C. Grifa, Multi-analytical and non-invasive characterization of the polychromy of wall paintings at the Domus of Octavius Quartio in Pompeii, Eur. Phys. J. Plus 133 (2018) 359.

[67] A.V.M. Samson, L.J. Wrapson, C.R. Cartwright, D. Sahy, R.J. Stacey, J. Cooper, Artists before Columbus: a multi-method characterization of the materials and practices of Caribbean cave art, J. Archaeol. Sci. 88 (2017) 24–36.

[68] E. Frahm, Ceramic studies using portable XRF: from experimental tempered ceramics to imports and imitations at tell Mozan, Syria, J. Archaeol. Sci. 90 (2018) 12–38.

[69] M.F. Bonomo, Ceramic production and provenance in the Yiluo Basin (Henan, China): geoarchaeological interpretations of utilitarian craft production in the Erlitou state, Archaeol. Res. Asia 14 (2018) 80–96.

[70] M.M. Burton, P.S. Quinn, A. Tamberino, T.E. Levy, Ceramic composition at Chalcolithic Shiqmim, northern Negev desert, Israel: investigating technology and provenance using thin section petrography, instrumental geochemistry and calcareous nannofossils, Levant (2019), https://doi.org/10.1080/00758914.2019.1625656.

[71] R.B. Scott, B. Neyt, C. Hofman, P. Degryse, Determining the provenance of Cayo pottery from Grenada, Lesser Antilles, using portable X-ray fluorescence spectrometry, Archaeometry 60 (5) (2018) 966–985.

[72] S. Gasanova, S. Pagès-Camagna, M. Andrioti, S. Hermon, Non-destructive in situ analysis of polychromy on ancient Cypriot sculptures, Archaeol. Anthropol. Sci. 10 (2018) 83–95.

[73] M. Ferretti, The investigation of ancient metal artefacts by portable X-ray fluorescence devices, J. Anal. At. Spectrom. 29 (2014) 1753.

[74] M.A. Roxburgh, B.J.H. Van Os, A comparative compositional study of 7th- to 11th-century copper-alloy pins from Sedgeford, England, and Domburg, the Netherlands, Mediev. Archaeol. 62 (2) (2018) 304–321.

[75] M.A. Roxburgh, S. Heeren, D.J. Huisman, B.J.H. Van Os, Non-destructive survey of early Roman copper-alloy brooches using portable X-ray fluorescence spectrometry, Archaeometry 61 (1) (2019) 55–69.

[76] C. Pelosi, G. Agresti, G.B. Gianni, S. De Angeli, P. Holmes, U. Santamaria, In situ investigation by X-ray fluorescence spectroscopy on Pian di Civita Etruscan lituus from the "monumental complex" of Tarquinia, Italy, Eur. Phys. J. Plus 133 (2018) 357.

[77] D. Magrini, D. Attanasio, S. Bracci, E. Cantisani, W. Prochaska, Innovative application of portable X-ray fluorescence (XRF) to identify Göktepe white marble artifacts, Archaeol. Anthropol. Sci. 10 (2018) 1141–1152.

[78] L. Van de Voorde, M. Vandevijvere, B. Vekemans, J. Van Pevenage, J. Caen, P. Vandenabeele, P. Van Espen, L. Vincze, Study of a unique 16th century Antwerp majolica floor in the Rameyenhof castle's chapel by means of X-ray fluorescence and portable Raman analytical instrumentation, Spectrochim. Acta Part B At. Spectrosc. 102 (2014) 28.

[79] M.S. Tite, The production technology of Italian maiolica: a reassessment, J. Archaeol. Sci. 36 (10) (2009) 2065–2080.

[80] M. Alfeld, K. Janssens, J. Dik, W. de Nolf, G. van der Snickt, Optimization of mobile scanning macro-XRF systems for the in situ investigation of historical paintings, J. Anal. At. Spectrom. 26 (5) (2011) 899–909.

[81] J. Dik, K. Janssens, G. Van der Snickt, L. van der Loeff, K. Rickers, M. Cotte, Visualization of a lost painting by Vincent van Gogh using synchrotron radiation based X-ray fluorescence elemental mapping, Anal. Chem. 80 (16) (2008) 6436–6442.

[82] Bruker M6 Jetstream. https://www.bruker.com/products/x-ray-diffraction-and-elemental-analysis/micro-xrf-and-txrf/m6-jetstream/overview.html. Surveyed on 23/12/2020.

[83] L. Van Tilborg, M. Vellekoop, Vincent van Gogh: Paintings. Volume 1 Dutch Period: 1881–1885, Van Gogh Museum, Amsterdam, The Netherlands, 1999, p. 84.

[84] S. Nash, Northern Renaissance Art. Oxford History of Art, Oxford University Press, Oxford, 2008.

[85] G. Van der Snickt, H. Dubois, J. Sanyova, S. Legrand, A. Coudray, C. Glaude, M. Postec, P. Van Espen, K. Janssens, Large-area elemental imaging reveals Van Eyck's original paint layers on the Ghent altarpiece (1432), rescoping its conservation treatment, Angew. Chem. Int. Ed. 56 (17) (2017) 4797–4801.

[86] F.P. Romano, C. Caliri, P. Nicotra, S. Di Martino, L. Pappalardo, F. Rizzo, H.C. Santos, Real-time elemental imaging of large dimension paintings with a novel mobile macro X-ray fluorescence (MA-XRF) scanning technique, J. Anal. At. Spectrom. 32 (2017) 773–781.

[87] N. De Keyser, G. Van der Snickt, A. Van Loon, S. Legrand, A. Wallert, K. Janssens, J. Davidsz, de Heem (1606–1684): a technical examination of fruit and flower still lifes combining MA-XRF scanning, cross-section analysis and technical historical sources, Herit. Sci. 5 (2017) 38.

[88] R. Alberti, T. Frizzi, L. Bombelli, M. Gironda, N. Aresi, F. Rosi, C. Miliani, G. Tranquilli, F. Talarico, L. Cartechini, CRONO: a fast and reconfigurable macro X-ray fluorescence scanner for in-situ investigations of polychrome surfaces, X-Ray Spectrom. 46 (5) (2017) 297–302.

[89] M. Alfeld, M. Mulliez, P. Martinez, K. Cain, P. Jockey, P. Walter, The eye of the medusa: XRF imaging reveals unknown traces of antique polychromy, Anal. Chem. 89 (3) (2017) 1493–1500.

[90] M. Alfeld, L. de Viguerie, Recent developments in spectroscopic imaging techniques for historical paintings – a review, Spectrochim. Acta B 136 (2017) 81–105.

[91] W.M. Gibson, M.A. Kumakhov, Applications of X-ray and neutron capillary optics, in: R.B. Hoover (Ed.), X-Ray Detector Physics and Applications, Proceedings of The Society of Photo-Optical Instrumentation Engineers (SPIE), vol. 1736, The International Society for Optics and Photonics, 1993, pp. 172–189.

[92] G. Havrilla, N. Gao, Dual-polycapillary micro X-ray fluorescence instrument, in: Book of Abstracts, Denver X-Ray Conference, Colorado Springs, USA, 2002.

[93] B. Vekemans, K. Janssens, L. Vincze, F. Adams, P. Vanespen, Analysis of X-ray-spectra by iterative least-squares (Axil) – new developments, X-Ray Spectrom. 23 (6) (1994) 278–285.

[94] B. Vekemans, K. Janssens, L. Vincze, F. Adams, P. Vanespen, Comparison of several background compensation methods useful for evaluation of energy-dispersive X-ray-fluorescence spectra, Spectrochim. Acta Part B At. Spectrosc. 50 (2) (1995) 149–169.

[95] V.A. Sole, E. Papillon, M. Cotte, P. Walter, J. Susini, A multiplatform code for the analysis of energy-dispersive X-ray fluorescence spectra, Spectrochim. Acta Part B At. Spectrosc. 62 (1) (2004) 63–68.

[96] B. Vekemans, K. Janssens, L. Vincze, A. Aerts, F. Adams, J. Hertogen, Automated segmentation of μ-XRF image sets, X-Ray Spectrom. 26 (6) (1997) 333–346.

[97] B. Vekemans, L. Vincze, F.E. Brenker, F. Adams, Processing of three-dimensional microscopic X-ray fluorescence data, J. Anal. At. Spectrom. 19 (10) (2004) 1302–1308.

[98] M. Cotte, E. Pouyet, M. Salomé, C. Rivard, W. De Nolf, H. Castillo-Michel, T. Fabris, L. Monico, K. Janssens, T. Wang, P. Sciau, L. Verger, L. Cormier, O. Dargaud, E. Brun, D. Bugnazet, B. Fayard, B. Hesse, A.E.P. del Real, G. Veronesi, J. Langlois, N. Balcar, Y. Vandenberghe, V.A. Solé, J. Kieffer, R. Barrett, C. Cohen, C. Cornu, R. Baker, E. Gagliardini, E. Papillon, J. Susini, The ID21 X-ray and infrared microscopy beamline at the ESRF: status and recent applications to artistic materials, J. Anal. At. Spectrom. 32 (2017) 477–493.

[99] G. Martinez-Criado, J. Villanova, R. Tucoulou, D. Salomon, J.P. Suuronen, S. Laboure, C. Guilloud, V. Valls, R. Barrett, E. Gagliardini, Y. Dabin, R. Baker, S. Bohic, C. Cohen, J. Morse, ID16B: a hard X-ray nanoprobe beamline at the ESRF for nano-analysis, J. Synchrotron Radiat. 23 (1) (2016) 344–352.

[100] C.G. Schroer, C. Baumbach, R. Dohrmann, S. Klare, R. Hoppe, M. Kahnt, J. Patommel, J. Reinhardt, S. Ritter, D. Samberg, M. Scholz, A. Schropp, F. Seiboth, M. Seyrich, F. Wittwer, G. Falkenberg, Hard X-ray nanoprobe of beamline P06 at PETRA III, in: Q. Shen, C. Nelson (Eds.), Book Series: AIP Conference Proceedings, vol. 1741, 2016, p. 030007.

[101] A. Gianoncelli, S. Raneri, S. Schoeder, T. Okbinoglu, G. Barone, A. Santostefano, P. Mazzoleni, Synchrotron mu-XRF imaging and mu-XANES of black-glazed wares at the PUMA beamline: insights on technological markers for colonial productions, Microchem. J. 154 (2020) 104629.

[102] L. Bertrand, M.A. Languille, S.X. Cohen, L. Robinet, C. Gervais, S. Leroy, D. Bernard, E. Le Pennec, W. Josse, J. Doucet, S. Schoder, European research platform IPANEMA at the SOLEIL synchrotron for ancient and historical materials, J. Synchrotron Radiat. 18 (5) (2011) 765–772.

[103] P. Tack, M. Cotte, S. Bauters, E. Brun, D. Banerjee, W. Bras, C. Ferrero, D. Delattre, V. Mocella, L. Vincze, Tracking ink composition on Herculaneum papyrus scrolls: quantification and speciation of lead by X-ray based techniques and Monte Carlo simulations, Sci. Rep. 6 (2016) 20763.

[104] V. Mocella, E. Brun, C. Ferrero, D. Delattre, Revealing letters in rolled Herculaneum papyri by X-ray phase-contrast imaging, Nat. Commun. 6 (2015) 5895.

[105] E.J. Kim, J.E. Herrera, D. Huggins, J. Braam, S. Koshowski, Effect of pH on the concentrations of lead and trace contaminants in drinking water: a combined batch, pipe loop and sentinel home study, Water Res. 45 (2011) 2763–2774.

[106] H. Delile, J. Blichert-Toft, J.P. Goiran, S. Keay, F. Albarede, Lead in ancient Rome's city waters, Proc. Natl. Acad. Sci. U. S. A. 111 (2014) 6594–6599.

[107] P. Martinetto, et al., Synchrotron X-ray micro-beam studies of ancient Egyption make-up, Nucl. Instrum. Meth. Phys. Rev. B 181 (2001) 744–748.

[108] M. Cotte, P. Dumas, G. Richard, R. Breniaux, P. Walter, New insight on ancient cosmetic preparation by synchrotron-based infrared microscopy, Anal. Chim. Acta 553 (2005) 105–110.

[109] P. Walter, et al., Making make-up in ancient Egypt, Nature 397 (1999) 483–484.

[110] B. Wagner, Analytical approach to the conservation of the ancient Egyptian manuscript "Bakai Book of the Dead": a case study, Microchim. Acta 159 (2007) 101–108.

[111] P. Baraldi, G. Moscardi, P. Bensi, M. Aceto, L. Tassi, An investigation of the palette and techniques of some high medieval codices by Raman microscopy. Pigment and ink analysis of medieval codices, e-PS 6 (2009) 163–168.

[112] N. Vornicu, N. Melniciuc-Puică, E. Ardelean, Red pigments used for writing and illuminating manuscripts, Scientific Annals of the Alexandru Ioan Cuza University of Iasi Orthodox Theology 1 (2013) 75–87.

[113] J. Maroger, The Secret Formulas and Techniques of the Masters, Chapters 7–8, Hacker Art Books, 1948.

[114] M. Cotte, et al., Kinetics of oil saponification by lead salts in ancient preparations of pharmaceutical lead plasters and painting lead mediums, Talanta 70 (2006) 1136–1142.

[115] M.L. Kastens, F.R. Hansen, Drier soap manufacture, Ind. Eng. Chem. 41 (1949) 2080–2090.

[116] A.H. Sabin, Industrial and Artistic Technology of Paint and Varnish, Chapters 4, 36, J. Wiley & Sons, Inc., New York, 1927. London, Chapman & Hall, limited.

[117] C. Canevali, et al., A multi-analytical approach for the characterization of powders from the Pompeii archaeological site, Anal. Bioanal. Chem. 401 (2011) 1801–1814.

[118] T. Christiansen, M. Cotte, W. de Nolf, E. Mouro, J. Reyes-Herrera, S. de Meyer, F. Vanmeert, N. Salvadó, V. Gonzalez, P.E. Lindelof, K. Mortensen, K. Ryholt, K. Janssens, S. Larsen, Insights into the composition of ancient Egyptian red and black inks on papyri achieved by synchrotron-based microanalyses, PNAS 117 (45) (2020) 27829.

[119] O. Hahn, B. Kanngiesser, W. Malzer, X-ray fluorescence analysis of iron gall inks, pencils and coloured crayons, Stud. Conserv. 50 (1) (2005) 23–32.

[120] B. Kanngiesser, O. Hahn, M. Wilke, B. Nekat, W. Malzer, A. Erko, Investigation of oxidation and migration processes of inorganic compounds in ink-corroded manuscripts, Spectrochim. Acta Part B At. Spectrosc. 59 (10–11) (2004) 1511–1516.

[121] W. Malzer, O. Hahn, B. Kanngiesser, A fingerprint model for inhomogeneous ink-paper layer systems measured with micro-x-ray fluorescence analysis, X-Ray Spectrom. 33 (4) (2004) 229–233.

[122] O. Hahn, W. Malzer, B. Kanngiesser, B. Beckhoff, Characterization of iron-gall inks in historical manuscripts and music compositions using x-ray fluorescence spectrometry, X-Ray Spectrom. 33 (4) (2004) 234–239.

[123] M. Wilke, O. Hahn, A.B. Woodland, K. Rickers, The oxidation state of iron determined by Fe K-edge XANES-application to iron gall ink in historical manuscripts, J. Anal. At. Spectrom. 24 (10) (2009) 1364–1372.

[124] V. Rouchon, M. Duranton, C. Burgaud, E. Pellizzi, B. Lavedrine, K. Janssens, W. de Nolf, G. Nuyts, F. Vanmeert, K. Hellemans, Room-temperature study of iron gall ink impregnated paper degradation under various oxygen and humidity conditions: time-dependent monitoring by viscosity and X-ray absorption near-edge spectrometry measurements, Anal. Chem. 83 (7) (2011) 2589–2597.

[125] V. Rouchon, E. Pellizzi, M. Duranton, F. Vanmeert, K. Janssens, Combining XANES, ICP-AES, and SEM/EDS for the study of phytate chelating treatments used on iron gall ink damaged manuscripts, J. Anal. At. Spectrom. 26 (12) (2011) 2434–2441.

[126] N.K. Turner, C.S. Patterson, D.K. MacLennan, K. Trentelman, Visualizing underdrawings in medieval manuscript illuminations with macro-X-ray fluorescence scanning, X-Ray Spectrom. 48 (4) (2019) 251–261.

[127] G.V. Fichera, M. Malagodi, P. Cofrancesco, M.L. Weththimuni, L. Maduka, C. Guglieri, L. Olivi, S. Ruffolo, M. Licchelli, Study of the copper effect in iron-gall inks after artificial ageing, Chem. Pap. 72 (8) (2018) 1905–1915.

[128] R.J.D. Hidalgo, R. Córdoba, P. Nabais, V. Silva, M.J. Melo, F. Pina, N. Teixeira, V. Freitas, New insights into iron-gall inks through the use of historically accurate reconstructions, Herit. Sci. 6 (2018) 63.

[129] J. Bostock, H.T. Riley, The Natural History by Pliny the Elder, Taylor and Francis, 1855.

[130] K.V. Klementev, Extraction of the fine structure from x-ray absorption spectra, J. Phys. D Appl. Phys. 34 (2001) 209–217.

[131] J. Rehr, J. Kas, F. Vila, M. Prange, K. Jorissen, Parameter-free calculations of X-ray spectra with FEFF9, Phys. Chem.Chem. Phys. 12 (2010) 5503–5513.

[132] H. Ma, J. Zhu, J. Henderson, N. Li, Provenance of Zhangzhou export blue-and-white and its clay source, J. Archaeol. Sci. 39 (2012) 1218–1226.

[133] M. Isabel Dias, M. Isabel Prudêncio, M. Pinto De Matos, A. Luisa Rodrigues, Tracing the origin of blue and white Chinese Porcelain ordered for the Portuguese market during the Ming dynasty using INAA, J. Archaeol. Sci. 40 (2013) 3046–3057.

[134] I. Reiche, S. Röhrs, J. Salomon, B. Kanngießer, Y. Höhn, W. Malzer, F. Voigt, Development of a nondestructive method for underglaze painted tiles—demonstrated by the analysis of Persian objects from the nineteenth century, Anal. Bioanal. Chem. 393 (2009) 1025–1041.

[135] M. Mantler, M. Schreiner, X-ray fluorescence specrometry in art and archaeology, X-Ray Spectrom. 29 (2000) 3–17.

[136] T. Zhu, Y. Zhang, H. Xiong, Z. Feng, Q. Li, B. Cao, Porcelain from jingdezhen in the yuan dynasty of china (ad 1271–1368) by micro x-ray fluorescence spectroscopy and microscopy, Archaeometry 58 (6) (2016) 966–978.

[137] P. Colomban, G. Sagon, L. Huy, N. Liem, L. Mazerolles, Vietnamese (15th century) blue-and-white, tam thai and lustre porcelains/stonewares: glaze composition and decoration techniques, Archaeometry 46 (2004) 125–136.

[138] H. Cheng, Z. Zhang, H. Xia, J. Jiang, F. Yang, Non-destructive analysis and appraisal of ancient Chinese porcelain by PIXE, Nucl. Inst. Methods Phys. Res. B 190 (2002) 488–491.

[139] L.V. Ferreira, D. Ferreira, D. Conceicao, L. Santos, M. Pereira, T. Casimiro, F. Machado, Portuguese tin-glazed earthenware from the 17th century. Part 2: A spectroscopic characterization of pigments, glazes and pastes of the three main production centers, Spectrochim. Acta A Mol. Biomol. Spectrosc. 149 (2015) 285–294.

[140] R. Wen, C.S. Wang, Z.W. Mao, Y.Y. Huang, A.M. Pollard, The chemical composition of blue pigment on Chinese blue-and-white porcelain of the Yuan and Ming Dynasties (ad 1271–1644), Archaeometry 49 (1) (2007) 101–115.

[141] L. Wang, C. Wang, Co speciation in blue decorations of blue-and-white porcelains from Jingdezhen kiln by using XAFS spectroscopy, J. Anal. At. Spectrom. 26 (2011) 1796–1801.

[142] E. De Pauw, P. Tack, E. Verhaeven, S. Bauters, L. Acke, B. Vekemans, L. Vincze, Microbeam X-ray fluorescence and X-ray absorption spectroscopic analysis of Chinese blue-and-white kraak porcelain dating from the Ming dynasty, Spectrochim. Acta B 149 (2018) 190–196.

[143] S.G. Valenstein, A Handbook of Chinese Ceramics, Metropolitan Museum of Art, 1988.

[144] J. Garrevoet, B. Vekemans, P. Tack, B. De Samber, S. Schmitz, F.E. Brenker, G. Falkenberg, L. Vincze, Methodology toward 3D micro X-ray fluorescence imaging using an energy dispersive charge-coupled device detector, Anal. Chem. 86 (23) (2014) 11826–11832.

[145] O. Scharf, S. Ihle, I. Ordavo, V. Arkadiev, A. Bjeoumikhov, S. Bjeoumikhova, G. Buzanich, R. Gubzhokov, A. Gunther, R. Hartmann, M. Kuhbacher, M. Lang, N. Langhoff, A. Liebel, M. Radtke, U. Reinholz, H. Riesemeier, H. Soltau, L. Struder, F. Thunemann, R. Wedell, Anal. Chem. 83 (2011) 2532–2538.

[146] M. Dowsett, M. Hand, P.J. Sabbe, P. Thompson, A. Adriaens, XEOM 1 – a novel microscopy system for the chemical imaging of heritage metal surfaces, Herit. Sci. 3 (2015) 14.

[147] M. Moini, C.M. Rollman, L. Bertrand, Assessing the impact of synchrotron X-ray irradiation on proteinaceous specimens at macro and molecular levels, Anal. Chem. 86 (2014) 9417–9422.

[148] L. Bertrand, S. Schöeder, D. Anglos, M.B.H. Breese, K. Janssens, M. Moini, A. Simon, Mitigation strategies for radiation damage in the analysis of ancient materials, Trends Anal. Chem. 66 (2015) 128–145.

[149] L. Monico, M. Cotte, F. Vanmeert, L. Amidani, K. Janssens, G. Nuyts, J. Garrevoet, G. Falkenberg, P. Glatzel, A. Romani, C. Miliani, Damages induced by synchrotron radiation-based X-ray microanalysis in chrome yellow paints and related Cr-compounds: assessment, quantification, and mitigation strategies, Anal. Chem. 92 (2020) 14164–14173.

Ion beam analysis for cultural heritage

10

Chris Jeynes

University of Surrey Ion Beam Centre, Guildford, United Kingdom

1 Overview

1.1 Why ion beam analysis?

Ion beam analysis (IBA) (see *Glossary*, Section 2) is a powerful modern analytical method that has enabled a very wide variety of measurements critically important in the development of the advanced materials underlying all modern technology (including the semiconductor industry); also in many other fields from Archaeology to Zoology! It has been used since the 1970s so could be thought to be a "mature" technique, but the last decade has seen dramatic improvements allowing much greater versatility in a much wider range of applications, especially including cultural heritage (CH) ones. IBA for CH application has been reviewed several times in the technical literature from different points-of-view (e.g., [1–3]): here we aim to make a comprehensible nontechnical overview for CH professionals.

We hope that this chapter will help users appreciate when IBA might solve their problems and what its limitations are, as well as encouraging newcomers to take advantage of its capabilities and informing the "old hands" of new possibilities they may not have been aware of.

Fig. 1 shows a recent example [4] of elemental imaging of a precious sample of prehistoric cave art carried out at the AGLAE accelerator laboratory sited at the Louvre Museum in Paris. The purpose was to determine the trace element concentrations characteristic of the pigments used by the ancient artists. The problem is that only *traces* of pigment remain, and before any measurements can be made the pigment must be found! Moreover, the artefact is priceless and cannot be damaged in any way by the scientists. However, used appropriately, IBA is nondestructive, and has high-resolution imaging of large areas. From the elemental maps obtained, the regions containing the pigment could be recognised and the analysis successfully completed. We will discuss this and other examples in more detail later, but it is important to note at the outset that the analysis is fully quantitative, and that minor elements (in this case Mn, Ti, K) and trace elements (As, V, Cr, Mo) are reliably detected and quantified, as well as the major elements.

Spectroscopy, Diffraction and Tomography in Art and Heritage Science. https://doi.org/10.1016/B978-0-12-818860-6.00008-8

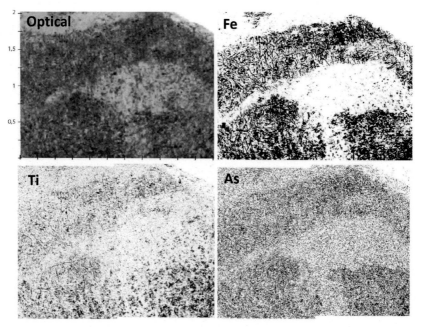

FIG. 1

PIXE of ancient cave art. Elemental images of part of a decorated bone (Scapula 2, Ref: V76B-5741) from the Abri Pataud rock-shelter (Dordogne, France) dated >26,000 years BP. Images are 2×2.75 mm using a 3 MeV external proton beam of size 40 μm.

Reproduced with permission from Fig. 2 of M. Lebon, L. Pichon, L. Beck, Enhanced identification of trace element fingerprint of prehistoric pigments by PIXE mapping, Nucl. Instrum. Meth. B 417 (2018) 91–95 (© 2017 Elsevier B.V.).

1.2 What is IBA?

IBA needs an electrostatic accelerator, typically running up to 2 million volts (2 MV). This is a high voltage that cannot be run in air (it will spark!): it requires a pressure vessel typically several meters long and containing something like 10 bar of SF_6 (sulfur hexafluoride: an insulating gas widely used in the electrical power industry). Such an installation is large and intricate, involving pressurised gas handling, heavy lifting gear, lots of high vacuum equipment, extensive control electronics, all together with the associated sophisticated spectrometry instrumentation. This is expensive both to build and run (of the order of €4 M, with annual running costs around 10% of the capital cost), and IBA labs are typically run as central facilities (national or regional). Almost all are based in universities since IBA is a cluster of techniques, not a single analytical method implementable as industrial equipment. It is more similar to synchrotron installations that cater for a wide variety of users, than (for example) a scanning electron microscope (SEM) that can be bought off the shelf, sits on the bench, and has a standard set of uses. Of course, the SEM is an essential equipment for any analytical laboratory, and modern SEMs are now very versatile.

Cultural heritage professionals use a variety of analytical methods. The closest to IBA and widely used in CH is X-ray fluorescence (XRF). This is a nondestructive technique capable of giving an immediate accurate and detailed elemental analysis of any object: it is valuable because it is so general, as well as being portable. This is not the place for an XRF-IBA comparison, except to say that if you can solve your problem by XRF (without going to one of the synchrotron laboratories) then you definitely should!

Using a synchrotron beamline (for example, *Bessy II* in Berlin, which has a dedicated metrology line), XRF can be used as a *reference method* to determine the basic physics parameters (*Fundamental Parameters*) needed to interpret XRF spectra [5], but RBS (an IBA technique, see *Glossary*) has also been demonstrated a *primary reference method* [6] and standard samples used to determine XRF fundamental parameters have been certified by RBS [7]. Standard samples used to calibrate industrial XRF instruments have also recently been certified by elastic backscattering spectrometry (EBS) [8]. This chapter shows why IBA is more powerful than XRF. The extent of this advantage has only recently become clear, as a consequence of two decades of intensive development of IBA as analytical method. It is true that IBA is normally a remote (inconvenient) and rather difficult technique used only where more convenient complementary methods fail, but the user will be supported by the accelerator facility staff not only to design and run measurements but also to interpret the resulting datasets.

We will walk through the accelerator laboratory (Section 3) and introduce readers to some of the basic physics ideas needed to make sense of what is going on (Section 4), explaining and illustrating the variety of types of IBA measurements that can be made (Section 5). This is all quite involved but I think it is essential, to provide to users the brief account they need to get started with their own intelligent and creative use of these versatile facilities. In Section 6 we describe how to get the best out of IBA, discussing Total-IBA examples. We conclude by summarising (Section 7) how users can make best use of IBA facilities, all of which have scientific liaison staff to help users take advantage of the facilities, and this chapter will enable users to quickly appreciate what the facility staff tell them.

2 Glossary

Atoms are one of the 92 chemical elements in the periodic table. Atoms are electrically neutral, with a heavy *nucleus* (made of *protons* and *neutrons*) surrounded by the <u>same number</u> of *electrons* (with a negative charge) as there are <u>protons</u> in the nucleus (with a positive charge).

Atomic number: How many protons the nucleus has. Atomic number is usually called Z. For example, the nuclei of {H, He, C, O, Pb} have atomic number $Z = \{1, 2, 6, 8, 82\}$.

Ions are atoms with the *wrong* number of electrons. Mostly ions are positively charged, since it is usually fairly easy to detach one or more of the electrons. But sometimes we want to make *negative ions*, where we persuade extra electrons to hitch a ride on the atom. The atom does not usually like this, and tries to get rid of the extra electrons at the first opportunity.

Charge state of the ion is how many electrons it has, too many or too few. For example, helium has two protons (and usually two neutrons), so the helium atom has two electrons. A helium ion with only one electron is called *singly charged* (written He^+), but a helium ion with no electrons is called *doubly charged* (He^{++}; this is also known as an *alpha particle*). A helium ion with three electrons is also singly charged (He^-). He^+ and He^{++} are *positive ions*, and He^- is a *negative ion*. A *proton* (H^+) is a hydrogen atom (with a nucleus of only one proton) without its electron. You cannot get doubly charged hydrogen! A *fully stripped* ion does not have any electrons left. So protons and alphas are fully stripped. So are Li^{3+}, C^{6+}, etc.

Electrostatics is how you think about apparatuses where some bits (the *electrodes*) are kept in a stable state of being charged up (i.e., with the *wrong number* of electrons) thus creating the electric fields used to form ions into beams and guide them according to the principles of ion optics.

Ion optics involves specially shaped electrodes such that you can think of them together as a *lens* for ions.

Ion beams are obtained when lots of ions are all travelling at the same speed in the same direction. This can happen when you grab ions out of a pot of ions (using electrostatics, i.e., an appropriate voltage on a suitable electrode) and squirt them into a particular direction using the right ion lens.

Ion energy is a very convenient way of speaking about how fast the ion is going. And this is given simply and unambiguously by the voltage through which the ion has fallen and the charge state of the ion. So if a singly charged ion has travelled through 1000 V (say), that is 1 kV, we say it has an energy of 1 keV ("kilo-electron-volt"). If He^+ is accelerated through 1,000,000 V (1000 kV), we say it has an energy of 1000 keV or 1 MeV ("mega-electron-volt"). But if He^{++} (an alpha particle) is accelerated through 1 MV, it has an energy of 2 MeV (double, since it is doubly charged).

Accelerator: the central equipment in an "accelerator laboratory." It accelerates ions from being essentially at rest (and going every which way with their thermal energy) to going very fast, all at the same speed and in the same direction. To have an "ion beam" you need an accelerator.

Sample: Analysts habitually refer to the "sample" when they really mean the "artefact." IBA is nondestructive: that means that the artefact to be measured is *not* required to be "sampled." Generally, artefacts to be measured by IBA require no preparatory treatments: they are analysed "as-received."

IBA: Ion beam analysis is a cluster of methods of handling the radiations produced when bombarding a *sample* with an *ion beam*. These include IBIL, PIXE, RBS, EBS, ERD, NRA, PIGE (*q.v.*).

IBIL: Ion beam induced luminescence. Visible light produced in the sample by the ion beam. IBIL is exactly equivalent to the better known *cathodoluminescence* (produced by an electron beam), and also to *X-ray excited optical luminescence* (XEOL).

PIXE: Particle-induced X-ray emission. The same as *IBIL* except that X-rays (not visible light) results. PIXE is exactly equivalent to X-ray fluorescence (XRF), except that XRF is produced by photon bombardment.

RBS: Rutherford backscattering spectrometry. Elastic backscattering (EBS *q.v.*) of incident ions from target atomic nuclei is interpreted as point charges in a Coulomb field. This is a special case of EBS. It is always an approximation, but frequently a very useful one since the RBS scattering cross-sections are known exactly.

EBS: Elastic backscattering spectrometry. Commonly, and incorrectly, called "RBS" in the literature. The scattering cross-sections must be measured, and for accurate work interpreted by an explicit nuclear model of the scattering event.

ERD: Elastic recoil detection. Elastic scattering is a two-body problem: EBS follows the incident ion (which is scattered from the target nucleus), ERD follows the recoiled target nucleus.

NRA: Nuclear reaction analysis. The incident ion interacts ***inelastically*** with the nucleus, with three or more particles resulting, any (or all) of which can be measured. The number of nucleons in the nucleus is changed in such an event, and the atomic number may also change.

PIGE: Particle-induced gamma-ray emission. A variety of NRA where one of the resulting particles is a (γ-ray) photon which is measured.

Total-IBA: An analysis that uses multiple IBA signals *synergistically.* Typically these are PIXE/EBS, but any combination is possible (and there are examples of all of these).

3 Surveying the accelerator laboratory

3.1 Quick walk through the lab

Fig. 2 shows a schematic of a typical IBA lab. Central is the accelerator which essentially is a fancy high-voltage power supply. We show a "tandem" accelerator, where the ion source is external to the accelerator. In this case *negative* ions are extracted from the ion source (using an "extraction lens") and injected into the accelerator. The high-voltage (positive) terminal is in the centre of the accelerator so that the negative ions are attracted *towards* it. When they arrive (at high speed) they are made to pass through a "stripper gas" which efficiently strips electrons from the ions. Only negative ions enter the high-voltage terminal, but only positive ions (or neutral atoms) exit it: negative ions are very easily stripped of their excess electrons.

Now the positive ions see the (positive) charge on the terminal and are accelerated *away* from it. So when the ion exits the accelerator it has been accelerated *twice* by the voltage (hence the name "tandem"). So if the extraction voltage on the ion source is 20 kV, extracting a helium beam (He^-), and the accelerator terminal voltage

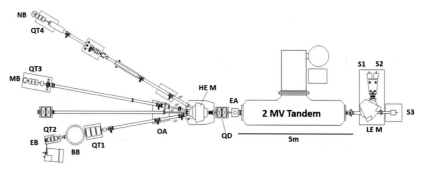

FIG. 2

IBA laboratory of the Surrey Ion Beam Centre (schematic). Central is the 2 MV HVEE Tandetron linear electrostatic accelerator (note scale bar). S1, S2, S3 are ion sources: S1 is a He$^+$ duoplasmatron, with a Li charge exchange canal to generate He$^-$ ions; S2 is a general purpose sputter source with Cs ion sputtering producing negative ions; S3 is a H$^-$ duoplasmatron with a Wien filter. LE M and HE M are the low-energy and high-energy switching magnets (LE M is not used for S3). EA and OA are the exit and object apertures. QD and QT are a quadrupole doublet and quadrupole triplets. The scattering chambers are: *BB*, broad beam; *EB*, external beam (in air); *MB*, microbeam; *NB*, nanobeam (a test system).

Figure courtesy of V.V. Palitsin.

is 1 MV, we will have exiting the accelerator *two* positive ion beams (He$^+$ and He^{++} respectively at 2020 and 3020 keV) and a neutral atom beam (at 1020 keV).

"Single-ended" accelerators are also often used: these seem simpler, but the ion source is now in the high-voltage terminal itself which can be rather inconvenient if the source goes wrong. This is an important consideration since even "simple" ion sources have plenty to go wrong. Tandems are very popular because they are actually easier to use and far more versatile. All of these fun and games are *very* much easier to say than to do: they have a very long history, and a very large technical literature which happily we do not need to discuss.

So we have an ion beam. The thing about ion beams is that they are very well defined (ideally!). All the ions have the same charge, the same speed, and the same direction. So when we put our sample in the way it will be "simple" to calculate what will happen. We know what the ions are (given by the ion source), what their energy is (given by the accelerator voltage), and how many there are (given by the beam current). All the rest is (quite complicated, but well known) physics! Typical might be a 3 MeV proton beam of 1 nA. Clearly, the ammeters have to be quite good to measure nanoamps reliably, but this sort of instrumentation is now standard.

What happens now is that the ion beam is directed down a beam line, using a big magnet to turn it (a charge moving through a magnetic field experiences the *Lorentz force*: this is all perfectly understood and for us counts only as "practical details"). In the old days we needed an expensive magnet effectively to define the energy of the ion beam, but on modern accelerators the beam energy is given very accurately by the *terminal voltage* (the voltage of the high-voltage terminal of the accelerator) and the magnet is only there to switch between beam lines. And different beam lines have different equipment used for different purposes.

Of course, ion beams are scattered readily by any matter they encounter, so to keep all the ions going at the same speed in the same direction the beam lines must be at a *high vacuum*, that is, pressures of a billionth of an atmosphere (10^{-6} mb, i.e., a millionth of a millibar) or so. This needs high vacuum pumps, today usually *turbo-pumps* which have compression ratios of around a million but cannot have either the input or the output at atmospheric pressure: they will break, explosively! This is frowned on, since they are very expensive ($\sim$£10K). So the turbos have *backing* pumps, usually *rotary pumps*: these have compression ratios of 10,000 or so but are designed to accept atmospheric pressure at either input or output (or both).

Cultural heritage samples are usually priceless and fragile, we do not want to put them in vacuum! Instead, we want to bring the ion beam out of the vacuum system and on to the sample: the so-called *external beam*. Fortunately, these fast ion beams are very penetrating and will travel many millimeters in air before stopping. So the last element of the beamline is the airtight "exit window" which is thin enough to let the fast ions out without too much scattering or energy loss, and strong enough to withstand atmospheric pressure. External beam installations have become so much better in the last decade or so with the introduction of thin silicon nitride exit windows. These are now available with diameter 5 mm and thickness 200 nm (yes, a fifth of a micron!), and will last for months. But you must not touch them (!!) and the vacuum system must be protected against the window breaking.

Finally there is the spectrometry equipment (not indicated in the figure). The whole point of using this complex apparatus for directing an energetic ion beam onto the sample is to see what happens next: how the sample responds. The purpose of the ion bombardment is to catch and interpret the resulting particles and photons. This is what the rest of the chapter will describe.

3.2 More leisurely walk through the lab

The curious reader will have lots of questions stimulated by the details in the figure. Why are there *three* ion sources? Why are *four* sample "chambers" shown? What is the difference between quadrupole *doublets* and *triplets*? What is the function of the *apertures*? How much detail do I need to know?

If you visit the laboratory to make measurements the lab staff will talk this strange language involving where the beam is, what it is doing, and how the various instruments are behaving: this present summary is a sort of "tourist's phrasebook" to help you make some sense of it. The more you understand in a foreign city the more valuable the experience will have been to you. So also here: the more you understand— even at a sketchy, superficial level—the better you will appreciate the results.

The duoplasmatron ion source is essentially very simple: it ionises a gas by electron bombardment from a hot filament (the *cathode*: the *anode* is an internal component of the source), then the ions are sucked out through the *aperture* by an appropriate "extraction lens." We want negative ions: both H^- and He^-. But helium being a noble gas, it does not like making a negative ion, and sufficiently large He^- beams cannot be extracted from these sources. Instead, a He^+ beam is extracted and then passed through a hot lithium (or rubidium) vapor (the Li or Rb *charge exchange canal*) to press electrons onto the He atoms and give a He^- beam. A subsequent magnet (the *LE magnet*) is then essential to filter out the (large) He^+ beam that also exits the charge exchange canal.

It turns out that the gas ionisation behaviour inside the source is quite complex, and placing the aperture in different geometries allows the extraction of either negative or positive ions (but not both!). Hydrogen is very happy to make a negative ion, and large H^- beam currents can be extracted from the duoplasmatron without any need for a charge exchange canal. One can use the same source for both H^- and He^+ beams, but only if the source is rebuilt which is very time consuming. So Surrey has two sources for the different beams. We should mention that in an ion source *anything* in the source is ionised, including any impurities. This is particularly problematic for the He^+ source since helium is hard to ionise, and therefore any impurities are ionised first. So however clean the source is, there will be other parasitic beams, which can themselves be quite intense. But we do not want to inject these parasitic beams into the accelerator, and the *LE magnet* is there to filter them out. But clearly this magnet must be switched off when the H^- source is in use. Instead a *Wien filter* (with crossed electric and magnetic fields as a velocity selector: not shown) can be used for the same purpose of cleaning up the beam injected into the accelerator.

The sputter source is a general purpose source from which negative ions of almost any element can be extracted. It is actually rather good at generating H^- beams, but the accelerator can satisfy a variety of other purposes (which do not concern us here) for which this flexibility is very useful.

At the high-energy end of the accelerator, the *HE magnet* has the same purpose as the LE magnet: to clean up the beam injected into the *beam line*. In particular there will always be a neutral beam coming out of the accelerator: such a beam obviously cannot be controlled with ion lenses, and its *beam current* cannot be measured: it is dumped in the "straight-through" line, which has no *sample chamber* for just this reason. The beam does need some steering into the beamline, and the *quadrupole doublet* before the HE magnet in Fig. 1 is a special and rather elegant device developed at Surrey to do just that (see Ref. [9]).

Cultural heritage applications almost always need imaging, that is, the beam must be focused and scanned over the area of interest on the sample. The ion lenses used for focusing have particular characteristics, most obviously that they have a fixed focal length and a fixed demagnification. That is, the *object aperture* position is determined by the optical properties of the ion lens, and the final spot size is determined by the aperture size (and the lens quality). The beam line length is required by the ion-optical properties of the focusing lenses.

The four sample "chambers" in Fig. 1 include the external beam (see Refs. [10–12]); and also three other vacuum chambers intended for a wide range of analyses: a *microbeam* chamber and a *broadbeam* chamber which includes a large 6-axis goniometer [13]; and a *nanobeam* chamber [14].

4 Ion beams

Before looking at examples of IBA for cultural heritage applications, it is necessary to gallop through some elementary physics, without which none of the spectrometry will make any sense.

4.1 Energy loss (elastic and inelastic)

Atoms have Ångström sizes (i.e., of order 0.1 nm) but atomic nuclei are measured in fm (10^{-6} nm). Thus ions travelling through matter spend almost all their time travelling through an electron soup. It is very easy to see that a charged particle moving through electrons is going to be dragged back by the attraction of all the electrons it passes, like running through water. This is called the *electronic energy loss* process which has been studied intensively for over a century (starting with William Bragg's classical work of 1905 [15]) and is now well understood even though it is quite complicated. See Ref. [16] for a recent technical comparison between experiment and various calculations, including a summary of the extensive literature.

Electronic energy loss is an *inelastic* process, and as the (monoenergetic and unidirectional) beam slows down in the sample there will be an increasing spread both in the beam energy and in the beam directions. The physics of all this is intricate, but now is well known: Fig. 3 shows that compared to electrons, protons do not scatter very much either in energy or direction (as expected, since they are nearly 2000 times heavier). Of course, the enormous scattering of the electrons is mostly *elastic* scattering, which is also the origin of the few large-angle proton scattering events. But it is important to mention that the inelastic electronic energy loss process is almost entirely an atomic (not a molecular) one. That is, we only (!!) need to know how the energy loss behaves for each ion in pure elemental materials. It turns out that we can almost always simply treat compound materials as mixtures of elemental

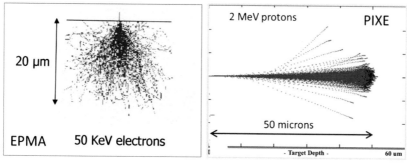

FIG. 3

Energy loss of electrons and protons compared. Monte Carlo simulations of 50 keV electrons (common in electron-probe microanalysis, EPMA) and 2 MeV protons in silicon (common in particle-induced X-ray emission, PIXE).

materials to a remarkable degree of accuracy. This is known as *Bragg's rule*, after William Bragg's original work.

If the particle were a neutron (i.e., uncharged) the electrons would be essentially invisible to it so there would be no inelastic energy loss. But the neutron could still scatter elastically from the nuclei, and in this case conservation of energy and momentum mean that the struck nucleus would recoil with a recoil momentum given by (i) the ratio of the masses and (ii) how directly the nucleus was struck. That is, the neutron will share its energy (as *elastic* energy loss) with all the recoils it creates. For further on this please see the scattering of thermal neutrons described in Chapter 8 by Winfried Kocklemann and Evelyne Godfrey.

A proton will have elastic energy losses in the same way as the neutron (although the nuclear interaction details are different) but it will also have the inelastic energy losses due to its being a charged particle. We will see that this distinction between *elastic* and *inelastic* energy losses is central to the interpretation of IBA spectra. This might seem terribly arcane, but actually things are much simpler at this level than in normal experience. Nuclear elastic scattering is just like billiards—except that it is simpler since you do not have to worry about rotational or frictional effects, nor about how flat the billiard table is, nor how bouncy the cushions! Elastic scattering is like bouncing a bouncy ball on a nice hard floor—except it is simpler since with elastic scattering there really is *no* energy loss and the ball will carry on bouncing forever. Of course, there is also *in*elastic (nuclear) scattering, which we will describe under *nuclear reaction analysis*.

4.2 Ionising radiation hazard

The high-energy ion beam is always stopped by *something*, either by the sample you want to analyse or by the walls of the vacuum system, or by the air if you are using an external beam. Where does all that energy go? It turns out that the great bulk of it goes into the *electronic energy loss* processes we described in Section 4.1: that is to say, into *ionising events*. Of course, the whole point of using this equipment is to create ionising events (X-rays, etc.!) that inform us about the artefact under investigation.

These fast beams are therefore classed by the radiological protection authorities as *ionising radiation*, which is always very closely regulated. Actually, these installations are capable of being very dangerous, but the lab staff will know all the proper procedures and the *local rules* thoroughly: you only have to follow instructions! It should be added that normally the hazard is rather small, even for the external beam.

4.3 Imaging

The great virtue of ion beams is that ions are charged particles that you can grab hold of with electric or magnetic fields. The scanning electron microscope (SEM) works because the electron is a charged particle and it is easy to scan it over the screen (this is how television sets used to work). The ion beam can be made to work just the same way: the ion lenses needed are rather different from the lenses you use in the SEM, but for our purposes these are details which have been sorted out. Interested readers can look at such details in a recent "nuclear microprobe" installation [17].

Modern scanning ion microbeams can focus the ion beam to submicron dimensions and can scan over areas of millimeter dimension. This means that precious cultural heritage artefacts can be imaged easily at high spatial resolution. This is very valuable, as Fig. 1 shows and as Grassi et al. [18] emphasise; they say: *scanning-mode measurements of ancient inks, Roman glass and metal point drawings clearly demonstrate [scanning to be] fundamental to avoid deceptive information and to obtain more reliable quantitative results.* But we can point out here that cultural heritage applications do not usually need submicron spatial resolution (the resolution in Fig. 1 is given by a proton beam size of $40\,\mu m$), and this is very convenient for us since there are lots of details to understand if such high resolution is needed, which for our purposes here we can overlook.

4.4 Sample damage

But we must say something about sample damage. We are using nanoampere (nA) beam currents of MeV ions: this is a power in the beam of milliwatts (mW) all of which is dumped in the surface $50\,\mu m$ or so of the sample. If the beam has a size of say $10\,\mu m$ this is an energy density of some MW/cm^3! Obviously this looks somewhat hazardous. These ion accelerators can pack a pretty punch: it has been known for incautious operators to blow holes in their vacuum systems—an embarrassing (and expensive) mistake! When you visit the accelerator labs you should ask to see their "chamber of horrors" where they keep relics of such mistakes to keep reminding themselves of the dangers.

However, ion beam analysis is not *deliberately* destructive. We do not dissolve the sample as for ICP-MS, or cut it up as for XTEM, or dig holes in it as for (dynamic) SIMS or LA-ICP-MS[a]. We aim to give the sample back to the curator unchanged, and curators tend to be rather fussy!

How is this managed? First, we can make the beam spot larger. Fig. 1 was made with a spot size of $40\,\mu m$ which immediately reduces the instantaneous power density

---

[a]*MS*, mass spectrometry; *ICP-MS*, inductively-coupled-plasma MS; *XTEM*, cross-sectional transmission electron microscopy; *SIMS*, secondary ion MS; *LA-ICP-MS*, laser ablation ICP-MS.

to $<100\,\text{kW/cm}^3$. Then the area scanned is about $5\,\text{mm}^2$ which reduces the average power density to about $4\,\text{W/cm}^3$. But perhaps the most important thing is that these delicate and precious samples are analysed *in air* (using the *external beam*). The fact that the surface of the sample is bathed in gas means that the heat deposited has an easy way to get away. After all, a vacuum is a very good thermal insulator.

In any case, the IBA scientists will carefully evaluate how the particular sample responds to the beam. An example from the AGLAE laboratory sited at the Louvre Museum is given by Beck et al. [19] for precisely how "zinc white" (zinc oxide) behaves under a 3 MeV proton beam. Calusi [20] gives a more general treatment of other examples (from the LABEC laboratory in Florence), acknowledging the issue of sample damage but without specifying the damage mitigation protocols in detail. A systematic and very useful treatment of a variety of damage mechanisms has been provided by the laboratory in Madrid [21], and details have been given of the potential damage to historical paintings [22] and modern pigments and papers [23].

5 IBA methods

A favourite game of children at the swimming pool is to jump in trying to make the biggest splash. A 2 MeV proton hitting a sample makes a big splash! Although IBA is not *deliberately* destructive, there is still a lot of energy in the beam (which may well alter the sample, as we have seen in the last section).

And all that energy is dumped into the sample extremely quickly: where does it all go? The thing about IBA which can be very confusing is that there are lots of analytical possibilities: you can surround the sample with a variety of different detectors and obtain a variety of spectroscopic information. In this section we try to describe this variety, starting with the lowest energy interactions and ending with the highest. That is, we will start with atomic interactions which have energies from electron-volts up to kilo-electron-volts, and we will end with the high-energy (MeV) interactions which all involve the atomic nuclei. We will see that the atomic physics and the nuclear physics are quite different in their details.

5.1 Ion beam-induced luminescence

What is the *electronic energy loss* we described in Section 4.1? The energetic ion stops in the sample—where does its energy go? Of course, it *ends up* as heat, but that is not very interesting! What are the intermediate steps? The first thing that happens is that the fast particle *scatters electrons* left right and centre. It is these scattered electrons that spread the energy of the particle around (ending up as heat, eventually). How fast is a 3 MeV proton? We can get to this via Newton's equation for kinetic energy $KE = \frac{1}{2}mv^2$, where m is the proton mass ($1.67 \times 10^{-27}\,\text{kg}$) and $1\,\text{MeV} = 1.60218 \times 10^{-13}\,\text{J}$: in this case the proton speed $v = 8 \times 10^6\,\text{m/s}$, or nearly 3% of the speed of light. At this fairly low speed Newton's equation is quite a good approximation and we do not have to use the exact relativistic equation. The point is that in most materials a 3 MeV proton stops in the first 100 µm or so, and everything is over in a few picoseconds.

When an atom loses an electron (perhaps due to some ion barging past, or perhaps simply because it has got hot) it is said to be in an *excited state*. It does not like it and wants to *relax* back to its nice neutral condition. When schoolchildren drop salt into the flame of a Bunsen burner and see the yellow sodium colour, this is what is happening: outer electrons of the sodium atom can be thought to "boil" off. It is the next part of the story that is quantum magic, bursting with information.

The atom is not allowed to just grab another electron to neutralise itself! Why? All the electrons flying around the atom have too much energy, and for an electron to stick to the atom it has to lose this extra energy. The rules of quantum mechanics apply to fundamental particles (like electrons) and the electron must lose its excess energy the proper way. The *energy levels* of each atom are very well defined, and for the free electron to stick to its atom a *photon* of the proper colour (equivalent to energy: yellow in the case of sodium) must carry the excess energy away. These characteristic colours give *characteristic lines* in the energy spectrum. Characteristic lines of the elements were recognised in the first half of the 19th century, and this is precisely why "helium" has its name (from the Greek *helios*, sun): Janssen and Lockyer independently observed the characteristic lines of helium during the solar eclipse of 1868 and recognised it as an unknown element (not observed on Earth until 1895).

Fast ions travelling through a solid will ionise the atoms of the solid, and then the excited atoms have to relax. Outer-shell electronic relaxation will give visible photons: this is called *ionoluminescence* (ion beam-induced luminescence, IBIL) when produced by ion bombardment. It is entirely comparable to *cathodoluminescence*, which is produced by electron bombardment.

The trouble with IBIL (like cathodoluminescence) is that the visible (low-energy) photons produced are very sensitive both to the *molecular* structure, and especially to defects. So the signals are usually hard to interpret, and IBIL is not yet very widely used. But there has been significant interest for certain applications: for example, determining provenance of lapis lazuli artefacts [24]. Lapis lazuli from Chile is *the easiest to distinguish due to the widespread presence of wollastonite ($CaSiO_3$), which … can be recognised [from] the ionoluminescence spectrum*. It is important to note that in this major study IBIL was used in conjunction with particle-induced X-ray emission (PIXE).

5.2 Particle-induced X-ray emission

High-energy particles bombarding the sample leave the atoms in the sample highly excited affecting not only the outer electrons but also the inner electrons. The atoms relax by electrons filling the empty shells, emitting characteristic photons to lose enough energy. Fig. 4 [25] shows K-shell photons (X-rays) from P, S, K, Ca, Mn, Fe, Cu, Zn atoms in the paint and L-shell X-rays from Pb. X-rays have a high-energy

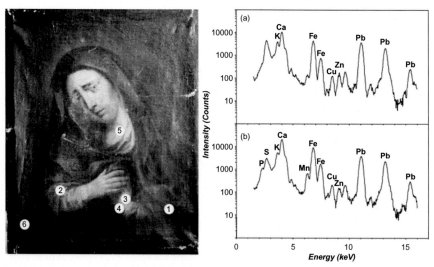

FIG. 4

PIXE spectra from the *Virgin of Sorrows* oil painting (Mexico, 1750–1830). The upper PIXE spectrum is from position #2 on the painting, the lower is from #6.

Reproduced with permission from Figs. ##1, 2 of M. Ortega-Avilés, P. Vandenabeele, D. Tenorio, G. Murillo, M. Jiménez-Reyes, N. Gutiérrez, Spectroscopic investigation of a 'Virgin of Sorrows' canvas painting: a multi-method approach, Anal. Chim. Acta 550 (2005) 164–172 (© 2005 Elsevier B.V.).

(keV: kilo-electron-volts) because they are a result of electron transitions in the inner electron shells (closest to the nucleus). But IBIL involves visible photons from the outer shells which have a much lower energy. For example, the yellow lines of sodium come from M-shell transitions and only have about 2 eV (electron-volts).

Fig. 4 underlines that the PIXE spectra are quite complicated. For example, two lines are shown for Fe: these are known as the Kα and Kβ lines. These two lines are also visible for Ca (but only one is labelled); they are also present for the other elements but obscured by other lines. So, for example, the Cu Kβ overlaps the Zn Kα. In turn, the Zn Kβ overlaps one of the (unlabelled) Pb lines. Three Pb lines are labelled: these are only the most prominent of the many L-shell transitions. It is plain that as the atomic number goes up, so do the energies of the characteristic photons. This is because there are more protons in the nucleus so that the electrons are more tightly bound. For comparison, Pb (lead) is very heavy ($Z = 82$) and Pb Kα has energy 72.9 keV, which is too high to be seen with the usual detectors. So we use the Pb L lines instead.

The other important thing to notice about the PIXE spectra in Fig. 4 is the logarithmic scale of the ordinate axis. "Counts" is the number of X-rays detected at that energy. The X-rays enter the detector almost always one-by-one, and the detector very cleverly works out what its colour is, incrementing the appropriate counter in memory. We say, *the spectrum is accumulating*. Obviously, the detector needs some time to do this, but modern detectors can handle 100 kHz and more without much loss of performance. But if two photons arrive together the detector counts them as one: this is called *pileup*.

So a logarithmic scale means that PIXE is sensitive to *all* the elements in the sample: major, minor and trace elements. Of course, there is always a limit, but provided there are more than a few parts-per-million of an element present in the sample than PIXE will probably see it. Moreover, PIXE now has a vast scientific literature, being more than a century old (it was discovered by James Chadwick in 1913 [26]) and these spectra can be treated entirely quantitatively.

Such a quantitative and highly detailed treatment of imaging data obtained non-destructively from priceless samples has been found to be of immense value. Some recent impressive examples include: provenancing of materials used for Persian glass from the 10th and 11th centuries [27]; an exploratory analysis of gilded leathers used as luxury "wallpapers" in *c.*17th century Europe [28]; a study of weathering and deterioration of Sicilian archaeological glasses of 4th to 7th centuries [29]; a study of how potters spread technology into Northern Europe in prehistory (4000–5000 years BP [30]); characterising Byzantine pottery from Romania [31]; comparing golden thread embroidery threads from Romanian Byzantium [32]; a study of an obsidian artefact reputed to be an "Inca Mirror" [33]; and provenancing postmediaeval grisaille glasswork [34].

5.3 Elastic backscattering spectrometry

In the discussion of PIXE we have ignored them so far, but of course atoms do have *nuclei*, and the incident ion beam can scatter from these nuclei. This scattering is elastic and (like billiard balls) can be directly backwards (180 degrees scattering angle) for a head-on collision. Of course, detecting such events is impossible since then the detector would have to be *in* the incident beam! But you can get quite close to 180 degrees, and it turns out that there are good analytical reasons to use large scattering angles. Fig. 5 shows a scattering angle of 150 degrees: this sample (from the recent restoration of the Rosslyn Chapel near Edinburgh [35]) is a 19th century soda-lime glass whose composition was not known beforehand.

Even though the glass is uniform in composition, Fig. 5 is clearly a complicated spectrum: there are no less than 16 measurable elements present in the glass! How are such spectra to be interpreted? And when would we need such data?

At the beginning of the 20th century, the big questions in physics included trying to work out what the mysterious α, β, and γ rays were. Ernest Rutherford's team in Manchester were watching how α rays passed through gold foil, and they noticed α particles scattering at all angles including more or less straight backwards! They were greatly surprised since by that time they already knew that an α particle was He^{++} (Rutherford got the Chemistry Nobel Prize for this in 1908), and they knew that there were protons and electrons in atoms (they did not know about neutrons then), but they thought these positive and negative charges were all mixed up in the atom, since *unlike charges attract* (J.J. Thompson's "plum pudding" model

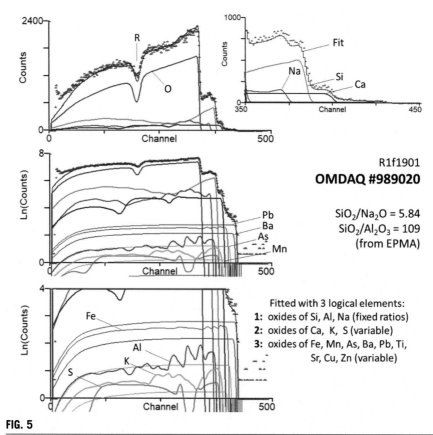

FIG. 5

Elastic backscattering spectrometry of Rosslyn glass. Top (linear scale): EBS total spectrum from a 19th century soda-glass using 3080 keV protons and a backscattering angle of 150 degrees. The O non-Rutherford signal dominates the spectrum. The feature labelled "R" shows the effect in the spectrum of the 2663 keV resonance in the O signal. Partial spectra are shown for O, Na, Si, Ca (respectively *purple, blue, orange, green*). Top (inset): detail of the EBS signal above the O edge. Centre & bottom (log scale): EBS of minor elements Al, K, S (*purple, green, dark green*); RBS of minor elements Pb, Ba, As, Mn (*red, blue, orange, magenta*).

From Fig. 5A of C. Jeynes, V.V. Palitsin, M. Kokkoris, A. Hamilton, G.W. Grime, On the accuracy of Total-IBA, Nucl. Instrum. Methods Phys. Res. B 465 (2020) 85–100 (© Elsevier B.V., 2020, by permission).

was proposed in 1904). But Rutherford recognised that this newly observed scattering behaviour meant that the atomic nucleus (including *all* the protons) had to be both very small and also the same polarity as the α particle: by simply pretending that the electrostatic force (*like charges repel*) was responsible he was able to account entirely for the observations. The Rutherford model of the atom was published in

1911 [36], and *Rutherford backscattering spectrometry* (RBS) refers to this analytical model. To sum up: RBS assumes *point charges in a Coulomb field*.

But this must be an approximation! Neither the α particle nor the atomic nuclei of the sample are "point charges": the α particle has a radius 1.68 fm and the O nucleus has a radius 2.71 fm [37]. These values may be small, but they are not nothing! And the ion beam does have a high energy! It turns out that when the incident particle is going fast enough, it comes close enough to the scattering nucleus for the nonzero sizes to become important. In this case you cannot pretend that Rutherford's approximation is good enough anymore, and you have to do a proper quantum mechanics calculation to work out what happens. This gets complicated: an overview with references to the technical literature can be found in Ref. [38].

Fig. 5 shows this complicated non-Rutherford behaviour: there is a pronounced "dip" in the O signal, and the Al signal has got lots of peculiar wiggles in it. Actually in this case, only the signals from the heaviest elements, including lead (Pb), are Rutherford. Looking at the literature, elastic scattering spectra are quite often (but incorrectly) called "RBS" when actually EBS (non-Rutherford) spectra are in view. In fact, especially for cultural heritage applications where proton beams are usually used, the particle spectra cannot be correctly interpreted *unless* they are treated as EBS (not RBS) spectra. Fig. 9 of Jeynes and Colaux [38] shows that for 2 MeV protons, elastic scattering is *not* Rutherford for all elements lighter than Fe.

To interpret EBS spectra we need to know what the probability is of a certain energy particle scattering from a given nucleus into a given direction. This probability is known as the "differential scattering cross-section" σ', which for RBS and the centre of mass frame of reference is a very simple formula (Eq. 1 in [39]):

$$\sigma' = \left\{ Z_1 Z_2 e^2 \, \mathrm{cosec}^2(\theta/2)/(4E) \right\}^2$$

where Z_1 and Z_2 are the atomic numbers of the incident and target nuclei, e is the charge on the electron, θ is the scattering angle and E is the beam energy at the scattering event. But for EBS σ' must be *measured*: it is not possible to calculate it ab initio sufficiently accurately for analytical purposes (and in any case, the calculations are very difficult).

The good news is that in the last couple of decades an enormous amount of work has been done both to catalogue all the scattering cross-section measurements done over the last 50 years or so (and make new measurements) [40, 41], and to fit these data to nuclear scattering models [42]. This means that although Fig. 5 shows that EBS for glass is complicated by the need for extensive and detailed scattering cross-section data, these data now exist in a form convenient for analysts. Therefore, EBS is now feasible in a way that simply was not available last century.

However, given the difficulties, why should curators want EBS spectra from their precious samples when they already have detailed information from PIXE? The first and trivial answer is that they are there for free! When the proton beam strikes, the sample responds in multiple ways: IBIL, PIXE, EBS (and the other responses including PIGE we will discuss below). You only have to put in the detector to collect the data! And given that the sample is precious, so are the data.

But in fact, we want the EBS data specifically when we suspect interesting things going on at the sample surface (such as corrosion effects, or surface coatings—gilding, etc.: we will discuss various such examples in Section 6). To understand this, consider the differences between PIXE and EBS spectra (compare Figs. 3 and 5). The abscissa on the PIXE spectra is in units of "keV," where for the EBS spectra it is in units of "channel number". This is because PIXE intrinsically gives an *energy* spectrum, but EBS intrinsically gives an *energy loss* spectrum. For PIXE the characteristic lines have their own characteristic energies which automatically calibrate their own energy scale. But for EBS the energy scale has to be calibrated externally. The signal for PIXE is the energy (colour) of X-ray photons, and photons are either absorbed or not as they pass through the material. If they are not absorbed their energy does not change. But for EBS the signal is the energy of the scattered particle, which steadily loses its energy as it passes through matter. That means that EBS is sensitive to the *elemental depth profile* in a way the PIXE spectrum is not.

This means that the EBS spectrum is more complicated than the PIXE one. If we ask the spectrum, *what sample generated you?* there is an analytical answer for the PIXE spectrum, but the EBS spectrum will answer: *this question is mathematically ill-posed*. Programs to automatically handle PIXE data (and the very similar XRF, SEM-EDX, SEM-WDX, EPMA[b] spectra) have been standard for over 20 years. But *fitting* EBS spectra in general (including fitting the elemental depth profile) must use advanced numerical methods (see the "Topical Review" of Jeynes et al. [43] describing the DataFurnace code: to date there is no other code capable of *fitting* PIXE+EBS datasets self-consistently; the use of DataFurnace for such Total-IBA was reviewed in 2008 [44], in 2012 [58], and in 2020 [12]).

Note that *fitting a spectrum* (i.e., extracting from the spectrum the composition of the sample that generated it) is the inverse operation to *simulating the spectrum* one would get from that sample. Of course, simulating the EBS spectrum is "easy" (provided one knows the scattering cross-sections) and suitable programs have been available since the 1970s, but the *inverse operation*—fitting it—is actually ill-posed mathematically.

We now need to look in more detail at the EBS spectrum in Fig. 5. Remember that this is the spectrum of a simple piece of uniform glass, for which the only interesting question from the ion beam analysis point-of-view is, what is my (elemental) composition? Therefore, the complexities of the spectrum are entirely due to the

---

[b]*XRF*, X-ray fluorescence; *SEM-EDX*, energy-dispersive X-ray spectrometry (XRS) on the scanning electron microscope; *SEM-WDX*, wavelength-dispersive XRS on the SEM; *EPMA*, electron probe microanalysis.

technique not the sample, which is pedagogically very helpful. There are four crucial aspects to understand:

1. ***Elemental edges.*** Looking at the logarithmic plot of the whole spectrum it is clear that there are four main steps, due to (i) the heavy group of elements (mostly Pb), (ii) the intermediate group of elements (mainly Ca), (iii) the matrix elements (mainly Si), and (iv) oxygen. In an elastic collision the recoiled particle takes more energy from the incident particle the more closely their masses match. Think of a ping pong ball bouncing off a billiard ball. The little ball will bounce very fast and the big ball will hardly move. But if the little ball is a bit heavier—say a tennis ball—it will not come back so fast and the big ball will roll away more. So here, the incident proton bounces back from the lead atom with almost all its initial speed, but it bounces off the oxygen atom much slower.

2. ***Mass-depth ambiguity.*** The spectrum is continuous! Compare channels 341 and 421 (the O and Pb edges). If you detect a backscattered proton with an energy that puts it in channel 421, it must have scattered from a Pb atom at the surface of the glass. But if you detect a backscattered proton with an energy that puts it in channel 341, it could have scattered from a O atom at the surface of the glass, or it could have scattered from a Pb (or other) atom buried in the sample. The incident proton would have had to go through some glass to get to it, losing energy on the way, and then having scattered it would have to come back along a (slightly different) path, losing more energy. This is why we say that the EBS spectrum is an *energy loss* spectrum. It is this mass-depth ambiguity that makes inverting the spectrum an ill-posed problem. And what makes interpreting EBS spectra notoriously difficult.

3. ***Mass closure properties.*** Glass is always essentially a mixture of oxides. In this case there is no boron or lithium so that the EBS spectrum sees *all the elements in the sample*—especially including the oxygen! Note that this is *not* the case for RBS spectra, for which the scattering cross-section goes with Z^2—the square of the atomic number. So for RBS the step height ratio of the O:Si signals for this sample (with an O:Si ratio of 2.4) would be $(2.4) * (8/14)^2 = 0.8$. But for this EBS spectrum the O:Si step height ratio is about 3. This is because the scattering cross-section for 3 MeV protons on O at this angle is about seven times what it would be if it were Rutherford scattering (protons on Si is not Rutherford either). The sensitivity to the light elements may be greatly enhanced in EBS.

    But now, with all the signals visible, the spectrum has a *mass closure* property that PIXE spectra do not have (since PIXE does not usually detect elements lighter than about Na). That is, the composition of the sample is determined unambiguously by the spectrum. It turns out (see [14] for details) that this mass closure property allows the analyst to correct the spectrum for over- or underestimated scattering cross-section functions. So the inset of Fig. 5 showing the details of the Si signal also shows that the fit is systematically *below* the data. That is, the spectrum is illegal, strictly speaking: it cannot be fitted properly by any composition under the given assumptions—which of course include the scattering cross-section functions used. Putting this another way, the spectrum allows one to extract correct cross-sections (this has been known and used since the 1980s: see [12] for details). In this case, even though it is clear that the scattering cross-sections for Si are underestimated by about 5%, the sample composition is obtained with a much lower uncertainty of about 2%.

4. **Synergy with PIXE.** In this case it is clear that the trace elements are undetectable by EBS, although they are easily measured by PIXE. Actually, in this case EPMA (i.e., SEM-WDX) was used to obtain the minor elements Al and Na (as well as the matrix elements—but not O). In principle both are obtainable with PIXE but were not in this case. Na can be measured successfully by EBS, but not so accurately as EPMA. It is important to stress the synergy between PIXE and EBS: each is strong where the other is weak.

The value of the backscattering spectrum is obvious where one wishes to measure (for example) gilding layer thicknesses, as Ortega-Feliu et al. [45] have done recently. But EBS is invaluable wherever there is compositional variation near the surface (corrosion layers for example), as we will discuss in the following text. Such cases have been regularly ignored up to now because the spectra are difficult, requiring the analyst to make use of the EBS-PIXE synergy. This has become possible only quite recently (see the review of IBA depth profiling by Jeynes and Colaux [38]).

5.4 Elastic recoil detection

In an elastic scattering event we can follow the incident particle (RBS or EBS) or we can follow the recoiled particle (or both!). Now, if the beam is incident normally onto a thick target, any recoiled atoms will just be pushed further into the target, and if the target is sufficiently thin the recoils can exit from its back, as was done in a beautiful measurement of H at the grain boundaries of a polycrystalline diamond sample [46]. But most cultural heritage samples are thick, and then recoils can only be detected out of the front of the sample—that is, if the beam is incident at a glancing angle and if the detector is in a forward direction. Typical is a scattering angle of 30 degrees (this is forward scattering, not backscattering!) and a beam incident at 15 degrees to the sample surface. This requires the sample to be quite flat, and is geometrically fiddly especially for the external beam where path lengths have to be kept very short (because of the energy loss in air). These difficulties mean that elastic recoil detection (ERD) is rarely used on the external beam, even though there are very few feasible ways of measuring hydrogen in samples.

ERD is mentioned in the 2004 review of Dran et al. [47], and was used by Calligaro et al. [48] to date two rock crystal artefacts by their hydrogen profile from the surface (quartz hydration dating). Both were alleged to be pre-Columbian meso-American artefacts, but only one was: the other was no earlier than the 18th century.

5.5 Particle-induced γ-ray emission

In a high-energy collision between two nuclei a variety of things may happen. The simplest is an elastic scattering event. For example, consider a proton striking a sample containing sodium. The elastic event is written ^{23}Na(p,p$_0$)^{23}Na. But other reactions are possible, such as ^{23}Na(p,p′γ)^{23}Na where a gamma ray is emitted together with a proton that has a much lower energy; or ^{23}Na(p,α$_0$)^{20}Ne where the sodium nucleus is transmuted to a neon one and an alpha particle is emitted rather than a proton. There is also the similar reaction ^{23}Na(p,α′γ)^{20}Ne where a gamma ray is emitted as well as the alpha (which will have a lower energy); or even the "proton capture" reaction ^{23}Na(p,γ)^{24}Mg where the sodium is transmuted to magnesium.

These are all *nuclear reactions*, and they can all be utilised for analytical purposes giving the IBA method: "nuclear reaction analysis" (NRA). The reactions yielding gamma rays are useful since these γ-rays have energies *characteristic* of the nuclei from which they came, just as the X-rays are characteristic of the atoms from which they come. When we do NRA and measure the γ-rays it is called particle induced γ-ray emission (PIGE).

The other really useful thing about PIGE is that γ-rays have very high energies: for example, the most probable one characteristic of Na has energy 440 keV (this reaction has recently been accurately characterised by Gurbich [49]) and such high-energy photons are very penetrating so that absorption is negligible and the detector geometry is not critical.

PIGE is used systematically (in conjunction with PIXE) for cultural heritage applications since it allows the convenient detection and quantisation of light elements in samples. So for example, Balvanović and Šmit [50] measured Na, Mg, Al in the 6th century Serbian glass by PIGE. These elements can usually also be measured by PIXE, but the K-shell X-rays are low in energy and very easily absorbed, meaning that both the information depth and the sensitivity are low. PIGE therefore gives both better sensitivity and more representative (reliable) results. The same methods were also applied by Šmit et al. [51] to analyse prehistoric Slovenian glass, by Gueli et al. [29] for the weathering of 3rd century Sicilian glass, and by Maggetti [52] in a review of mass production 19th century European earthenware. The method is standard for reliable measurements of these important elements of glass: we have here cited only some recent examples.

PIGE was used by Reiche et al. [53] to measure fluorine in Paleolithic mammoth ivory, and by Bouquillon et al. [54] to measure boron in 16th century glazed ceramics. Lithium can also be measured by PIGE [55].

6 Total-IBA

PIXE and EBS are beautifully synergistic: where one is weak the other is strong. PIXE is a spectroscopic technique directly sensitive to atomic number: that is, we directly see which elements are present in a sample. EBS on the other hand is an energy loss (spectrometric) technique, which is directly sensitive to the elemental depth profile near the surface. Both are sensitive to the first 10 microns or so near the sample surface. Roughly speaking, PIXE says which elements are present, and EBS says where they are.

Fig. 6 shows a spot analysis by Pascual-Izarra et al. [56] of a site of corrosion on a very famous artefact—the first ever photograph! The question was basic to any restoration: what sort of corrosion was it? This work is the *first example* of strictly synergistic analysis of PIXE/EBS data where neither the PIXE nor the EBS on their own allowed the structure of the sample to be resolved, and where the spectra are *fitted* (as opposed to being roughly simulated with a trial-and-error method), and where both the particle scattering and the X-ray spectra are handled with a fully general code. A simplified synergistic code is widely available (see Ref. [57], and references therein) but could not have handled this case.

In Fig. 6 the number of counts in the characteristic lines of the PIXE spectrum is shown, not the spectrum itself. This is because it is the line areas that are fitted, not the full spectrum itself. Conversely, the energy-loss EBS/RBS spectra must be fitted as such, and Fig. 6 shows the fits together with the partial spectra of the various elements present.

The PIXE data shows line areas {for respectively Sn K, Sn L, Pb L, Mn K, Cu K, As K, Bi L, Fe K}, with the grey bars showing the experiment and the black bars the fit. Note that the ordinate is on a logarithmic scale. So the PIXE tells us that the sample is largely Pb and Sn (lead and tin) with small amounts of other elements present. But it gives no useful hint of the corrosion mechanism. Now, if we knew that all these elements were uniformly distributed in the depth of the sample then we could easily calculate the sample composition. But we suspect that corrosion starts at the surface and changes the composition in the sample depth. Here the PIXE is of no help.

The He RBS spectrum very clearly gives a useful boundary condition: what the Sn/Pb ratio is at the surface of the sample. However, it is the H EBS spectrum that solves the puzzle: this spectrum makes O visible (this light element is completely invisible in RBS: note that C is also present), and of course, corrosion is (usually) an oxidation process. But the EBS cannot disentangle the Sn from the Pb—we need the PIXE for that, together with the boundary condition supplied by the RBS. We can see that the few % of Pb present in the sample does not participate in the corrosion, which is an oxidation of the tin extending to a depth of about 10^{20} at/cm^2 (equivalent to about 27 μm of tin assuming a density of 7.3 g/cm^3 or 3.7×10^{22} at/cm^3).

Jeynes et al. [58] gave a general account of Total-IBA in the technical literature in 2012; Lucile Beck said of this in a 2014 review [59]:

> *Analytical challenge can be achieved by coupling IBA techniques. Multiple combinations enable various types of ancient materials to be characterised with low limits of detection, non-destructivity, depth resolution and information on organic phase. For the last two points, data processing can now benefit of the progress in self-consistent analysis of simultaneously collected IBA data by using the "Total IBA" approach. This improvement may resolve complex structures requiring information in depth and composed of numerous mixed compounds. It could be very useful for the direct analysis of paintings by collecting a maximum of data with a minimum number of experiments. This procedure has also the advantage to reduce the possible irradiation damage pointed out by several studies …*

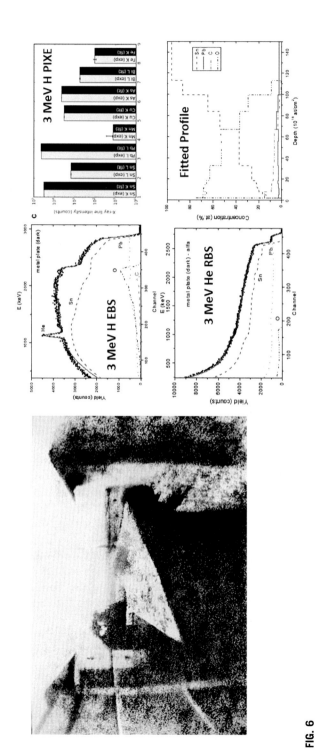

FIG. 6

Corrosion of Niépce's heliograph. See text. The external beam proton PIXE & EBS are obtained simultaneously, the external beam alpha RBS subsequently. All three are needed to obtain the profile by fitting. Image created by Joseph Nicéphore Niépce in 1827, Point de vue du Gras ("View from the window"), enhanced by Helmut Gersheim *c.*1952 (Wikimedia Commons, public domain).

IBA results from Figs. 2, 3 of C. Pascual-Izarra, N.P. Barradas, M.A. Reis, C. Jeynes, M. Menu, B. Lavedrine, J.J. Ezrati, S. Röhrs, Towards truly simultaneous PIXE and RBS analysis of layered objects in cultural heritage, Nucl. Instrum. Methods Phys. Res. B 261 (2007) 426–429 (© Elsevier 2007, reproduced by permission).

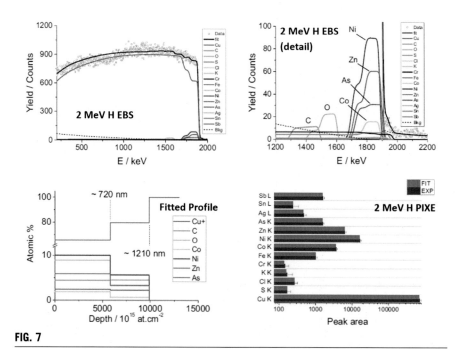

FIG. 7

Corrosion of Portuguese copper coins. Example of the analysis of corrosion products extending ~1 μm into copper coins from XV and XVI centuries.

Reproduced with permission from Fig. 12 of J. Cruz, M. Manso, V. Corregidor, R.J.C. Silva, E. Figueiredo, M.L. Carvalho, L.C. Alves, Surface analysis of corroded XV–XVI century copper coins by μ-XRF and μ-PIXE/μ-EBS self-consistent analysis, Mater. Charact. 161 (2020) 110170 (© Elsevier B.V. 2020).

Cruz et al. [60] have made a detailed study of the corrosion products on two historic and rare Portuguese copper coins which are representative of important collections. Fig. 7 gives a flavour of the analysis.

Similar methods have been used by Beck et al. [61] to unambiguously characterise the layer structure of paint (including a partial analysis of a 19th-century reproduction of Frans Hals' "La Bohémienne"), and also by Corregidor et al. [62] to measure the thickness and composition of the mercury gilding of sacred art objects from the 16th and 18th centuries. Cultural heritage artefacts commonly have "severe" surface roughness (i.e., from the point-of-view of IBA, which is sensitive to the condition of surfaces). Gurbich et al. [63] have shown that even though such effects as observed by Beck et al. cannot be ignored, they can be simply (and now routinely) corrected.

It has of course long been known that PIXE and EBS are highly synergistic, but it has not been feasible to handle them self-consistently until recently. In simple cases, or where rough results are adequate, it is possible to approximate a *Total-IBA* approach by manually fitting the PIXE spectra and simulating the EBS spectra sequentially, iterating as required. Cruz et al. [60] point out that in such cases the data analysis *performed step by step by combining X-ray spectrum fitting and backscattered particle spectrum simulation ... is time consuming and not automatically self-consistent*. This approximate "Total-IBA" has been used by de Viguerie et al. [64] in an application very similar to that of Beck et al. [61].

---

7 Summary

Ion beam analysis is an off-line spectrometric method requiring a high-energy (MeV) ion beam from a large accelerator. It is nondestructive, very versatile, and capable of very high spatial resolution while being fully quantitative. Measurement times can be very short: even multitechnique high-resolution images over large areas can be completed in a few hours. Cultural heritage samples (which are usually large, fragile and priceless) are regularly analysed in air as received (without any sample preparation). Measurement campaigns are regularly of large sample sets.

Imaging modes may be (and regularly are) any or all of optical or X-ray characteristic photons induced by atomic excitation of the sample (IBIL/PIXE), or of elastic nuclear scattering (EBS/RBS), or of nuclear excitations yielding γ-ray photons (PIGE). The information depth extends only to 10–20 μm or so (depending on the sample and the analysis conditions), but detailed information is also available for the surface submicron layer.

In the last two decades IBA has grown dramatically in power. Previously the IBA community was effectively divided between PIXE and EBS practitioners, both for practical reasons and because there is a large class of samples where the interest is simply the average composition, and another distinct large class where the interest is largely in the surface modifications. Of course, both aspects are important for most real samples, and this is what has driven the development of Total-IBA methods. But these have been enabled by:

1. a synergistic handling of multimodal PIXE/EBS/PIGE datasets, which has only become feasible in the last decade;
2. an explosion in the availability and quality of the scattering cross-section data essential to EBS (also valuable technical developments in PIGE), thanks partly to the substantial support of the International Atomic Energy Agency (IAEA);
3. the Fundamental Parameters programme of the European X-ray Spectrometry Association (EXSA) supporting developments in PIXE (and XRF), although this has not yet attracted significant European Union funding;
4. substantial European Union funding for various programs supporting IBA: SPIRIT (Mar. 2009-Aug. 2013), CHARISMA (Oct. 2009-Mar. 2014), RADIATE (Jan. 2019-Dec. 2022). The European Research Infrastructure for Heritage Science (*E*-RIHS) is ongoing;
5. dramatic advances both in X-ray detectors and in digital spectroscopic electronics revolutionising imaging applications;
6. greatly improved performance of ion accelerators and related equipment.

Heritage scientists should today not overlook the enormous value of well-designed research campaigns making full use of the power of IBA. They will find that it is well worth the effort.

References

[1] A. Adriaens, Review: Non-destructive analysis and testing of museum objects: an overview of 5 years of research, Spectrochim. Acta Part B At. Spectrosc. 60 (2005) 1503–1516, https://doi.org/10.1016/j.sab.2005.10.006.

[2] J. Salomon, J.-C. Dran, T. Guillou, B. Moignard, L. Pichon, P. Walter, F. Mathis, Present and future role of ion beam analysis in the study of cultural heritage materials: the example of the AGLAE facility, Nucl. Instrum. Methods Phys. Res. B 266 (2008) 2273–2278, https://doi.org/10.1016/j.nimb.2008.03.076.

[3] A. Zucchiatti, Ion beam analysis for the study of our cultural heritage. A short history and its milestones, Nucl. Instrum. Methods Phys. Res. B 452 (2019) 48–54, https://doi.org/10.1016/j.nimb.2019.05.059.

[4] M. Lebon, L. Pichon, L. Beck, Enhanced identification of trace element fingerprint of prehistoric pigments by PIXE mapping, Nucl. Instrum. Methods Phys. Res. B 417 (2018) 91–95, https://doi.org/10.1016/j.nimb.2017.10.010.

[5] B. Beckhoff, Reference-free X-ray spectrometry based on metrology using synchrotron radiation, J. Anal. Atom. Spectrom. 23 (2008) 845–853, https://doi.org/10.1039/b718355k.

[6] C. Jeynes, RBS as a new primary direct reference method for measuring quantity of material, Nucl. Instrum. Methods Phys. Res. B 406 (2017) 30–31, https://doi.org/10.1016/j.nimb.2016.11.041.

[7] R. Unterumsberger, P. Hönicke, J.L. Colaux, C. Jeynes, M. Wansleben, M. Müller, B. Beckhoff, Accurate experimental determination of gallium K- and L3-shell XRF fundamental parameters, J. Anal. Atom. Spectrom. 33 (2018) 1003–1013, https://doi.org/10.1039/c8ja00046h.

[8] C. Jeynes, E. Nolot, C. Costa, C. Sabbione, W. Pessoa, F. Pierre, A. Roule, G. Navarro, M. Mantler, Quantifying nitrogen in GeSbTe:N alloys, J. Anal. Atom. Spectrom. 35 (2020) 701–712, https://doi.org/10.1039/C9JA00382G.

[9] G.W. Grime, A compact beam focusing and steering element using quadrupoles with independently excited poles, Nucl. Instrum. Methods Phys. Res. B 306 (2013) 12–16, https://doi.org/10.1016/j.nimb.2012.10.041.

[10] M.J. Merchant, P. Mistry, M. Browton, A.S. Clough, F.E. Gauntlett, C. Jeynes, K.J. Kirkby, G.W. Grime, Characterisation of the University of Surrey Ion Beam Centre in-air scanning microbeam, Nucl. Instrum. Methods Phys. Res. B 231 (2005) 26–31, https://doi.org/10.1016/j.nimb.2005.01.029.

[11] L. Matjačić, V. Palitsin, G.W. Grime, N. Abdul-Karim, R.P. Webb, Simultaneous molecular and elemental mapping under ambient conditions by coupling AP MeV SIMS and HIPIXE, Nucl. Instrum. Methods Phys. Res. B 450 (2019) 353–356, https://doi.org/10.1016/j.nimb.2018.08.007.

[12] C. Jeynes, V.V. Palitsin, G.W. Grime, C. Pascual-Izarra, A. Taborda, M.A. Reis, N.P. Barradas, External beam total-IBA using data furnace, Nucl. Instrum. Methods Phys. Res. B 481 (2020) 47–61, https://doi.org/10.1016/j.nimb.2020.08.002.

[13] A. Simon, C. Jeynes, R.P. Webb, R. Finnis, Z. Tabatabian, P.J. Sellin, M.B.H. Breese, D. F. Fellows, R. van den Broek, R.M. Gwilliam, The new Surrey ion beam analysis facility, Nucl. Instrum. Methods Phys. Res. B 219 (2004) 405–409, https://doi.org/10.1016/j.nimb.2004.01.091.

[14] C. Jeynes, V.V. Palitsin, M. Kokkoris, A. Hamilton, G.W. Grime, On the accuracy of Total-IBA, Nucl. Instrum. Methods Phys. Res. B 465 (2020) 85–100, https://doi.org/10.1016/j.nimb.2019.12.019.

[15] W.H. Bragg, M.A. Elder, On the alpha particles of radium, and their loss of range in passing through various atoms and molecules, Philos. Mag. (Series 6) 10 (1905) 318–340, https://doi.org/10.1080/14786440509463378.

[16] E. Vagena, E.G. Androulakaki, M. Kokkoris, N. Patronis, M.E. Stamati, A comparative study of stopping power calculations implemented in Monte Carlo codes and compilations with experimental data, Nucl. Instrum. Methods Phys. Res. B 467 (2020) 44–52, https://doi.org/10.1016/j.nimb.2020.02.003.

[17] I. Rajta, G.U.L. Nagy, I. Vajda, S.Z. Szilasi, G.W. Grime, F. Watt, First resolution test results of the Atomki nuclear nanoprobe, Nucl. Instrum. Methods Phys. Res. B 449 (2019) 94–98, https://doi.org/10.1016/j.nimb.2019.03.056.

[18] N. Grassi, L. Giuntini, P.A. Mandò, M. Massi, Advantages of scanning-mode ion beam analysis for the study of Cultural Heritage, Nucl. Instrum. Methods Phys. Res. B 256 (2007) 712–718, https://doi.org/10.1016/j.nimb.2006.12.196.

[19] L. Beck, P.C. Gutiérrez, S. Miro, F. Miserque, Ion beam modification of zinc white pigment characterized by *ex situ* and *in situ* micro-Raman and XPS, Nucl. Instrum. Methods Phys. Res. B 409 (2017) 96–101, https://doi.org/10.1016/j.nimb.2017.04.071.

[20] S. Calusi, The external ion microbeam of the LABEC laboratory in Florence: some applications to cultural heritage, Microsc. Microanal. 17 (2011) 661–666, https://doi.org/10.1017/S1431927611000092.

[21] A. Zucchiatti, F. Agulló-Lopez, Potential consequences of ion beam analysis on objects from our cultural heritage: an appraisal, Nucl. Instrum. Methods Phys. Res. B 278 (2012) 106–114, https://doi.org/10.1016/j.nimb.2012.02.016.

[22] T. Calligaro, V. Gonzalez, L. Pichon, PIXE analysis of historical paintings: is the gain worth the risk? Nucl. Instrum. Methods Phys. Res. B 363 (2015) 135–143, https://doi.org/10.1016/j.nimb.2015.08.072.

[23] A. Zucchiatti, S. Martina, IBA analysis of modern pigments and papers: new data on color alteration and limits for safe analysis, Eur. Phys. J. Appl. Phys. 86 (2019) 10701 (7 pp) https://doi.org/10.1051/epjap/2019190060.

[24] A.L. Giudice, D. Angelici, A. Re, G. Gariani, A. Borghi, S. Calusi, L. Giuntini, M. Massi, L. Castelli, F. Taccetti, T. Calligaro, C. Pacheco, Q. Lemasson, L. Pichon, B. Moignard, G. Pratesi, M.C. Guidotti, Protocol for lapis lazuli provenance determination: evidence for an Afghan origin of the stones used for ancient carved artefacts kept at the Egyptian Museum of Florence (Italy), Archaeol. Anthropol. Sci. 9 (2017) 637–651, https://doi.org/10.1007/s12520-016-0430-0.

[25] M. Ortega-Avilés, P. Vandenabeele, D. Tenorio, G. Murillo, M. Jiménez-Reyes, N. Gutiérrez, Spectroscopic investigation of a 'Virgin of Sorrows' canvas painting: a multi-method approach, Anal. Chim. Acta 550 (2005) 164–172, https://doi.org/10.1016/j.aca.2005.06.059.

[26] J. Chadwick, The excitation of γ rays by α rays, Philos. Mag. Series 6 25 (1913) 703–713, https://doi.org/10.1080/14786440108634324.

[27] N. Salehvand, D. Agha-Aligol, A. Shishegar, M.L. Rachti, The study of chemical composition of Persian glass vessels of the early Islamic centuries (10th – 11th centuries AD) by micro-PIXE; Case Study: Islamic collection in the National Museum of Iran, J. Archaeol. Sci. Rep. 29 (2020) 102034 (10 pp) https://doi.org/10.1016/j.jasrep.2019.102034.

[28] M. Radepont, L. Robinet, C. Bonnot-Diconne, C. Pacheco, L. Pichon, Q. Lemasson, B. Moignard, Ion beam analysis of silver leaves in gilt leather wall coverings, Talanta 206 (2020) 120191 (7 pp) https://doi.org/10.1016/j.talanta.2019.120191.

[29] A.M. Gueli, S. Pasquale, D. Tanasi, S. Hassam, Q. Lemasson, B. Moignard, C. Pacheco, L. Pichon, G. Stella, G. Politi, Weathering and deterioration of archeological glasses from late Roman Sicily, Int. J. Appl. Glas. Sci. 11 (2020) 215–225, https://doi.org/10.1111/ijag.14076.

[30] E. Holmqvist, Å.M. Larsson, A. Kriiska, V. Palonen, P. Pesonen, K. Mizohata, P. Kouki, J. Räisänen, Tracing grog and pots to reveal Neolithic Corded Ware Culture contacts in the Baltic Sea region (SEM-EDS, PIXE), J. Archaeol. Sci. 91 (2018) 77–91, https://doi.org/10.1016/j.jas.2017.12.009.

[31] R. Bugoi, C. Talmaṭchi, C. Haiṭă, D. Ceccato, Archaeometric characterization of Byzantine pottery from Păcuiul lui Soare, Herit. Sci. 7 (2019) 55 (16 pp) https://doi.org/10.1186/s40494-019-0298-2.

[32] Z.I. Balta, I. Demetrescu, I. Petroviciu, M. Lupu, Advanced micro-chemical investigation of golden threads from Romanian Byzantine embroideries by micro-particle induced X-ray emission (micro-PIXE), Rev. de Chim. 70 (2019) 1956–1959, https://doi.org/10.37358/RC.19.6.7253.

[33] T. Calligaro, P.-J. Chiappero, F. Gendron, G. Poupeau, New clues on the origin of the "Inca Mirror" at the Museum National d'Histoire Naturelle in Paris, Lat. Am. Antiq. 30 (2019) 422–428, https://doi.org/10.1017/laq.2019.3.

[34] M. Vilarigues, I. Coutinho, T. Medici, L.C. Alves, B. Gratuze, A. Machado, From beams to glass: determining compositions to study provenance and production techniques, Phys. Sci. Rev. (2019) 20180019. (24 pp); Also Chapter 12 in L. Sabbatini, I. Dorothé van der Werf (Eds.) Chemical Analysis in Cultural Heritage, De Gruyter, Berlin, 2020 https://doi.org/10.1515/9783110457537-012.

[35] M. Bambrough, Rosslyn Chapel: innovation in stained glass conservation, J. Stained Glass 42 (2018) 122–138.

[36] E. Rutherford, The scattering of α and β particles by matter and the structure of the atom, Philos. Mag. Series 6 21 (1911) 669–688, https://doi.org/10.1080/14786435.2011.617037.

[37] M.C. Parker, C. Jeynes, W.N. Catford, On the geometric (holographic) entropy of helium isotopes, Scientific Reports (2021). (in review) Submitted for publication https://doi.org/10.21203/rs.3.rs-112066/v3.

[38] C. Jeynes, J.L. Colaux, Thin film depth profiling by ion beam analysis (Tutorial Review), Analyst 121 (2016) 5944–5985, https://doi.org/10.1039/c6an01167e.

[39] C. Jeynes, N.P. Barradas, E. Szilágyi, Accurate determination of quantity of material in thin films by Rutherford backscattering spectrometry, Anal. Chem. 84 (2012) 6061–6069, https://doi.org/10.1021/ac300904c.

[40] D. Abriola, N.P. Barradas, I. Bogdanović-Radović, M. Chiari, A.F. Gurbich, C. Jeynes, M. Kokkoris, M. Mayer, A.R. Ramos, L. Shi, I. Vickridge, Development of a reference database for Ion Beam Analysis and future perspectives, Nucl. Instrum. Methods Phys. Res. B 269 (2011) 2972–2978, https://doi.org/10.1016/j.nimb.2011.04.056.

[41] A. Abriola, P. Dimitriou, Development of a reference database for ion beam analysis, Report of a Coordinated Research Project (CRP) on Reference Database for Ion Beam Analysis (IAEA-TECDOC-1780), List of participants: N. P. Barradas, I. B. Radović, M. Chiari, A. F. Gurbich, C. Jeynes, M. Kokkoris, M. Mayer, A. R. Lopes Ramos, E. Rauhala, I. Vickridge, International Atomic Energy Agency, Vienna, 2015.

[42] A.F. Gurbich, Evaluated differential cross-sections for IBA, Nucl. Instrum. Methods Phys. Res. B 268 (2010) 1703–1710, https://doi.org/10.1016/j.nimb.2010.02.011.

[43] C. Jeynes, N.P. Barradas, P.K. Marriott, G. Boudreault, M. Jenkin, E. Wendler, R.P. Webb, Elemental thin film depth profiles by ion beam analysis using simulated annealing – a new tool (Topical Review), J. Phys. D 36 (2003) R97–R126, https://doi.org/10.1088/0022-3727/36/7/201.

[44] N.P. Barradas, C. Jeynes, Advanced physics and algorithms in the IBA DataFurnace, Nucl. Instrum. Methods Phys. Res. B 266 (2008) 1875–1879, https://doi.org/10.1016/j.nimb.2007.10.044.

[45] I. Ortega-Feliu, F.J. Ager, C. Roldán, M. Ferretti, D. Juanes, S. Scrivano, M.A. Respaldiza, L. Ferrazza, I. Traver, M.L. Grilli, Multi-technique characterization of gold electroplating on silver substrates for cultural heritage applications, Nucl. Instrum. Methods Phys. Res. B 406 (2017) 318–323, https://doi.org/10.1016/j.nimb.2017.02.016.

[46] P. Reichart, G. Datzmann, A. Hauptner, R. Hertenberger, C. Wild, G. Dollinger, Three-dimensional hydrogen microscopy in diamond, Science 306 (2004) 1537–1540, https://doi.org/10.1126/science.1102910.

[47] J.-C. Dran, J. Salomon, T. Calligaro, P. Walter, Ion beam analysis of art works: 14 years of use in the Louvre, Nucl. Instrum. Methods Phys. Res. B 219–220 (2004) 7–15, https://doi.org/10.1016/j.nimb.2004.01.019.

[48] T. Calligaro, Y. Coquinot, I. Reiche, J. Castaing, J. Salomon, G. Ferrand, Y. Le Fur, Dating study of two rock crystal carvings by surface microtopography and by ion beam analyses of hydrogen, Appl. Phys. A Mater. Sci. Process. 94 (2009) 871–878, https://doi.org/10.1007/s00339-008-5018-9.

[49] A.F. Gurbich, Evaluation of cross-sections for particle induced gamma-ray emission (PIGE) spectroscopy, Nucl. Instrum. Methods Phys. Res. B 331 (2014) 31–33, https://doi.org/10.1016/j.nimb.2013.11.031.

[50] R. Balvanović, Ž. Šmit, Sixth-century AD glassware from Jelica, Serbia—an increasingly complex picture of late antiquity glass composition, Archaeol. Anthropol. Sci. 12 (2020) 94 (17 pp.) https://doi.org/10.1007/s12520-020-01031-3.

[51] Ž. Šmit, B. Laharnar, P. Turk, Analysis of prehistoric glass from Slovenia, J. Archaeol. Sci. Rep. 29 (2020) 102114 (13 pp.) https://doi.org/10.1016/j.jasrep.2019.102114.

[52] M. Maggett, Archaeometric analyses of European 18th–20th century white earthenware—a review, Minerals 8 (2018) 269 (38 pp.) https://doi.org/10.3390/min8070269.

[53] I. Reiche, C. Heckel, K. Müller, O. Jöris, T. Matthies, N.J. Conard, H. Floss, R. White, Combined non-invasive PIXE/PIGE analyses of mammoth ivory from Aurignacian archaeological sites, Angew. Chem. Int. Ed. 57 (2018) 7428–7432, https://doi.org/10.1002/anie.201712911.

[54] A. Bouquillon, J. Castaing, F. Barbe, S.R. Paine, B. Christman, T. Crepin-Leblond, A.H. Heuer, Lead-glazed Rustiques Figulines (Rustic Ceramics) of Bernard Palissy (1510–90) and his followers, Archaeometry 59 (2017) 69–83, https://doi.org/10.1111/arcm.12247.

[55] T. Calligaro, Y. Coquinot, L. Pichon, B. Moignard, Advances in elemental imaging of rocks using the AGLAE external microbeam, Nucl. Instrum. Methods Phys. Res. B 269 (2011) 2364–2372, https://doi.org/10.1016/j.nimb.2011.02.074.

[56] C. Pascual-Izarra, N.P. Barradas, M.A. Reis, C. Jeynes, M. Menu, B. Lavedrine, J.J. Ezrati, S. Röhrs, Towards truly simultaneous PIXE and RBS analysis of layered objects in cultural heritage, Nucl. Instrum. Methods Phys. Res. B 261 (2007) 426–429, https://doi.org/10.1016/j.nimb.2007.04.259.

[57] G.W. Grime, O.B. Zeldin, M.E. Snell, E.D. Lowe, J.F. Hunt, G.T. Montelione, L. Tong, E.H. Snell, E.F. Garman, High-throughput PIXE as an essential quantitative assay for accurate metalloprotein structural analysis: development and application, J. Am. Chem. Soc. 142 (2020) 185–197, https://doi.org/10.1021/jacs.9b09186.

[58] C. Jeynes, M.J. Bailey, N.J. Bright, M.E. Christopher, G.W. Grime, B.N. Jones, V.V. Palitsin, R.P. Webb, Total IBA—where are we? Nucl. Instrum. Methods Phys. Res. B 271 (2012) 107–118, https://doi.org/10.1016/j.nimb.2011.09.020.

[59] L. Beck, Recent trends in IBA for cultural heritage studies, Nucl. Instrum. Methods Phys. Res. B 332 (2014) 439–444, https://doi.org/10.1016/j.nimb.2014.02.113.

[60] J. Cruz, M. Manso, V. Corregidor, R.J.C. Silva, E. Figueiredo, M.L. Carvalho, L.C. Alves, Surface analysis of corroded XV–XVI century copper coins by μ-XRF and μ-PIXE/μ-EBS self-consistent analysis, Mater. Charact. 161 (2020) 110170 (16 pp) https://doi.org/10.1016/j.matchar.2020.110170.

[61] L. Beck, C. Jeynes, N.P. Barradas, Characterization of paint layers by simultaneous self-consistent fitting of RBS/PIXE spectra using simulated annealing, Nucl. Instrum. Methods Phys. Res. B 266 (2008) 1871–1874, https://doi.org/10.1016/j.nimb.2007.12.091.

[62] V. Corregidor, L.C. Alves, N.P. Barradas, M.A. Reis, M.T. Marques, J.A. Ribeiro, Characterization of mercury gilding art objects by external proton beam, Nucl. Instrum. Methods Phys. Res. B 269 (2011) 3049–3053, https://doi.org/10.1016/j.nimb.2011.04.070.

[63] S.L. Molodtsov, A.F. Gurbich, C. Jeynes, Accurate ion beam analysis in the presence of surface roughness, J. Phys. D 41 (2008) 205303 (8 pp.) https://doi.org/10.1088/0022-3727/41/20/205303.

[64] L. de Viguerie, L. Beck, J. Salomon, L. Pichon, P. Walter, Composition of renaissance paint layers: simultaneous particle induced X-ray emission and backscattering spectrometry, Anal. Chem. 81 (2009) 7960–7966, https://doi.org/10.1021/ac901141v.

High-energy particle analysis

Andrea Denker
*Protons for Therapy, Helmholtz-Zentrum Berlin, Berlin, Germany University of Applied Sciences,
Beuth Hochschule für Technik Berlin, Berlin, Germany*

1 Introduction and motivation

Research on cultural heritage objects comprises, among others, questions concerning dating, provenance, attribution to an artist or a workshop, and conservation. Stylistic and historical considerations, combined with comprehensive studies of historical sources, provide essential insights. However, in many cases additional information of utmost importance can be obtained by natural science analyses. Information on which elements are present in an object may already allow indirect dating or the detection of forgeries using the chronology of pigments [1] (see Fig. 1). If quantification is possible, the composition of alloys, for example, can be determined as discussed later.

As sampling is in most cases restricted in view of the value or the uniqueness of the object, nondestructive and noninvasive investigations are mandatory. Even in the case sampling is allowed, nondestructive testing offers the possibility of obtaining more information on one specific sample as complementary techniques may be applied.

Several nondestructive and noninvasive methods rely on the emission of characteristic X-rays. They are characteristic for the emitting element, thus allowing its identification. To create excited atoms which hereupon emit these X-rays, one can use either X-rays (X-ray fluorescence—XRF [2]), electrons (electron microprobe [3]), or ions. For the excitation with ions, mainly hydrogen ions are used; therefore, this technique is called proton-induced X-ray emission (PIXE).

2 Proton-induced X-ray emission
2.1 Basic principles

When material is irradiated with ions with an energy of some MeV, these can interact either with the atomic nuclei of the sample or with the electrons. The abbreviation MeV stands for mega-electron-volt. One MeV is the kinetic energy which a singly

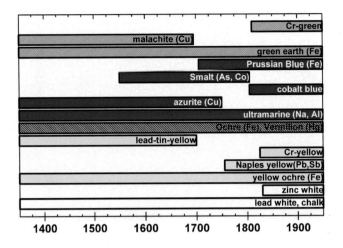

FIG. 1

Chronology of pigments. Many pigments, like earth pigments, have been known since antiquity. Others have been invented later or became unfashionable [1].

charged particle obtains by traveling through a potential difference of one million volts. For comparison, the X-ray photons used for medical radiography each carry between 2% and 15% of an MeV depending on the application. The ions collide with the inner-shell electrons, thus ejecting them. This effect was observed for the first time by Chadwick [4]. This excitation process with ions is similar to excitation using electrons; however, due to the smaller energy loss of the ions, there is less bremsstrahlung in the spectra. The vacancies are filled with electrons from higher shells. The energy difference is released either by the emission of X-rays or the energy is transferred to another electron which leaves the atom as a so-called Auger electron. The chance that an X-ray is emitted instead of an Auger electron increases with the atomic number Z and is described by the fluorescence yield ω which can be calculated by the semiempirical formula of Bambynek et al. [5]:

$$\left(\frac{\omega}{1-\omega}\right)^{\frac{1}{4}} = \sum_{i=0}^{3} B_i Z^i \tag{1}$$

The coefficients B_i are listed in Table 1 [6] and Fig. 2 shows the fluorescence yields for the K- and L-shell.

As the arrangement of the shells is different for each element, the energy of the emitted X-rays is characteristic for the element in question. Moseley [7] found the following empirical correlation between the frequency $\nu_{K\alpha}$ of the Kα line and Z:

$$\sqrt{\nu_{K\alpha}} = \kappa(Z-1) \tag{2}$$

with $\kappa^2 = 2.48 \times 10^{15}$ Hz. The determination of the X-ray energy allows the identification of the emitting element. Fig. 3 shows the dependence of the X-ray energy as a function of Z [8].

Table 1 Coefficients for the calculation of the K- and L-shell fluorescence yields [6].

	K	L
B_0	3.7×10^{-2}	0.17765
B_1	3.112×10^{-2}	2.98937×10^{-3}
B_2	5.44×10^{-5}	8.91297×10^{-5}
B_3	-1.25×10^{-6}	-2.67184×10^{-7}

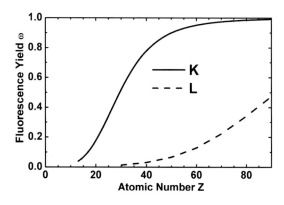

FIG. 2

Fluorescence yield as a function of the atomic number. For heavy elements, the emission of a characteristic X-ray is the most probable process.

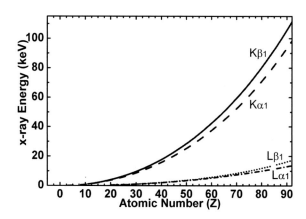

FIG. 3

Energy of the characteristic X-rays as a function of the atomic number.

The electrons around an atom are arranged in the main shells K, L, M, ... corresponding to the main quantum numbers $n = 1, 2, 3, \ldots$. The orbital angular momentum l and the spin s of the electrons interact, leading to the fine structure of the electron shells (see Fig. 4). The total angular momentum j is given by:

$$\vec{j} = \vec{l} + \vec{s} \tag{3}$$

Electrons filling a vacancy in the K- or L-shell have to fulfil the rules $\Delta l = \pm 1$ and $\Delta j = 0, \pm 1$. The transitions from L to K are called Kα-lines, from M to K Kβ-lines, and so forth. For the K-lines, the relative intensity of the various lines is quite well known [9, 10]. For the L-lines, the situation is more complicated, because a vacancy in the L-shell may, before direct emission of an X-ray or Auger-electron, be transferred via a nonradiation transfer to a higher shell. This effect is called the Coster-Kronig effect.

The probability for the production of an inner shell vacancy is described by the cross-section. The highest probability is achieved when the velocity of the proton is

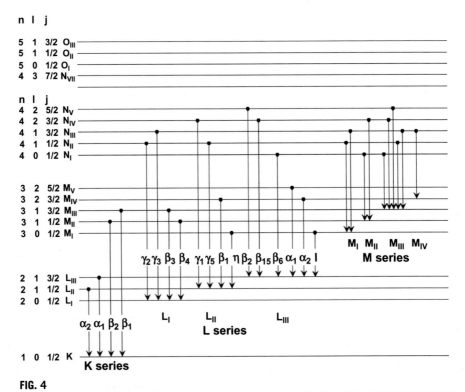

FIG. 4

Schematic diagram of the excited atomic states and the possible transitions leading to the emission of characteristic X-rays. Main quantum number n, angular momentum l, and total angular momentum j are indicated.

the same as the classical velocity of the bound electron. (The classical velocity is the orbital velocity which the electron would have if it were a quasi-classical particle with the same angular momentum $v = Z\hbar/m_e a_0 n$ where $\hbar$ is Planck's constant/2π, m_e is the electron mass, a_0 is the Bohr radius (0.0529 nm), and n is the principal quantum number.) Due to the different binding energies of the electrons in the various shells and atoms with different Z these velocities vary. Hence, the projectile energy with maximum production probability is different for each element (Fig. 5).

The first theoretical calculations used the plane wave Born approximation (PWBA), where perturbation theory is applied. The start system describes the projectile by a plane wave and includes a bound electron; the final state is a plane wave projectile and an electron in the continuum. This theory has been improved by taking into account the energy loss (E), the deviation/slowing down of the projectile in the Coulomb field of the atom (C), the perturbation of the stationary status of the atom (PSS), and relativistic effects (R), which gave the name ECPSSR theory [11]. In addition to theoretical calculations, a vast amount of experimental data is available. Paul and Sacher analyzed the data for the K-shell [12], resulting in so-called reference values. At 3-MeV proton energy, experimental values and theoretical values of the cross-sections for K X-rays agree within 2% (Fig. 6).

Orlic et al. [13] provide for the three L-subshells a compilation of experimental cross-section measurements using the data of Krause [14] for the L-subshell fluorescence as well as the Coster-Kronig probabilities. Although the precision for the L-shells is not as good as for the K-shells, these data can be considered as sufficient for most analytical purposes.

The characteristic X-rays are produced along the total flight path of the projectile ion through the sample. Therefore due to the energy dependence of the cross-section for the production rate of the X-rays in a given depth, the energy loss of the ion has to

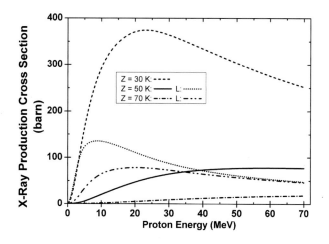

FIG. 5

Total X-ray production cross-sections as a function of proton energy for various Z.

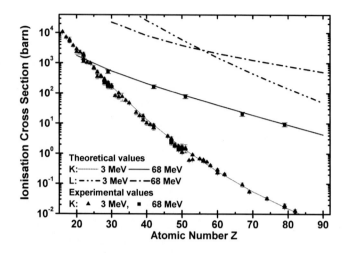

FIG. 6

Total ionization cross-sections for 3- and 68-MeV protons as a function of Z. The experimental values are from Paul and Sacher [12] for 3-MeV protons, and from Denker et al. for 68-MeV protons [13]. Note: For heavy elements, at a proton energy of 68 MeV the cross-section for the K-lines is several orders of magnitude larger than for 3-MeV protons.

be taken into account. Data for the energy loss can be found in the book of Ziegler et al. [15].

In addition, for the detection of the X-rays, their attenuation in matter has to be considered. In thick samples, the attenuation in the sample will determine the detection limits and the possible analytical depth. The intensity I from the original X-ray intensity I_0 after the transition of a material with a thickness d is given by

$$I = I_0 \exp(-\mu d) \qquad (4)$$

where μ is the attenuation coefficient [16]. Table 2 gives the thicknesses of different materials which reduce the intensity of X-rays to one-tenth of the original intensity.

For quantitative analysis, the following equation has to be solved:

Table 2 Thickness of glass, silver, and gold which will attenuate particular X-rays by 90% (i.e., leaving 10% of the original intensity).

X-ray	Glass (SiO₂) (mm)	Silver (mm)	Gold (mm)
Copper Kα (8 keV)	0.32	0.01	0.006
Lead Lα (10.5 keV)	0.7	0.02	0.01
Lead Kα (74.3 keV)	52	0.63	0.4

$$N_t = \frac{Y(Z)}{N_p \omega_Z b_Z \varepsilon_{\text{abs}} \displaystyle\int\limits_0^{x_{\text{max}}} \sigma(x) \exp\left(\frac{-\mu x}{\sin\theta}\right) \mathrm{d}x} \tag{5}$$

where x is the depth in the sample, N_t is the number of analyte atoms in the sample, $\sigma(x)$ is the X-ray production cross-section (a function of depth as the protons lose energy along their flight path), θ is the angle between the sample-normal and the detector, $Y(Z)$ is the X-ray yield (in counts) from peak area of the corresponding spectral line, N_p is the number of protons, ω_Z is the fluorescence yield, b_Z is the fraction of X-rays in the line, ε_{abs} is the absolute detector efficiency, and absorption of X-rays between target material and detector crystal is the absorption of X-rays between target material and detector crystal.

To solve this equation, various computer codes are available, e.g., Geopixe [17], Gupix [18], Pixan [19], Pixeklm [20], Sapix [21], and Winaxil [22].

The first application of the PIXE technique, as we know it today, was performed by Johansson et al. [23]. They used proton beams of 2–3-MeV energy from a Van-de-Graaff accelerator and samples consisting of carbon foils with subnanogram depositions of titanium and copper. Since that time, PIXE has evolved tremendously, being applied in archaeometry, biology, geosciences, medicine, and environmental studies. Under favorable conditions, detection limits below 1 ppm can be achieved. The interested reader will find detailed descriptions of PIXE in the books of Campbell and Tesmer [6, 24, 25]. Worldwide, there are many laboratories applying the PIXE technique to art objects (e.g. [26–32]). The materials investigated comprise paintings and drawings, porcelain, ceramics, and gems, as well as all kinds of metals. At the International Conference on PIXE and its applications held in Mexico City in 2007, laboratories from Bordeaux, Bucharest, Florence, Lecce, Liege, Ljubljana, Paris, Sacavem, Seville, Surrey, Mexico, and Denton presented work on art and archaeometry [33]. And this list is, by far, not complete.

2.2 High-energy PIXE

In PIXE, the achievable analytical depth is defined by two factors: the range of the protons in the investigated material and the absorption of the X-rays.

The range of protons with a given energy depends on the composition and density of the material. Table 3 gives the ranges of 3-MeV protons in air, glass, and

Table 3 Ranges of 3- and 68-MeV protons in air, glass, and gold.

	3 MeV	68 MeV
Air	136 mm	34 m
Glass	85 μm	21 mm
Gold	27 μm	4 mm

The values have been calculated using the code SRIM (stopping and ranges of ions in matter [34]).

gold compared to the ranges of 68-MeV protons, calculated by SRIM (stopping and ranges of ions in matter) [34].

The achievable analytical depth is smaller than the proton range, as the protons lose energy in the sample: the decreasing energy leads to decreasing cross-sections. As mentioned previously, the cross-section reaches a maximum if the velocity of the protons matches the velocity of the electrons. For 3-MeV protons, this is the case for the K electrons of fluorine, for higher Z, the cross-section drops tremendously. Therefore for $Z > 50$, usually the L-lines are used to detect heavy elements. For higher proton energies, the maximum of the cross-section is shifted to higher Z, so that heavy elements may also be detected via the higher energy K-lines.

The second parameter defining the analytical depth is the absorption of the X-rays in material lying above the interaction. For proton energies above 50 MeV, the achievable analytical depth depends solely on the absorption. For low-Z elements only the K-lines are visible in the spectrum. For heavy elements, also the L-lines have sufficient energy to be detected. As the energy of the K-lines is much higher compared to the L-lines, they will suffer less absorption. Fig. 7 shows, as an example, that it is possible to identify gold even behind thick layers of lead glass.

At proton energies above 3 MeV, nuclear reactions may occur and γ-rays can be produced. These γ-rays may be used to identify elements otherwise not detectable with the PIXE technique, like silicon or fluorine whose X-ray energy is too low. This is called proton induced γ-ray emission (PIGE).

Fig. 8 shows an X-ray spectrum of a 0.1-μm gold layer on a silicon substrate obtained by the irradiation with 68-MeV protons. The four major K-lines are clearly distinguishable, and also the four major groups of L-lines. The lead K-lines are a background contribution of the lead shielding. The silicon γ-line originates from

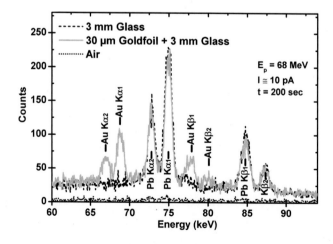

FIG. 7

High-energy PIXE spectra of air, lead glass, and a thin gold foil behind the lead glass. Even behind 3 mm of lead glass, the K-X-rays of gold are clearly visible.

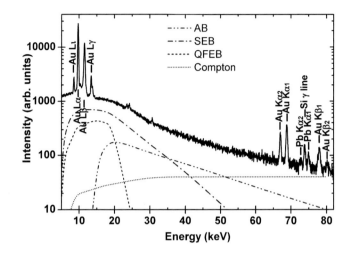

FIG. 8

High-energy PIXE spectrum of a 0.1-μm gold layer on silicon substrate using a 68-MeV proton beam of about 5 nA, measuring time about 25 min. All four K-lines of Au are clearly visible, also the major L-lines. The various background contributions are marked (see text).

nuclear reactions of the high-energy protons with the substrate. As the X-rays always appear in groups, single γ-lines cannot be confused with X-rays. In addition to the peaks, there is a background which consists of:

- Atomic bremsstrahlung (AB) which is due to the deceleration of bound target electrons in the Coulomb field of the projectile.
- Secondary electron bremsstrahlung (SEB) is created by the deceleration of electrons emitted from atoms in the ionization processes. The maximum energy E_{max} which can be transferred to an electron in a central collision is given by

$$E_{max} = \frac{4m_e}{M_p} E_p \qquad (6)$$

where m_e and M_p are the mass of electron and projectile and E_p is the projectile energy. If the collision is not head-on, an energy smaller than E_{max} will be transferred by Coulomb interaction. In the spectrum we observe an intense contribution below E_{max}, decreasing rapidly with increasing energy.

- Quasi free electron bremsstrahlung (QFEB) originates from the quasi-free electrons in the solid, and the maximum energy transfer to those electrons is

$$T_r = \frac{m_e}{M_p} E_p \qquad (7)$$

increasing the low-energy part of the background.

- Compton background is due to the Compton scattering of γ-rays created in nuclear reactions. Comparing this background for protons and heavy ions at the same velocity, the Compton background is much larger for heavy ions.

Both AB and SEB have an anisotropic behavior, and are maximal at 90 degrees with respect to the incoming particle beam. Therefore for most PIXE set-ups, a backward geometry of the detector is preferred.

2.3 Experimental set-up

The protons are produced in an ion source and accelerated to the desired energy. A beam-line system guides the proton beam to the experiment. The beam-line is under vacuum to avoid interactions of the protons with air. For the analysis of art objects, the proton beam is extracted via a thin foil from the vacuum of the beam-line. Thus the object under study is kept in normal atmospheric conditions. The size of the proton beam spot on the sample depends on the beam-line: in most cases, it has a diameter of 0.5–1 mm, which is suitable for many purposes. The size is controlled by looking at a quartz plate which emits visible light when irradiated. At special beam-lines, a size of about 10 μm has been achieved. Fine structures, like metal point drawings, require such excellent spatial resolution [35]. The object to be studied is mounted on an x–y table which allows positioning of the object. The X-rays are measured by solid-state detectors made from single crystals. For the X-ray energy range of 1–25 keV, usually lithium drifted silicon detectors (Si(Li)) are employed. Above 25 keV, high-purity Germanium (HPGe) detectors are used. The signals of the detector are amplified, passed to an analog/digital converter, and then processed by a multichannel analyzer, usually combined with a computer to allow on-line display of the spectra. The identification of the elements present in the sample can be done during the measurements, whereas the final deconvolution of peak heights to concentrations is done off-line.

The intensity of the proton beam can be determined by measuring the current on rotating metal fingers, X-rays from the exit window or from the Argon in the air, or by using transmission ionization chambers. Typically the proton beam intensity is kept below 1 nA for a beam spot of 1 mm. This is a sufficient intensity to obtain good count rates in the X-ray detectors, and will not damage the object. For very radiation-sensitive material, like glass and porcelain, usually test irradiations are performed on modern material. If there are no observable changes in the modern material, the beam intensity is then reduced even further to ensure the safety of the art object.

Fig. 9 shows a photograph of the high-energy PIXE set-up at the Helmholtz-Zentrum Berlin, Germany, with silver coins as samples.

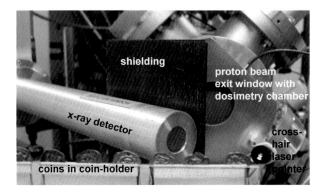

FIG. 9

High-energy PIXE set-up at the Helmholtz-Zentrum Berlin, with coins from the hoard of Tulln mounted in front of the beam. The X-ray detector is shielded against radiation from the exit foil and ionization chamber. The laser is equipped with cross-hair optics for precise positioning of the samples.

3 Paintings

The PIXE technique is a valuable tool for the nondestructive analysis of paintings. For a mere detection of the elements present in a painting, a portable XRF-system would be the choice of today, thus avoiding the transport of the painting to an accelerator.

However, complementary information can be obtained through the use of protons. One example is the analysis of "Madonna and Child Enthroned" from the Finnish National Gallery [36]. T. Tuurnala and A. Hautojärvi show the simultaneous analysis of the blue element: typical "X-ray elements" such as calcium, copper, lead, as well as elements not visible in X-ray spectra like fluorine, sodium, and aluminum can be detected in one experimental run on the same spot.

A further advantage is that the PIXE technique provides a depth-resolved analysis, when varying the incident proton energy [37]. For instance, if increasing the proton energy leads to a strong increase of the lead Lα/mercury Lα ratio, this indicates a mercury-containing layer on top of lead. The feasibilities and limitations have been investigated by a European collaboration on various paint test samples [38].

When very thick layers, above 100 μm, come into play, high-energy PIXE allows a nondestructive analysis. The excitation conditions are the same throughout the paint-layers, therefore, the intensity ratios of the lines of the same element, e.g., copper Kα/copper Kβ, give an indication of the depth of the element. On paint mock-ups of varying complexity, it could be shown that paint sequences, e.g., white ground and then layers containing pure pigments, can be clearly distinguished and identified, whereas limitations occur when the same pigment is present in various layers [39].

The painting shown in Fig. 10 belongs to a trust collection in Berlin. It was considered to be a 19th century copy created by an unknown artist of the famous painting by Christofano Allori: Judith with the head of Holofernes. Only slight changes in Judith's clothes distinguish the paintings, e.g., the girdle in the copy has blue stripes. A recent art historical examination of the painting resulted in doubts about the age. The fineness of the brush strokes and the overall high quality of the painting lead to the assumption that the painting might be much older, possibly even dating back to the times of Allori himself [40]. As mentioned earlier, indirect dating is possible by the determination of the pigments used: some pigments like natural earth pigments and mineral pigments have been known since antiquity, others were discovered in the 18th or 19th century or became unfashionable at a certain time [41]. For example, the use of titanium in a white pigment would give a maximum age of the painting of about 100 years, as TiO_2 as white pigment was discovered in 1908.

Fourteen different spots have been analyzed on this painting. Just three elements were detected: iron, lead, and on some spots calcium. Chalk and gypsum contain calcium and both were widely used as ground. The thickness of organic material, like

FIG. 10

The Allori copy in front of the high-energy PIXE set-up.

binding media, which reduces the 3.6-keV K X-rays of calcium to half of their original intensity is of the order of 50 µm. Paint-layers are not uniform in thickness; they vary within the brushstrokes and, therefore, the presence of calcium in the spectra indicates thin or abraded paint-layers at these spots. In the following, three colors will be discussed in more detail.

White: On the white color, only lines from lead and some small amounts of iron are visible in the spectra, denoting the use of lead-white ($2PbCO_3 \cdot Pb(OH)_2$). Lead-white has been used since antiquity, and it was the sole white pigment until 1835. Neither zinc, barium, nor titanium—elements used for recent white pigments—was detected in the spectra.

Yellow: The yellow color shows lead and iron in different amounts. Tin, arsenic, or antimony is absent; hence, lead-tin-yellow, Auripigmentum (As_2S_3), and Naples yellow ($Pb(SbO_3)_2$) can be excluded. The Kα-line of arsenic overlaps with the Lα-line of lead, having nearly the same energy, which makes it difficult to estimate the amount of arsenic. However, the arsenic Kβ-line is also absent. Therefore arsenic is not present in the painting in the amount used for a pigment. The modern pigments can be excluded as well, as neither chromium nor cadmium was detected. Chromate of lead may be a possible pigment. However, as iron is at this spot on top of the paint sequence, the yellow coat was painted using a mixture of lead-white and yellow ochre, the latter containing Fe_2O_3 (Fig. 11).

Blue: This is the only spot where no iron was present in the spectrum (see Fig. 12). Hence, Prussian blue ($Fe_4[Fe(CN)_6]_3$), which came widely into use after its discovery in 1725, was not used. Cobalt blue, smalt, and azurite can be excluded, as neither cobalt nor copper was detected. Possible blue pigments are either organic indigo or ultramarine ($Na_{8...10}Al_6Si_6O_{24}S_{2...4}$). The strong absorption of the

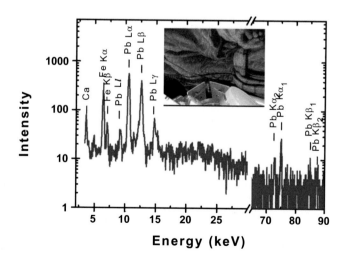

FIG. 11

Spectrum of Judith's *yellow* coat. The inset shows the corresponding spot marked by the laser cross hair. Calcium, iron, and lead lines are present in the spectrum. None of the elements typical of modern yellow pigments is visible.

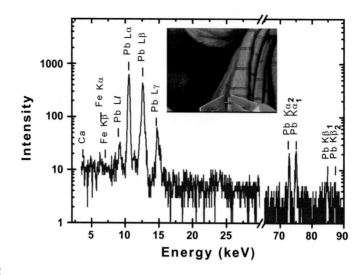

FIG. 12

Spectrum of the *blue stripes* in the girdle. The inset shows the corresponding spot marked by the laser cross hair. Only lead lines are visible in the spectrum. Hence the *blue color* is either an organic pigment (indigo) or ultramarine, both not detectable in the high-energy PIXE set-up.

Table 4 The intensity ratios for Fe Kα/Kβ, Pb Lα/Kα$_1$, and Lα/Lβ for the analyzed painting and their corresponding depth in chalk.

	Judith's cloth		Holofernes' hair	Dark background
	Yellow coat	White sleeve		
Fe Kα/Kβ (7.0)	5.42 40 μm		3.98 88 μm	6.78 5 μm
Pb Lα/Kα$_1$ (44)	22.26 98 μm	23.15 93 μm	21.2 104 μm	20.67 109 μm
Pb Lα/Lβ (1.66)	1.23 109 μm	1.26 99 μm	1.43 53 μm	1.19 119 μm

The values in brackets in the first column give the ratios measured on thin foils with the same set-up.

low-energy X-rays of organic elements, sodium or aluminum makes it impossible to detect them behind the varnish layer.

Table 4 shows the intensity ratios for K- and L-lines for iron and lead as measured on different spots. The ratios have been converted to depth in chalk using Eq. (4). For Judith's cloth and the dark background, iron has the smallest depth. Therefore the paint sequence consists of an iron containing pigment on top of a lead-based pigment.

For the white sleeve, both lead intensity ratios yield the same depth, so one can assume that only one lead-containing layer is present. In contrast to that, the spot measured on Holofernes' hair gives very different depths, if converting the intensity ratios to thickness. This fact excludes the assumption of one single thin paint-layer containing lead. Either there are two different layers (ground and highlight, with iron in between) or there is a very thick layer.

Using high-energy protons, not only the elements used for the pigments could be identified, but also information about the paint sequence could be obtained. However, although no modern pigments were found, this is not an unambiguous proof that the painting dates back to before the 19th century. It cannot be excluded that, for whatsoever reasons, an artist in the 19th century used "ancient" pigments. Therefore further investigations on the painting are necessary.

4 Metals

The fact that the PIXE technique allows quantitative analysis by solving Eq. (5), making it an important tool for the analysis of metal artifacts. Not only major elements, but also trace elements can be detected in a nondestructive way. The investigations range from the distinction between silver from the "old world" and America to the identification of soldering procedures in gold jewelry items [42–44]. However, the structure of the artifact has to be considered: corrosion layers, enriched or depleted layers may yield misleading results if a homogenous material is assumed.

4.1 Egyptian scarab

The Egyptian scarab depicted in Fig. 13 appeared during World War II, and the question arose whether it was genuine or a modern fake [45]. Hence, it was analyzed by high-energy PIXE. The aim of the PIXE analysis was the determination of the elemental composition and especially the distinction between massive and gilded parts. A distinction between gilded and massive parts by using the density of the metal was not possible due to the complicated shape of the object.

Fig. 14 shows the spectrum of scarab shown in Fig. 13 compared to the results of a thin gold foil measured in the same set-up with the same experimental conditions using a measuring time of 200 s. The intensity ratio of the gold $L\alpha/K\alpha_1$-line of the scarab in this spectrum is 1.25, whereas the thin gold foil yields a ratio of 44. For comparison, a 1-cm thick lead sample gives a ratio of $L\alpha/K\alpha_1$ of 1.3. Hence, a thin gold layer, i.e., gilding, on top of a different material can be excluded. Similar ratios were observed on all three investigated spots, showing that scarab was made of a massive gold alloy. A precise quantitative analysis of the metal alloy was difficult due to the fact that the irregular shape of the scarab changed the geometry of the set-up. For the scarab, the composition was estimated to be 95% gold and 5% silver. This composition comes close to the composition of river gold used in Egyptian times. Thus the PIXE measurements provided an indication that the object might be ancient.

FIG. 13

Egyptian scarab during the PIXE investigation.

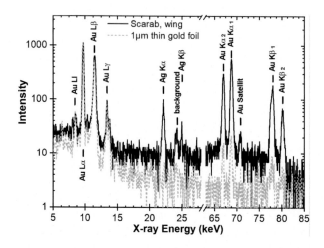

FIG. 14

Spectra of the scarab *(solid)* and a thin gold foil *(dashed)*. The large intensity of the gold K-lines implies the use of a massive gold alloy, and a gilding can be excluded.

4.2 Silver coins

The analysis of the copper content in silver using X-ray techniques will only allow a determination of the copper content close to the surface due to the strong absorption of the copper X-rays in the silver matrix. Fig. 15 shows the transmission of the copper Kα-line in that matrix: already after 10 μm, the transmission is reduced to 10%. However, ancient silver objects are often not homogenous [46]. The use of high-energy protons will give two pieces of information: the normal PIXE technique will provide

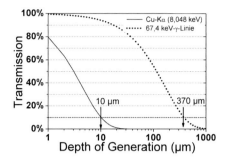

FIG. 15

Transmission of the Cu Kα X-ray *(solid line)* and the 67.4-keV γ-ray *(dotted line)* in a silver matrix as a function of depth. Only 10% of the Cu X-rays generated at a depth of 10 μm actually reach the sample surface.

the copper content close to the surface. In addition, 68-MeV protons also undergo nuclear reactions, leading to a single line at 67.4 keV (see Fig. 15). Copper exists in two isotopes, one containing 34 neutrons, the other one has 36 neutrons. The notation of the different isotopes is done by preceding the elemental abbreviation with the total mass of the atom, the sum of proton and neutron numbers. If a copper atomic nucleus with a mass of 63 incorporates a proton and emits three neutrons, thus forming the zinc isotope of mass 61 (^{61}Zn), this can be written as:

$$^{63}Cu\,(p, 3n)^{61}Zn$$

The isotope ^{61}Zn (an atom with 30 protons and 31 neutrons) is not stable, and decays by two consecutive beta emissions. In this process, a proton in the atomic nucleus transforms to a neutron, and, as the charge has to be maintained, a positive β^+ particle is emitted (the antiparticle of the electron, i.e., a positron). Thus the ^{61}Zn is transformed into ^{61}Ni, a nickel isotope. This nickel atomic nucleus is still in an excited state, relaxing to the ground state by emitting the excess energy as γ-radiation with an energy of 67.4 keV. This nickel isotope is a stable isotope, so no further transitions occur. The energy of this γ-line is high enough to be detected from large depths. Normalizing the intensity of the γ-line by using silver material of known compositions provides the copper concentration for large depths.

The disadvantage is the transformation of copper to nickel, however, only about 100,000 atoms undergo this reaction under experimental conditions. Compared to the about 3 000 000 000 000 000 000 000 copper atoms in a 1-g silver coin with about 30% copper, this can be neglected. The energy loss of 68-MeV protons in 400-μm silver is only 3 MeV, leading to the similar excitation conditions over the whole information depth. This technique has been applied to medieval silver coins, showing the depletion of copper close to the surface due to a long burial time and the restoration necessary as a consequence [47].

5 Conclusion

The whole field of proton beam analytical techniques cannot be presented in a single chapter. The methods presented, PIXE and PIGE, have their specific assets and drawbacks. PIXE is a very fast method when applied to thin samples, allowing determinations of trace elements on huge amounts of samples within a reasonable time ($\sim$100 s). The possibility to vary the incident proton beam energy provides additional information on the sample structure, especially for layered materials. Light elements such as silicon or fluorine, normally not visible in X-ray spectra, when the sample is measured under atmospheric conditions, may be detected via γ-rays created by nuclear reactions in the same experimental run. In addition, these γ-lines yield information about the composition of the sample not accessible with X-rays, like for copper in silver. PIXE and PIGE can be applied in a nondestructive way; therefore, measurements on very unique and precious objects are possible.

Acknowledgment

The author is indebted to K. Ebert, M. Griesser, R. Denk, and H. Winter for fruitful discussions and collaborations.

References

[1] H.P. Schramm, B. Hering, Historische Malpigmente und ihre Identifizierung, Bücherei des Restaurators Band 1, Enke Verlag, Stuttgart, 1995.

[2] B. Beckhoff, B. Kanngießer, N. Langhoff, R. Wedell, H. Wolff, Handbook of Practical X-Ray Fluorescence Analysis, Springer, 2006.

[3] J.I. Goldstein, D.E. Newbury, P. Echlin, D.C. Joy, A.D. Romig, C.E. Lyman, C. Fiori, E. Lifshin, Scanning Electron Microscopy and X-Ray Microanalysis, Plenum Press, New York, 1992.

[4] J. Chadwick, The γ rays excited by the β rays of radium, Phil. Mag. 24 (1912) 594–600.

[5] W. Bambynek, B. Crasemann, R.W. Fink, H.U. Freund, H. Mark, C.D. Swift, R.E. Price, P. Veugopalo, X-Ray Fluorescence Yields, Auger, and Coster-Kronig Transition Probabilities, Rev. Mod. Phys. 44 (1972) 716.

[6] S.A.E. Johansson, J.L. Campbell, PIXE: A Novel Technique for Elemental Analysis, John Wiley and Sons Ltd, 1988, p. 12.

[7] H.G.J. Moseley, The high-frequency spectra of the elements, Phil. Mag. 26 (1913) 1024–1034.

[8] J.A. Bearden, X-ray wavelengths, Rev. Mod. Phys. 39 (1967) 78.

[9] J.H. Scofield, Exchange corrections of K X-ray emission rates, Phys. Rev. A9 (1974) 1041.

[10] A. Perujo, J.A. Maxwell, W.J. Teesdale, J.L. Campbell, Deviation of the K_β/K_α intensity ratio from theory observed in proton-induced X-ray spectra in the $22 \leq Z \leq 32$ region, J. Phys. B: At. Mol. Phys. 20 (1987) 4973.

[11] W. Brandt, G. Lapicki, Energy-loss effect in inner-shell Coulomb ionization by heavy charged particles, Phys. Rev. A 23 (1981) 1717.

[12] H. Paul, J. Sacher, Fitted empirical reference cross sections for K-shell ionization by protons, At. Data Nucl. Data Tables 42 (1989) 105.

[13] I. Orlic, C.H. Sow, S.M. Tang, Experimental L-shell X-ray production and ionization cross sections for proton impact, At. Data Nucl. Data Tables 56 (1994) 159.

[14] M.O. Krause, Atomic radiative and radiationless yields for K and L shells, J. Phys. Chem. Ref. Data 8 (1979) 307.

[15] J.F. Ziegler, J.P. Biersack, U. Littmark, The Stopping and Range of Ions in Solids, Pergamon Press, 1985.

[16] H. Hubbell, S.M. Seltzer, Tables of X-Ray Mass Attenuation Coefficients and Mass Energy-Absorption Coefficients, Originally published as NISTIR 5632, National Institute of Standards and Technology, Gaithersburg, MD, 1995.

[17] C.G. Ryan, D.R. Cousens, S.H. Sie, W.L. Griffin, Quantitative analysis of PIXE spectra in geoscience applications, Nucl. Instrum. Methods Phys. Res. B 49 (1990) 271.

[18] J.L. Campbell, J.A. Maxwell, W.J. Teesdale, The Guelph PIXE software package II, Nucl. Instrum. Methods Phys. Res. B 95 (2005) 407–421.

[19] E. Clayton, PIXAN: The Lucas Heights PIXE analysis computer package, International Nuclear Information System (INIS), 1986, AAEC/M-113.

[20] G. Szabo, I. Borbely-Kiss, PIXYKLM computer package for PIXE analyses, Nucl. Instrum. Methods Phys. Res. B 75 (1993) 123.

[21] K. Sera, S. Futatsugawa, Personal computer aided data handling and analysis for PIXE, Nucl. Instrum. Methods Phys. Res. B 109 & 110 (1996) 99.

[22] B. Vekemans, K. Jensens, L. Vincze, F. Adams, P. Van Espen, Analysis of X-ray spectra by iterative least squares (AXIL): new developments, XRay Spectrom. 23 (1994) 278.

[23] T.B. Johansson, K.R. Akelsson, S.A.E. Johansson, X-ray analysis: elemental trace analysis at the $10 - 12$ g level, Nucl. Instrum. Meth. 84 (1970) 141.

[24] S.A.E. Johansson, J.L. Campbell, K. Malmqvist, Particle-Induced X-Ray Emission Spectrometry (PIXE), John Wiley & Sons Inc, 1995.

[25] J.R. Tesmer, M.A. Nastasi, Handbook of Modern Ion Beam Materials Analysis, Materials Research Society Handbook, Materials Research Society, 1995.

[26] J.L. Boutaine, The modern museum, in: J. Bradley, D. Creagh (Eds.), Physical Techniques in the Study of Art, Archaeology and Cultural Heritage, vol. 1, Elsevier, Amsterdam, 2006.

[27] M.A. Respaldiza, J. Gómez-Camacho, Applications of Ion Beam Analysis Techniques to Arts and Archaeometry, Universidad de Sevilla, 1997.

[28] G. Demortier, A. Adriaens (Eds.), Ion Beam Study of Art and Archaeological Objects, Office for Official Publications of the European Communities, Luxembourg, 2000.

[29] P.A. Mandó, Advantages and limitations of external beams in application to art & archaeology, geology and environmental problems, Nucl. Instrum. Methods Phys. Res. B 85 (1995) 815.

[30] A. Denker, A. Adriaens, M. Dowsett, A. Giumlia-Mair (Eds.), COST Action G8: Non-Destructive Testing and Analysis of Museum Objects, Fraunhofer IRB Verlag, Stuttgart, 2006.

[31] A. Zucchiatti, A. Bouquillon, G. Lanterna, F. Lucarelli, P.A. Mandò, P. Prati, J. Salomon, M.G. Vaccari, PIXE and μ-PIXE analysis of glazes from terracotta sculptures of the della Robbia workshop, Nucl. Instrum. Methods Phys. Res. B 189 (2002) 358.

[32] T. Calligaro, S. Colinart, J.P. Poirot, C. Sudres, Combined external-beam PIXE and μ-Raman characterisation of garnets used in Merovingian jewellery, Nucl. Instrum. Methods Phys. Res. B 189 (2002) 320.

[33] http:/www.fisica.unam.mx/pixe2007/.

[34] www.srim.org.

[35] A. Duval, H. Guicharnaud, J.C. Dran, Particle induced X-ray emission: a valuable tool for the analysis of metalpoint drawings, Nucl. Instrum. Methods Phys. Res. B 226 (2004) 60–74.

[36] T. Tuurnala, A. Hautojärvi, Original or forgery – pigment analysis of paintings using ion beams an ionising radiation, in: G. Demortier, A. Adriaens (Eds.), Ion Beam Study of Art and Archaeological Objects, Office for Official Publications of the European Communities, Luxembourg, 2000, p. 21.

[37] C. Neelmeijer, W. Wagner, H.P. Schramm, Depth resolved ion beam analysis of objects of art, Nucl. Instrum. Methods Phys. Res. B 118 (1996) 338.

[38] C. Neelmeijer, I. Brissaud, T. Calligaro, G. Demortier, A. Hautojärvi, R. Mäder, L. Martinot, M. Schreiner, T. Tuurnala, C. Welzer, Paintings – a challenge for XRF and PIXE analysis, X-Ray Spectrom. 29 (2000) 101.

[39] M. Griesser, A. Denker, H. Musner, K.H. Maier, Non-destructive investigation of paint layer sequences, in: A. Roy, P. Smith (Eds.), Tradition and Innovation – Advances in Conservations, International Institute for Conservation of Historic and Artistic Works, London, 2000, p. 82.

[40] A. Denker, J. Opitz-Coutureau, Paintings – high-energy protons detect pigments and paint layers, Nucl. Instrum. Methods Phys. Res. B 213C (2004) 677–682.

[41] H.P. Schramm, B. Hering, Historische Malmaterialien und ihre Identifizierung, Enke Verlag, Stuttgart, 1995.

[42] B. Constantinescu, J. Kennedy, G. Demortier, On relevant PIXE information for determining the compositional analysis of ancient silver and bronze coins, Int. J. PIXE 9 (1999) 487.

[43] M.F. Guerra, Fingerprinting ancient gold with proton beams of different energies, Nucl. Instrum. Methods Phys. Res. B 226 (2004) 185.

[44] G. Demortier, J.L. Ruvalcaba-Sil, Quantitative ion beam analysis of complex gold-based artefacts, Nucl. Instrum. Methods Phys. Res. B 239 (2005) 1–15.

[45] A. Grimm, Private communication, 2004.

[46] L. Beck, S. Reveillon, S. Bosonnet, D. Eliot, F. Pilon, Silver surface enrichment of silver–copper alloys: a limitation for the analysis of ancient silver coins by surface techniques, Nucl. Instrum. Methods Phys. Res. B 226 (2004) 153–162.

[47] A. Denker, J. Opitz-Coutureau, M. Griesser, R. Denk, H. Winter, Non-destructive analysis of coins using high-energy PIXE, Nucl. Instrum. Methods Phys. Res. B 226 (2004) 163–171.

Index

Note: Page numbers followed by *f* indicate figures and *t* indicate tables.

Printed in the United States
by Baker & Taylor Publisher Services